differential equations and control theory

PURE AND APPLIED MATHEMATICS

A Program of Monographs, Textbooks, and Lecture Notes

EXECUTIVE EDITORS

Earl J. Taft
Rutgers University
New Brunswick, New Jersey

Zuhair Nashed
University of Delaware
Newark, Delaware

EDITORIAL BOARD

M. S. Baouendi
University of California,
San Diego

Jane Cronin
Rutgers University

Jack K. Hale
Georgia Institute of Technology

S. Kobayashi
University of California,
Berkeley

Marvin Marcus
University of California,
Santa Barbara

W. S. Massey
Yale University

Anil Nerode
Cornell University

Donald Passman
University of Wisconsin,
Madison

Fred S. Roberts
Rutgers University

David L. Russell
Virginia Polytechnic Institute
and State University

Walter Schempp
Universität Siegen

Mark Teply
University of Wisconsin,
Milwaukee

LECTURE NOTES IN PURE AND APPLIED MATHEMATICS

1. N. Jacobson, Exceptional Lie Algebras
2. L.-Å. Lindahl and F. Poulsen, Thin Sets in Harmonic Analysis
3. I. Satake, Classification Theory of Semi-Simple Algebraic Groups
4. F. Hirzebruch et al., Differentiable Manifolds and Quadratic Forms
5. I. Chavel, Riemannian Symmetric Spaces of Rank One
6. R. B. Burckel, Characterization of C(X) Among Its Subalgebras
7. B. R. McDonald et al., Ring Theory
8. Y.-T. Siu, Techniques of Extension on Analytic Objects
9. S. R. Caradus et al., Calkin Algebras and Algebras of Operators on Banach Spaces
10. E. O. Roxin et al., Differential Games and Control Theory
11. M. Orzech and C. Small, The Brauer Group of Commutative Rings
12. S. Thomier, Topology and Its Applications
13. J. M. Lopez and K. A. Ross, Sidon Sets
14. W. W. Comfort and S. Negrepontis, Continuous Pseudometrics
15. K. McKennon and J. M. Robertson, Locally Convex Spaces
16. M. Carmeli and S. Malin, Representations of the Rotation and Lorentz Groups
17. G. B. Seligman, Rational Methods in Lie Algebras
18. D. G. de Figueiredo, Functional Analysis
19. L. Cesari et al., Nonlinear Functional Analysis and Differential Equations
20. J. J. Schäffer, Geometry of Spheres in Normed Spaces
21. K. Yano and M. Kon, Anti-Invariant Submanifolds
22. W. V. Vasconcelos, The Rings of Dimension Two
23. R. E. Chandler, Hausdorff Compactifications
24. S. P. Franklin and B. V. S. Thomas, Topology
25. S. K. Jain, Ring Theory
26. B. R. McDonald and R. A. Morris, Ring Theory II
27. R. B. Mura and A. Rhemtulla, Orderable Groups
28. J. R. Graef, Stability of Dynamical Systems
29. H.-C. Wang, Homogeneous Branch Algebras
30. E. O. Roxin et al., Differential Games and Control Theory II
31. R. D. Porter, Introduction to Fibre Bundles
32. M. Altman, Contractors and Contractor Directions Theory and Applications
33. J. S. Golan, Decomposition and Dimension in Module Categories
34. G. Fairweather, Finite Element Galerkin Methods for Differential Equations
35. J. D. Sally, Numbers of Generators of Ideals in Local Rings
36. S. S. Miller, Complex Analysis
37. R. Gordon, Representation Theory of Algebras
38. M. Goto and F. D. Grosshans, Semisimple Lie Algebras
39. A. I. Arruda et al., Mathematical Logic
40. F. Van Oystaeyen, Ring Theory
41. F. Van Oystaeyen and A. Verschoren, Reflectors and Localization
42. M. Satyanarayana, Positively Ordered Semigroups
43. D. L Russell, Mathematics of Finite-Dimensional Control Systems
44. P.-T. Liu and E. Roxin, Differential Games and Control Theory III
45. A. Geramita and J. Seberry, Orthogonal Designs
46. J. Cigler, V. Losert, and P. Michor, Banach Modules and Functors on Categories of Banach Spaces
47. P.-T. Liu and J. G. Sutinen, Control Theory in Mathematical Economics
48. C. Byrnes, Partial Differential Equations and Geometry
49. G. Klambauer, Problems and Propositions in Analysis
50. J. Knopfmacher, Analytic Arithmetic of Algebraic Function Fields
51. F. Van Oystaeyen, Ring Theory
52. B. Kadem, Binary Time Series
53. J. Barros-Neto and R. A. Artino, Hypoelliptic Boundary-Value Problems
54. R. L. Sternberg et al., Nonlinear Partial Differential Equations in Engineering and Applied Science
55. B. R. McDonald, Ring Theory and Algebra III
56. J. S. Golan, Structure Sheaves Over a Noncommutative Ring
57. T. V. Narayana et al., Combinatorics, Representation Theory and Statistical Methods in Groups
58. T. A. Burton, Modeling and Differential Equations in Biology
59. K. H. Kim and F. W. Roush, Introduction to Mathematical Consensus Theory

60. J. Banas and K. Goebel, Measures of Noncompactness in Banach Spaces
61. O. A. Nielson, Direct Integral Theory
62. J. E. Smith et al., Ordered Groups
63. J. Cronin, Mathematics of Cell Electrophysiology
64. J. W. Brewer, Power Series Over Commutative Rings
65. P. K. Kamthan and M. Gupta, Sequence Spaces and Series
66. T. G. McLaughlin, Regressive Sets and the Theory of Isols
67. T. L. Herdman et al., Integral and Functional Differential Equations
68. R. Draper, Commutative Algebra
69. W. G. McKay and J. Patera, Tables of Dimensions, Indices, and Branching Rules for Representations of Simple Lie Algebras
70. R. L. Devaney and Z. H. Nitecki, Classical Mechanics and Dynamical Systems
71. J. Van Geel, Places and Valuations in Noncommutative Ring Theory
72. C. Faith, Injective Modules and Injective Quotient Rings
73. A. Fiacco, Mathematical Programming with Data Perturbations I
74. P. Schultz et al., Algebraic Structures and Applications
75. L Bican et al., Rings, Modules, and Preradicals
76. D. C. Kay and M. Breen, Convexity and Related Combinatorial Geometry
77. P. Fletcher and W. F. Lindgren, Quasi-Uniform Spaces
78. C.-C. Yang, Factorization Theory of Meromorphic Functions
79. O. Taussky, Ternary Quadratic Forms and Norms
80. S. P. Singh and J. H. Burry, Nonlinear Analysis and Applications
81. K. B. Hannsgen et al., Volterra and Functional Differential Equations
82. N. L. Johnson et al., Finite Geometries
83. G. I. Zapata, Functional Analysis, Holomorphy, and Approximation Theory
84. S. Greco and G. Valla, Commutative Algebra
85. A. V. Fiacco, Mathematical Programming with Data Perturbations II
86. J.-B. Hiriart-Urruty et al., Optimization
87. A. Figa Talamanca and M. A. Picardello, Harmonic Analysis on Free Groups
88. M. Harada, Factor Categories with Applications to Direct Decomposition of Modules
89. V. I. Istrătescu, Strict Convexity and Complex Strict Convexity
90. V. Lakshmikantham, Trends in Theory and Practice of Nonlinear Differential Equations
91. H. L. Manocha and J. B. Srivastava, Algebra and Its Applications
92. D. V. Chudnovsky and G. V. Chudnovsky, Classical and Quantum Models and Arithmetic Problems
93. J. W. Longley, Least Squares Computations Using Orthogonalization Methods
94. L. P. de Alcantara, Mathematical Logic and Formal Systems
95. C. E. Aull, Rings of Continuous Functions
96. R. Chuaqui, Analysis, Geometry, and Probability
97. L. Fuchs and L. Salce, Modules Over Valuation Domains
98. P. Fischer and W. R. Smith, Chaos, Fractals, and Dynamics
99. W. B. Powell and C. Tsinakis, Ordered Algebraic Structures
100. G. M. Rassias and T. M. Rassias, Differential Geometry, Calculus of Variations, and Their Applications
101. R.-E. Hoffmann and K. H. Hofmann, Continuous Lattices and Their Applications
102. J. H. Lightbourne III and S. M. Rankin III, Physical Mathematics and Nonlinear Partial Differential Equations
103. C. A. Baker and L. M. Batten, Finite Geometrics
104. J. W. Brewer et al., Linear Systems Over Commutative Rings
105. C. McCrory and T. Shifrin, Geometry and Topology
106. D. W. Kueke et al., Mathematical Logic and Theoretical Computer Science
107. B.-L. Lin and S. Simons, Nonlinear and Convex Analysis
108. S. J. Lee, Operator Methods for Optimal Control Problems
109. V. Lakshmikantham, Nonlinear Analysis and Applications
110. S. F. McCormick, Multigrid Methods
111. M. C. Tangora, Computers in Algebra
112. D. V. Chudnovsky and G. V. Chudnovsky, Search Theory
113. D. V. Chudnovsky and R. D. Jenks, Computer Algebra
114. M. C. Tangora, Computers in Geometry and Topology
115. P. Nelson et al., Transport Theory, Invariant Imbedding, and Integral Equations
116. P. Clément et al., Semigroup Theory and Applications
117. J. Vinuesa, Orthogonal Polynomials and Their Applications
118. C. M. Dafermos et al., Differential Equations
119. E. O. Roxin, Modern Optimal Control
120. J. C. Díaz, Mathematics for Large Scale Computing

121. P. S. Milojević, Nonlinear Functional Analysis
122. C. Sadosky, Analysis and Partial Differential Equations
123. R. M. Shortt, General Topology and Applications
124. R. Wong, Asymptotic and Computational Analysis
125. D. V. Chudnovsky and R. D. Jenks, Computers in Mathematics
126. W. D. Wallis et al., Combinatorial Designs and Applications
127. S. Elaydi, Differential Equations
128. G. Chen et al., Distributed Parameter Control Systems
129. W. N. Everitt, Inequalities
130. H. G. Kaper and M. Garbey, Asymptotic Analysis and the Numerical Solution of Partial Differential Equations
131. O. Arino et al., Mathematical Population Dynamics
132. S. Coen, Geometry and Complex Variables
133. J. A. Goldstein et al., Differential Equations with Applications in Biology, Physics, and Engineering
134. S. J. Andima et al., General Topology and Applications
135. P Clément et al., Semigroup Theory and Evolution Equations
136. K. Jarosz, Function Spaces
137. J. M. Bayod et al., p-adic Functional Analysis
138. G. A. Anastassiou, Approximation Theory
139. R. S. Rees, Graphs, Matrices, and Designs
140. G. Abrams et al., Methods in Module Theory
141. G. L. Mullen and P. J.-S. Shiue, Finite Fields, Coding Theory, and Advances in Communications and Computing
142. M. C. Joshi and A. V. Balakrishnan, Mathematical Theory of Control
143. G. Komatsu and Y. Sakane, Complex Geometry
144. I. J. Bakelman, Geometric Analysis and Nonlinear Partial Differential Equations
145. T. Mabuchi and S. Mukai, Einstein Metrics and Yang–Mills Connections
146. L. Fuchs and R. Göbel, Abelian Groups
147. A. D. Pollington and W. Moran, Number Theory with an Emphasis on the Markoff Spectrum
148. G. Dore et al., Differential Equations in Banach Spaces
149. T. West, Continuum Theory and Dynamical Systems
150. K. D. Bierstedt et al., Functional Analysis
151. K. G. Fischer et al., Computational Algebra
152. K. D. Elworthy et al., Differential Equations, Dynamical Systems, and Control Science
153. P.-J. Cahen, et al., Commutative Ring Theory
154. S. C. Cooper and W. J. Thron, Continued Fractions and Orthogonal Functions
155. P. Clément and G. Lumer, Evolution Equations, Control Theory, and Biomathematics
156. M. Gyllenberg and L. Persson, Analysis, Algebra, and Computers in Mathematical Research
157. W. O. Bray et al., Fourier Analysis
158. J. Bergen and S. Montgomery, Advances in Hopf Algebras
159. A. R. Magid, Rings, Extensions, and Cohomology
160. N. H. Pavel, Optimal Control of Differential Equations
161. M. Ikawa, Spectral and Scattering Theory
162. X. Liu and D. Siegel, Comparison Methods and Stability Theory
163. J.-P. Zolésio, Boundary Control and Variation
164. M. Křížek et al., Finite Element Methods
165. G. Da Prato and L. Tubaro, Control of Partial Differential Equations
166. E. Ballico, Projective Geometry with Applications
167. M. Costabel et al., Boundary Value Problems and Integral Equations in Nonsmooth Domains
168. G. Ferreyra, G. R. Goldstein, and F. Neubrander, Evolution Equations
169. S. Huggett, Twistor Theory
170. H. Cook et al., Continua
171. D. F. Anderson and D. E. Dobbs, Zero-Dimensional Commutative Rings
172. K. Jarosz, Function Spaces
173. V. Ancona et al., Complex Analysis and Geometry
174. E. Casas, Control of Partial Differential Equations and Applications
175. N. Kalton et al., Interaction Between Functional Analysis, Harmonic Analysis, and Probability
176. Z. Deng et al., Differential Equations and Control Theory
177. P. Marcellini et al. Partial Differential Equations and Applications
178. A. Kartsatos, Theory and Applications of Nonlinear Operators of Accretive and Monotone Type
179. M. Maruyama, Moduli of Vector Bundles
180. A. Ursini and P. Aglianò, Logic and Algebra
181. X. H. Cao et al., Rings, Groups, and Algebras
182. D. Arnold and R. M. Rangaswamy, Abelian Groups and Modules
183. S. R. Chakravarthy and A. S. Alfa, Matrix-Analytic Methods in Stochastic Models

184. *J. E. Andersen et al.*, Geometry and Physics
185. *P.-J. Cahen et al.*, Commutative Ring Theory
186. *J. A. Goldstein et al.*, Stochastic Processes and Functional Analysis
187. *A. Sorbi*, Complexity, Logic, and Recursion Theory
188. *G. Da Prato and J.-P. Zolésio*, Partial Differential Equation Methods in Control and Shape Analysis
189. *D. D. Anderson*, Factorization in Integral Domains
190. *N. L. Johnson*, Mostly Finite Geometries
191. *D. Hinton and P. W. Schaefer*, Spectral Theory and Computational Methods of Sturm–Liouville Problems
192. *W. H. Schikhof et al.*, p-adic Functional Analysis
193. *S. Sertöz*, Algebraic Geometry
194. *G. Caristi and E. Mitidieri*, Reaction Diffusion Systems
195. *A. V. Fiacco*, Mathematical Programming with Data Perturbations
196. *M. Křížek et al.*, Finite Element Methods: Superconvergence, Post-Processing, and A Posteriori Estimates
197. *S. Caenepeel and A. Verschoren*, Rings, Hopf Algebras, and Brauer Groups
198. *V. Drensky et al.*, Methods in Ring Theory
199. *W. B. Jones and A. Sri Ranga*, Orthogonal Functions, Moment Theory, and Continued Fractions
200. *P. E. Newstead*, Algebraic Geometry
201. *D. Dikranjan and L. Salce*, Abelian Groups, Module Theory, and Topology
202. *Z. Chen et al.*, Advances in Computational Mathematics
203. *X. Caicedo and C. H. Montenegro*, Models, Algebras, and Proofs
204. *C. Y. Yıldırım and S. A. Stepanov*, Number Theory and Its Applications
205. *D. E. Dobbs et al.*, Advances in Commutative Ring Theory
206. *F. Van Oystaeyen*, Commutative Algebra and Algebraic Geometry
207. *J. Kakol et al.*, p-adic Functional Analysis
208. *M. Boulagouaz and J.-P. Tignol*, Algebra and Number Theory
209. *S. Caenepeel and F. Van Oystaeyen*, Hopf Algebras and Quantum Groups
210. *F. Van Oystaeyen and M. Saorin*, Interactions Between Ring Theory and Representations of Algebras
211. *R. Costa et al.*, Nonassociative Algebra and Its Applications
212. *T.-X. He*, Wavelet Analysis and Multiresolution Methods
213. *H. Hudzik and L. Skrzypczak*, Function Spaces: The Fifth Conference
214. *J. Kajiwara et al.*, Finite or Infinite Dimensional Complex Analysis
215. *G. Lumer and L. Weis*, Evolution Equations and Their Applications in Physical and Life Sciences
216. *J. Cagnol et al.*, Shape Optimization and Optimal Design
217. *J. Herzog and G. Restuccia*, Geometric and Combinatorial Aspects of Commutative Algebra
218. *G. Chen et al.*, Control of Nonlinear Distributed Parameter Systems
219. *F. Ali Mehmeti et al.*, Partial Differential Equations on Multistructures
220. *D. D. Anderson and I. J. Papick*, Ideal Theoretic Methods in Commutative Algebra
221. *Á. Granja et al.*, Ring Theory and Algebraic Geometry
222. *A. K. Katsaras et al.*, p-adic Functional Analysis
223. *R. Salvi*, The Navier-Stokes Equations
224. *F. U. Coelho and H. A. Merklen*, Representations of Algebras
225. *S. Aizicovici and N. H. Pavel*, Differential Equations and Control Theory
226. *G. Lyubeznik*, Local Cohomology and Its Applications

Additional Volumes in Preparation

differential equations and control theory

edited by

Sergiu Aizicovici
Nicolae H. Pavel
**Ohio University
Athens, Ohio**

MARCEL DEKKER, INC. NEW YORK · BASEL

ISBN: 0-8247-0681-1

This book is printed on acid-free paper.

Headquarters
Marcel Dekker, Inc.
270 Madison Avenue, New York, NY 10016
tel: 212-696-9000; fax: 212-685-4540

Eastern Hemisphere Distribution
Marcel Dekker AG
Hutgasse 4, Postfach 812, CH-4001 Basel, Switzerland
tel: 41-61-261-8482; fax: 41-61-261-8896

World Wide Web
http://www.dekker.com

The publisher offers discounts on this book when ordered in bulk quantities. For more information, write to Special Sales/Professional Marketing at the headquarters address above.

Copyright © 2002 by Marcel Dekker, Inc. All Rights Reserved.

Neither this book nor any part may be reproduced or transmitted in any form or by any means, electronic or mechanical, including photocopying, microfilming, and recording, or by any information storage and retrieval system, without permission in writing from the publisher.

Current printing (last digit):
10 9 8 7 6 5 4 3 2 1

PRINTED IN THE UNITED STATES OF AMERICA

Preface

This volume is based on papers presented at the International Workshop on Differential Equations and Optimal Control, held at the Department of Mathematics of Ohio University in Athens, Ohio. The main objective of this international meeting was to feature new trends in the theory and applications of partial differential and functional-differential equations and their optimal control. The workshop can be viewed as a follow-up to the March 1993 Ohio University International Conference on Optimal Control of Differential Equations, whose proceedings were edited by N. H. Pavel and also published by Marcel Dekker, Inc., as volume 160 of the series: Lecture Note in Pure and Applied Mathematics.

A large variety of related topics is covered in this volume, both theoretical and applied, deterministic and stochastic. The topics include: nonlinear programming and control with closed range operators, stabilization of the diffusion equations, flow-invariant sets with respect to the Navier-Stokes equations, numerical approximation of the Riccati equation, telegraph systems, dispersive equations, viable domains for differential equations, almost periodic solutions to neutral functional equations, Wentzell boundary conditions, parabolic phase-field models with memory, Kato classes of distributions, optimal control and algebraic Riccati equations, identification problems for wave equations via optimal control, integrodifferential and other functional equations, stochastic Navier-Stokes equations, Lavrentiev phenomena, volatility for American options via optimal control, obstacle problems, necessary conditions of optimality for semilinear problems, Lyapunov stability, least action for N-body problems, and more.

The workshop was sponsored by the College of Arts and Sciences, the Department of Mathematics, and the Research Office of Ohio University. We gratefully acknowledge their financial support, which made this workshop possible. We are also very indebted to all participants and contributors. Finally, our thanks go also to Marcel Dekker, Inc., for undertaking the publication of this volume.

Sergiu Aizicovici
Nicolae H. Pavel

Contents

Preface		*iii*
Contributors		*vii*

1. Existence and Uniqueness of Solutions to a Second Order Nonlinear Nonlocal Hyperbolic Equation
   Azmy S. Ackleh, Sergiu Aizicovici, Michael Demetriou, and Simeon Reich — 1

2. Fully Nonlinear Programming Problems with Closed Range Operators
   Sergiu Aizicovici, D. Motreanu, and Nicolae H. Pavel — 19

3. Internal Stabilization of the Diffusion Equation
   Laura-Iulia Aniţa and Sebastian Aniţa — 31

4. Flow-Invariant Sets with Respect to Navier-Stokes Equation
   V. Barbu and Nicolae H. Pavel — 39

5. Numerical Approximation of the Riccati Equation via Fractional Steps Method
   Tudor Barbu and Costică Moroşanu — 55

6. Asymptotic Analysis of the Telegraph System with Nonlinear Boundary Conditions
   L. Barbu, E. Cosma, Gh. Moroşanu, and W. L. Wendland — 63

7. Global Existence for a Class of Dispersive Equations
   Radu C. Cascaval — 77

8. Viable Domains for Differential Equations Governed by Carathéodory Perturbations of Nonlinear *m*-Accretive Operators
   Ovidiu Cârjă and Ioan I. Vrabie — 109

9. Almost Periodic Solutions to Neutral Functional Equations
   C. Corduneanu — 131

10. The One Dimensional Wave Equation with Wentzell Boundary Conditions
    Angelo Favini, Gisèle Ruiz Goldstein, Jerome A. Goldstein, and Silvia Romanelli — 139

11. On the Longterm Behaviour of a Parabolic Phase-Field Model with Memory
    Maurizio Grasselli and Vittorino Pata — 147

12. On the Kato Classes of Distributions and the *BMO*-Classes
    Archil Gulisashvili — 159

13. The Global Solution Set for a Class of Semilinear Problems
    Philip Korman — 177

14. Optimal Control and Algebraic Riccati Equations under Singular Estimates for $e^{At}B$ in the Absence of Analyticity. Part I: The Stable Case 193
    Irena Lasiecka and Roberto Triggiani

15. Solving Identification Problems for the Wave Equation by Optimal Control Methods 221
    Suzanne Lenhart and Vladimir Protopopescu

16. Singular Perturbations and Approximations for Integrodifferential Equations 233
    J. Liu, J. Sochacki, and P. Dostert

17. Remarks on Impulse Control Problems for the Stochastic Navier-Stokes Equations 245
    J. L. Menaldi and S. S. Sritharan

18. Recent Progress on the Lavrentiev Phenomenon with Applications 257
    Victor J. Mizel

19. Abstract Eigenvalue Problem for Monotone Operators and Applications to Differential Operators 263
    Silviu Sburlan

20. Implied Volatility for American Options via Optimal Control and Fast Numerical Solutions of Obstacle Problems 277
    Srdjan Stojanovic

21. First Order Necessary Conditions of Optimality for Semilinear Optimal Control Problems 295
    M. D. Voisei

22. Lyapunov Equation and the Stability of Nonautonomous Evolution Equations in Hilbert Spaces 309
    Quoc-Phong Vu and Siu Pang Yung

23. Least Action for *N*-Body Problems with Quasihomogeneous Potentials 319
    Shih-liang Wen and Shiqing Zhang

Contributors

Azmy S. Ackleh   University of Louisiana at Lafayette, Lafayette, Louisiana

Sergiu Aizicovici   Ohio University, Athens, Ohio

Laura-Iulia Aniţa   University "Al.I. Cuza", Iaşi, Romania

Sebastian Aniţa   University "Al.I. Cuza", Iaşi, Romania

L. Barbu   Ovidius University, Constanţa, Romania

Tudor Barbu   Institute of Mathematics of Romanian Academy, Iaşi, Romania

V. Barbu   University of Iaşi, Iaşi, Romania

Ovidiu Cârjă   "Al.I. Cuza" University of Iaşi, Iaşi, Romania

Radu C. Cascaval   University of Missouri, Columbia, Missouri

C. Corduneanu   University of Texas at Arlington, Arlington, Texas

Michael Demetriou   Worcester Polytechnic Institute, Worcester, Massachusetts

P. Dostert   James Madison University, Harrisonburg, Virginia

Angelo Favini   Università di Bologna, Bologna, Italy

Gisèle Ruiz Goldstein   University of Memphis, Memphis, Tennessee

Jerome A. Goldstein   University of Memphis, Memphis, Tennessee

Maurizio Grasselli   Politecnico di Milano, Milan, Italy

A. Gulisashvili   Ohio University, Athens, Ohio

Philip Korman   University of Cincinnati, Cincinnati, Ohio

Irena Lasiecka   University of Virginia, Charlottesville, Virginia

Suzanne Lenhart   University of Tennessee, Knoxville, Tennessee

J. Liu   James Madison University, Harrisonburg, Virginia

J. L. Menaldi   Wayne State University, Detroit, Michigan

Victor J. Mizel   Carnegie-Mellon University, Phladelphia, Pennsylvania

Costică Moroşanu   University "Al.I. Cuza", Iaşi, Romania

Gh. Moroşanu   Stuttgart University, Stuttgart, Germany

D. Motreanu   University of Iaşi, Iaşi, Romania

Vittorino Pata   Politecnico di Milano, Milan, Italy

Nicolae H. Pavel   Ohio University, Athens, Ohio

Vladimir Protopopescu   Oak Ridge National Laboratory, Oak Ridge, Tennessee

Simeon Reich   The Technion-Israel Institute of Technology, Haifa, Israel

Silvia Romanelli   Università di Bari, Bari, Italy

Silviu Sburlan   Ovidius University, Constantza, Romania

J. Sochacki   James Madison University, Harrisonburg, Virginia

S. S. Sritharan   U.S. Navy, San Diego, California

Srdjan Stojanovic   University of Cincinnati, Cincinnati, Ohio

Roberto Triggiani   University of Virginia, Charlottesville, Virginia

M. D. Voisei   Ohio University, Athens, Ohio

Ioan I. Vrabie   "Al.I. Cuza" University of Iaşi, Iaşi, Romania

Quoc-Phong Vu   Ohio University, Athens, Ohio

Shih-liang Wen   Ohio University, Athens, Ohio

W. L. Wendland   Stuttgart University, Stuttgart, Germany

Siu Pang Yung   University of Hong Kong, Hong Kong, China

Shiqing Zhang   Chongqing University, Chongqing, China

Existence and Uniqueness of Solutions to a Second Order Nonlinear Nonlocal Hyperbolic Equation

AZMY S. ACKLEH
Department of Mathematics, University of Louisiana at Lafayette, Lafayette, LA 70504, USA

SERGIU AIZICOVICI
Department of Mathematics, Ohio University, Athens, OH 45701, USA

MICHAEL DEMETRIOU
Department of Mechanical Engineering, Worcester Polytechnic Institute, Worcester, MA 01609, USA

SIMEON REICH
Department of Mathematics, The Technion-Israel Institute of Technology, 32000 Haifa, ISRAEL

We establish existence and uniqueness of weak solutions to a class of second order distributed parameter systems with sudden changes in the input term. Such systems are often encountered in flexible structures and structure-fluid interaction systems that utilize smart actuators. A Galerkin finite dimensional approximation scheme for computing the solution of these systems is developed and its strong convergence is proved. Numerical results are also presented.

1 Introduction

In this paper we consider the nonlinear, nonlocal partial differential equation

$$w_{tt} + \kappa_1 w_{xxxx} + \kappa_2 w_{xxxxt} = [\beta(x,t)g(y)]_{xx} + f(x,t), \tag{1.1}$$

with boundary and initial conditions given by

$$w_x(0,t) = w(0,t) = 0, \quad w_x(1,t) = w(1,t) = 0,$$
$$w(\cdot,0) = w_0 \in H_0^2(0,1), \quad w_t(\cdot,0) = w_1 \in L^2(0,1). \tag{1.2}$$

In equation (1.1) the function y satisfies

$$y(t) = \int_0^1 k_s \chi_{[x_1,x_2]}(x) w_{xxt}(x,t) dx$$

where $\chi_{[x_1,x_2]}$ denotes the characteristic function on the interval $[x_1, x_2]$, with $0 \leq x_1 < x_2 \leq 1$. The constants κ_1, κ_2 and k_s are positive and g is a Lipschitz continuous function.

There is an extensive literature on linear and semilinear second order (in time) evolution equations (e.g., [1, 2, 3, 4, 6, 7, 12, 13, 14, 16]). For example, the existence-uniqueness results presented in [4] apply to the system (1.1)-(1.2) for the case $g \equiv 0$. However, to our knowledge, no existence-uniqueness results for the system (1.1)-(1.2) (with a nontrivial nonlinear function g) are available.

Equation (1.1) is a general form of the model developed by Demetriou and Polycarpou [9, 10]. Indeed, in the context of the flexible structure encountered in Demetriou and Polycarpou [9], κ_1 denotes the *stiffness* parameter, κ_2 the *damping* parameter and k_s the sensor *piezoceramic* constant; see Banks et al. [5] and Dosch et al. [11]. When the actuator (input) failure term $\beta(x,t)g(y)$ is written as

$$\beta(x,t)g(y) = \beta_1(t)\left(k_a \chi_{[x_1,x_2]}(x)\epsilon(t)\right) g(y)$$

with the *time profile* (Polycarpou and Helmicki [15]) of the failure given by

$$\beta_1(t) = \begin{cases} 0 & \text{if } t < T_f \\ 1 - e^{-\lambda(t-T_f)} & \text{if } t \geq T_f \end{cases}, \quad \lambda > 0, \qquad (1.3)$$

and the nominal forcing (*actuator*) term given by

$$f(x,t) = [k_a \chi_{[x_1,x_2]}(x)\epsilon(t)]_{xx}, \quad k_a > 0,$$

then equation (1.1) has exactly the same form as the beam equation considered in Demetriou and Polycarpou [9]. The time T_f denotes the unknown instance of the failure occurrence and the signal ϵ denotes the input voltage to the patch. Similarly, k_a denotes the actuator *piezoceramic* constant; see Banks et al. [5]. Therefore, this model describes the dynamics of a flexible cantilevered beam before ($t < T_f$) and after ($t \geq T_f$) the occurrence of an anticipated actuator failure commencing at an unknown time T_f. In view of the above, the plant equation (1.1) can now be written as follows:

$$w_{tt} + \kappa_1 w_{xxxx} + \kappa_2 w_{xxxxt} = \left[k_a \chi_{[x_1,x_2]}(x)\epsilon(t)\right]_{xx} + \beta_1(t)\left[k_a \chi_{[x_1,x_2]}(x)\epsilon(t)g(y(t))\right]_{xx}.$$

Our efforts here are a continuation of an earlier work [8]. There, the following Galerkin approximations for solutions of the system (1.1)-(1.2) were considered:

$$w^m(t) = \sum_{i=1}^{m} C_i^m(t)\psi_i,$$

where $\{\psi_j\}_{j=1}^{\infty}$ are the eigenfunctions corresponding to the eigenvalues $\{\lambda_j\}_{j=1}^{\infty}$ of the strictly positive self adjoint operator $A = \dfrac{d^4}{dx^4}$ with the dense domain in $L^2(0,1)$ given by

$$\mathcal{D}(A) = \{\phi \in H^4(0,1) : \phi'(0) = \phi(0) = 0,\ \phi'(1) = \phi(1) = 0\}.$$

It is well known that the eigenvalues λ_j are simple and that the set of eigenfunctions $\{\psi_j\}$ forms a complete orthonormal system in $L^2(0,1)$. A priori bounds which are based on energy estimates were established for these Galerkin approximations. Furthermore, in order to detect, diagnose and accommodate the actuator failure, a *model-based* fault diagnosis scheme was presented. Our scheme consisted of a detection/diagnostic observer and an estimator of the actuator failure term. Since the proposed scheme is infinite dimensional, a finite dimensional approximation was considered for computational purposes.

The present paper is organized as follows. In Section 2 we give the definition of weak solutions to problem (1.1)-(1.2) and establish existence-uniqueness of such solutions using a Galerkin approximation technique. The strong convergence of this approximation is also proved. In Section 3 we use the Galerkin method to give a numerical solution to a model problem.

2 Existence and Uniqueness of Weak Solutions

We begin this section by letting $H = L^2(0,1)$ and $V = H_0^2(0,1)$, so we have the Gelfand triple $V \hookrightarrow H \hookrightarrow V^*$ with $V^* = H^{-2}(0,1)$. We denote by $\langle \cdot, \cdot \rangle$ the inner product in H, while $\langle \cdot, \cdot \rangle_{V^*,V}$ stands for the usual duality product. Let $\|\cdot\|, \|\cdot\|_V$, and $\|\cdot\|_{V^*}$ denote the norms of the spaces H, V, and V^*, respectively. Assume that the parameters in (1.1)-(1.2) satisfy the following conditions:

(A_β) The function $\beta \in L^\infty(0,T,H)$, with $\|\beta\|_{L^\infty(0,T;H)} \leq L$.

(A_g) The nonlinear function g satisfies the following Lipschitz condition:
$$|g(\xi_1) - g(\xi_2)| \leq \frac{\widetilde{C}_1}{k_s} |\xi_1 - \xi_2|, \quad \text{for all } \xi_1, \xi_2 \in \mathbb{R},$$
where $\widetilde{C}_1 < \kappa_2/L$.

(A_f) The forcing term $f \in L^2(0,T;V^*)$.

To establish the existence-uniqueness of solutions we use a Galerkin type method which is comparable to the one employed in the study of well-posedness for other second order (in time) evolution equations (see, e.g., [1, 2, 3, 4, 12, 13, 14]). To this end, we define the space of functions
$$\mathcal{U}_T = \{u : u \in W^{1,2}(0,T;V), \ u_{tt} \in L^2(0,T;V^*)\}$$
with norm
$$\|u\|_{\mathcal{U}_T} = (\|u\|^2_{W^{1,2}(0,T;V)} + \|u_{tt}\|^2_{L^2(0,T;V^*)})^{1/2}.$$

We now define the notion of a weak solution to the problem (1.1)-(1.2).

Definition 2.1 *We say that a function $w \in \mathcal{U}_T$ is a weak solution of (1.1)-(1.2) if it satisfies*

$$\langle w_{tt}(t), \phi \rangle_{V^*,V} + \kappa_1 \langle w_{xx}(t), \phi_{xx} \rangle + \kappa_2 \langle w_{xxt}(t), \phi_{xx} \rangle \\ = \langle \beta(t) g(y(t)), \phi_{xx} \rangle + \langle f(t), \phi \rangle_{V^*,V}, \quad \forall \phi \in V \tag{2.1}$$

and

$$w(0) = w_0 \in V, \qquad w_t(0) = w_1 \in H. \tag{2.2}$$

Next we state the existence-uniqueness theorem which is the main result of this paper.

Theorem 2.2 *The problem (1.1)-(1.2) has a unique weak solution.*

Proof. Let $\{\psi_i\}_{i=1}^\infty$ be any linearly independent total subset of V. For each m, let

$$V^m = \mathrm{span}\{\psi_1, \ldots, \psi_m\}$$

and let $w_0^m, w_1^m \in V^m$ be chosen so that $w_0^m \to w_0$ in V, $w_1^m \to w_1$ in H as $m \to \infty$. For each m we define an approximate solution to the problem (1.1)-(1.2) by $w^m(t) = \sum_{i=1}^m C_i^m(t) \psi_i$, where w^m is the unique solution to the m-dimensional system

$$\langle w_{tt}^m(t), \psi_j \rangle + \kappa_1 \langle w_{xx}^m(t), \psi_{jxx} \rangle + \kappa_2 \langle w_{xxt}^m(t), \psi_{jxx} \rangle \\ = \langle \beta(t) g(y^m(t)), \psi_{jxx} \rangle + \langle f(t), \psi_j \rangle_{V^*,V}, \quad j = 1, 2, \ldots, m, \tag{2.3}$$

with initial conditions

$$w^m(0) = w_0^m, \qquad w_t^m(0) = w_1^m. \tag{2.4}$$

The function y^m in equation (2.3) satisfies

$$y^m(t) = \int_0^1 k_s \chi_{[x_1,x_2]}(x) w_{xxt}^m(x,t) dx.$$

Multiplying the equation (2.3) by $\frac{d}{dt} C_j^m(t)$ and summing up over j we obtain

$$\langle w_{tt}^m(t), w_t^m(t) \rangle + \kappa_1 \langle w_{xx}^m(t), w_{xxt}^m(t) \rangle + \kappa_2 \langle w_{xxt}^m(t), w_{xxt}^m(t) \rangle \\ = \langle \beta(t) g(y^m(t)), w_{xxt}^m(t) \rangle + \langle f(t), w_t^m(t) \rangle_{V^*,V}.$$

Hence,
$$\frac{d}{dt}\left[\frac{1}{2}\|w_t^m(t)\|^2 + \frac{\kappa_1}{2}\|w_{xx}^m(t)\|^2\right] + \kappa_2 \|w_{xxt}^m(t)\|^2$$
$$= \langle \beta(t)g(y^m(t)), w_{xxt}^m(t)\rangle + \langle f(t), w_t^m(t)\rangle_{V^*, V}.$$

Upon integrating this equality we obtain

$$\|w_t^m(t)\|^2 + \kappa_1 \|w_{xx}^m(t)\|^2 + 2\kappa_2 \int_0^t \|w_{xx\tau}^m(\tau)\|^2 \, d\tau = \|w_1^m\|^2 + \kappa_1 \|w_{0xx}^m\|^2$$
$$+ 2\int_0^t \langle \beta(\tau)g(y^m(\tau)), w_{xx\tau}^m(\tau)\rangle d\tau + 2\int_0^t \langle f(\tau), w_\tau^m(\tau)\rangle_{V^*, V} d\tau. \qquad (2.5)$$

Now, using the assumption (A_f), the fourth term on the right hand side of (2.5) can be bounded as follows:

$$2\int_0^t \langle f(\tau), w_\tau^m(\tau)\rangle_{V^*, V} d\tau \le \delta \int_0^t \|w_{xx\tau}^m(\tau)\|^2 d\tau + \frac{1}{\delta}\int_0^t \|f(\tau)\|_{V^*}^2 \, d\tau,$$

for any $\delta > 0$. Furthermore, note that from assumption (A_g) it follows that

$$|g(y^m(t))| \le |g(y^m(t)) - g(0)| + |g(0)|$$
$$\le \frac{\widetilde{C}_1}{k_s}|y^m(t)| + \widetilde{C}_2 \qquad (2.6)$$

for some $\widetilde{C}_2 > 0$. Hence, the third term on the right hand side of (2.5) satisfies the following estimate:

$$2\int_0^t \langle \beta(\tau)g(y^m(\tau)), w_{xx\tau}^m(\tau)\rangle d\tau$$
$$\le 2\int_0^t \|\beta(\tau)\| \, |g(y^m(\tau))| \, \|w_{xx\tau}^m(\tau)\| \, d\tau$$
$$\le 2\int_0^t \left(\frac{\widetilde{C}_1}{k_s}|y^m(\tau)| + \widetilde{C}_2\right) \|\beta(\tau)\| \, \|w_{xx\tau}^m(\tau)\| \, d\tau$$
$$\le 2\int_0^t \left(\frac{\widetilde{C}_1}{k_s}k_s \|w_{xx\tau}^m(\tau)\| + \widetilde{C}_2\right) L \, \|w_{xx\tau}^m(\tau)\| \, d\tau$$
$$\le 2\int_0^t L\widetilde{C}_1 \|w_{xx\tau}^m(\tau)\|^2 \, d\tau + 2\int_0^t L\widetilde{C}_2 \|w_{xx\tau}^m(\tau)\| \, d\tau$$
$$\le 2L\widetilde{C}_1 \int_0^t \|w_{xx\tau}^m(\tau)\|^2 \, d\tau + \frac{1}{\delta}\int_0^t (L\widetilde{C}_2)^2 d\tau$$
$$+ \delta \int_0^t \|w_{xx\tau}^m(\tau)\|^2 \, d\tau.$$

Now choose δ such that
$$\delta = \frac{1}{2}(\kappa_2 - L\widetilde{C}_1).$$

Then
$$\|w_t^m(t)\|^2 + \kappa_1 \|w_{xx}^m(t)\|^2 + \left(\kappa_2 - L\widetilde{C}_1\right)\int_0^t \|w_{xx\tau}^m(\tau)\|^2 \, d\tau \leq \|w_1^m\|^2$$
$$+ \kappa_1 \|w_{0xx}^m\|^2 + \frac{2}{\kappa_2 - L\widetilde{C}_1}(L\widetilde{C}_2)^2 T + \frac{2}{\kappa_2 - L\widetilde{C}_1}\|f\|_{L^2(0,T;V^*)}.$$

Recalling that $w_0^m \to w_0$ in V, $w_1^m \to w_1$ in H as $m \to \infty$, we conclude that there exists a positive constant C independent of m such that
$$\|w_t^m(t)\|^2 + \kappa_1 \|w_{xx}^m(t)\|^2 + \left(\kappa_2 - L\widetilde{C}_1\right)\int_0^t \|w_{xx\tau}^m(\tau)\|^2 \, d\tau \leq C.$$

It follows that $\{w_t^m\}$ is bounded in $C([0,T];H)$ and in $L^2(0,T;V)$, and that $\{w^m\}$ is bounded in $C([0,T];V)$. Furthermore, from (2.6) and the boundedness of w_t^m in $L^2(0,T;V)$ it follows that there exists a positive constant M such that
$$\|g(y^m(\cdot))\|_{L^2(0,T)} \leq M.$$

Hence, there exists a subsequence $\{w^{m_k}\}$ of $\{w^m\}$ and limit functions $w \in W^{1,2}(0,T;V)$ and $\widetilde{g} \in L^2(0,T)$ such that
$$w^{m_k} \to w \text{ weakly in } W^{1,2}(0,T;V)$$
$$g(y^{m_k}) \to \widetilde{g} \text{ weakly in } L^2(0,T).$$

Note that $w(0) = w_0$.

Following [13] we fix $j < m$ and let $\eta \in C^1[0,T]$ with $\eta(T) = 0$ be arbitrarily chosen. Set $\eta_j(t) = \eta(t)\psi_j$ and multiply both sides of (2.3) by the function $\eta(t)$. Integrating over $[0,T]$ we obtain
$$\int_0^T \{\langle w_{tt}^m(t), \eta_j(t)\rangle + \kappa_1 \langle w_{xx}^m(t), \eta_{jxx}(t)\rangle + \kappa_2 \langle w_{xxt}^m(t), \eta_{jxx}(t)\rangle\} \, dt$$
$$= \int_0^T \{\langle \beta(t)g(y^m(t)), \eta_{jxx}(t)\rangle + \langle f(t), \eta_j(t)\rangle_{V^*,V}\} \, dt.$$

Using integration by parts for the first term and letting $\dot{\eta}_j = \frac{d}{dt}\eta_j$ we get

$$\int_0^T \{-\langle w_t^m(t), \dot{\eta}_j(t)\rangle + \kappa_1 \langle w_{xx}^m(t), \eta_{jxx}(t)\rangle + \kappa_2 \langle w_{xxt}^m(t), \eta_{jxx}(t)\rangle\} dt$$
$$= \int_0^T \{\langle \beta(t) g(y^m(t)), \eta_{jxx}(t)\rangle + \langle f(t), \eta_j(t)\rangle_{V^*, V}\} dt + \langle w_1^m, \eta_j(0)\rangle.$$

Using the above weak convergences and taking subsequential limits as $m = m_k \to \infty$ in the previous equation we get

$$\int_0^T \{\langle -w_t(t), \dot{\eta}_j(t)\rangle + \kappa_1 \langle w_{xx}(t), \eta_{jxx}(t)\rangle + \kappa_2 \langle w_{xxt}(t), \eta_{jxx}(t)\rangle\} dt$$
$$= \int_0^T \{\langle \beta(t) \tilde{g}(t), \eta_{jxx}(t)\rangle + \langle f(t), \eta_j(t)\rangle_{V^*, V}\} dt + \langle w_1, \eta_j(0)\rangle. \quad (2.7)$$

Recalling that $\eta_j(t) = \eta(t) w_j$ and further restricting η so that $\eta \in C_0^\infty(0, T)$, we get

$$\int_0^T \{\dot{\eta}(t) \langle -w_t(t), \psi_j\rangle + \kappa_1 \eta(t) \langle w_{xx}(t), \psi_{jxx}\rangle + \kappa_2 \eta(t) \langle w_{xxt}(t), \psi_{jxx}\rangle\} dt$$
$$= \int_0^T \eta(t) \{\langle \beta(t) \tilde{g}(t), \psi_{jxx}\rangle + \langle f(t), \psi_j\rangle_{V^*, V}\} dt.$$

This implies that for each ψ_j,

$$\frac{d}{dt}\langle w_t(t), \psi_j\rangle + \kappa_1 \langle w_{xx}(t), \psi_{jxx}\rangle + \kappa_2 \langle w_{xxt}(t), \psi_{jxx}\rangle = \langle \beta(t) \tilde{g}(t), \psi_{jxx}\rangle + \langle f(t), \psi_j\rangle_{V^*, V}.$$
$$(2.8)$$

Since ψ_j is total in V we thus have that $w_{tt} \in L^2(0, T; V^*)$ and for all $\phi \in V$,

$$\langle w_{tt}(t), \phi\rangle + \kappa_1 \langle w_{xx}(t), \phi_{xx}\rangle + \kappa_2 \langle w_{xxt}(t), \phi_{xx}\rangle = \langle \beta(t) \tilde{g}(t), \phi_{xx}\rangle + \langle f(t), \phi\rangle_{V^*, V}. \quad (2.9)$$

We already have $w(0) = w_0$ and to argue that $w_t(0) = w_1$, we return to (2.7) which holds for all $\eta_j(t) = \eta(t)\psi_j$, $\eta \in C^1[0, T]$, $\eta(T) = 0$. Integrating by parts the first term in (2.7) and using (2.8) we obtain

$$\langle -w_t(t), \eta_j(t)\rangle|_{t=0}^{t=T} = \langle w_1, \eta_j(0)\rangle.$$

From this it follows that $w_t(0) = w_1$. To prove that the limit function is indeed a weak solution left to be shown that $\tilde{g}(t) = g(y(t))$ for a.e. $t \in [0, T]$. Recall that we already proved that $g(y^m) \to \widetilde{g}$ weakly in $L^2(0,T)$ (along a subsequence). Our next goal is to show that this weak convergence is actually a strong one.

To achieve this step we follow ideas developed in [7] for linear second order (in time) evolution equations, and adopted for other nonlinear second order problems in [1, 12]. we let $z^m(t) = w^m(t) - w(t)$, where w^m is the unique solution to the finite dimensional system (2.3)-(2.4) and w is the limit function which solves the linear problem (2.9) with $w(0) = w_0$ and $w_t(0) = w_1$. Now, use the test function w^m in (2.3) and the test function w in (2.9) and add and subtract terms to obtain

$$\begin{aligned}
&\|z_t^m(t)\|^2 + \kappa_1 \|z_{xx}^m(t)\|^2 + 2\kappa_2 \int_0^t \|z_{txx}^m(\tau)\|^2 d\tau \\
&= \|w_1^m - w_1\|^2 + \kappa_1 \|w_{0xx}^m - w_{0xx}\|^2 \\
&\quad + 2\int_0^t \langle \beta(\tau)(g(y^m(\tau)) - g(y(\tau))), z_{txx}^m(\tau)\rangle d\tau \\
&\quad + 2\int_0^t \langle \beta(\tau)(g(y(\tau)) - \widetilde{g}(\tau)), z_{txx}^m(\tau)\rangle d\tau \\
&\quad + 2\int_0^t \langle f(\tau), z_\tau^m(\tau)\rangle_{V^*,V} d\tau + \Gamma^m(t).
\end{aligned} \qquad (2.10)$$

Here,

$$\begin{aligned}
\Gamma^m(t) = 2\Big[&-\langle w_t(t), w_t^m(t)\rangle - \kappa_1 \langle w_{xx}(t), w_{xx}^m(t)\rangle - 2\kappa_2 \int_0^t \langle w_{xx\tau}(\tau), w_{xx\tau}^m(\tau)\rangle d\tau \\
&+\langle w_1, w_1^m\rangle + \kappa_1 \langle w_{0xx}, w_{0xx}^m\rangle + \int_0^t \langle \beta(\tau)g(y^m(\tau)), w_{xx\tau}(\tau)\rangle d\tau \\
&+\int_0^t \langle \beta(\tau)\widetilde{g}(\tau), w_{xx\tau}^m(\tau)\rangle + 2\int_0^t \langle f(\tau), w_\tau(\tau)\rangle_{V^*,V} d\tau\Big].
\end{aligned}$$

The third term on the right-hand side of (2.10) satisfies the following estimate:

$$\begin{aligned}
2\int_0^t &\langle \beta(\tau)(g(y^m(\tau)) - g(y(\tau))), z_{txx}^m(\tau)\rangle d\tau \\
&\leq 2\int_0^t \|\beta(\tau)\| \, |g(y^m(\tau)) - g(y(\tau))| \, \|z_{\tau xx}^m(\tau)\| \, d\tau \\
&\leq 2\int_0^t \frac{\widetilde{C}_1}{k_s} |y^m(\tau) - y(\tau)| \, \|\beta(\tau)\| \, \|z_{\tau xx}^m(\tau)\| \, d\tau \\
&\leq 2\int_0^t \frac{\widetilde{C}_1}{k_s} k_s \|z_{\tau xx}^m(\tau)\| \, L \, \|z_{\tau xx}^m(\tau)\| \, d\tau \\
&\leq 2L\widetilde{C}_1 \int_0^t \|z_{\tau xx}^m(\tau)\|^2 \, d\tau.
\end{aligned}$$

From this estimate we get the following inequality:

$$\|z_t^m(t)\|^2 + \kappa_1 \|z_{xx}^m(t)\|^2 + \left(2\kappa_2 - 2L\widetilde{C}_1\right) \int_0^t \|z_{txx}^m(\tau)\|^2 d\tau \leq \|w_1^m - w_1\|^2$$
$$+ \kappa_1 \|w_{0xx}^m - w_{0xx}\|^2 + 2 \int_0^t \langle \beta(\tau)(g(y(\tau)) - \widetilde{g}(\tau)), z_{txx}^m(\tau)\rangle d\tau$$
$$+ 2 \int_0^t \langle f(\tau), z_\tau^m(\tau)\rangle_{V^*,V} d\tau + \Gamma^m(t).$$

Letting $m = m_k \to \infty$, we clearly get $\|w_1^{m_k} - w_1\|^2 + \kappa_1 \|w_{0xx}^{m_k} - w_{0xx}\|^2 \to 0$. Recalling that $z^{m_k} = w^{m_k} - w \to 0$ weakly in $W^{1,2}(0,T;V)$ and that $g(y^{m_k}) \to \widetilde{g}$ weakly in $L^2(0,T)$, we see that $\Gamma^{m_k}(t) \to 0$ because w satisfies the integrated form of (2.9). Furthermore, we also see that the third and the fourth terms on the right-hand side of the above inequality converges to 0. Hence, we have that $w^{m_k} \to w$ strongly in $C([0,T];V)$ and that $w_t^{m_k} \to w_t$ strongly in $C([0,T];H) \cap L^2(0,T;V)$. This implies that

$$\int_0^T |g(y^{m_k}(t)) - g(y(t))|^2 dt \leq \left(\frac{\widetilde{C}_1}{k_s}\right)^2 \int_0^T |y^{m_k}(t) - y(t)|^2 dt$$
$$\leq (\widetilde{C}_1)^2 \int_0^T \|w_{xxt}^{m_k} - w_{xxt}\|^2 dt \to 0.$$

Since $g(y^{m_k}) \to \widetilde{g}$ weakly in $L^2(0,T)$ also, we get that $g(y(t)) = \widetilde{g}(t)$ for a.e. $t \in (0,T)$.

Now that we have proved the existence of a weak solution to problem (1.1)-(1.2) we would like to point out that a solution to this problem satisfies the additional regularity $w \in C([0,T];V)$ and $w_t \in C([0,T];H)$ (cf. [13, p. 273], [14, Chap. 3]). To see this, observe that for any $\widetilde{g} \in L^2(0,T)$ the linear problem (2.9) has a unique solution $\widetilde{w}$ with $\widetilde{w}(0) = w_0$, $\widetilde{w}_t(0) = w_1$ satisfying $\widetilde{w} \in C([0,T];V)$ and $\widetilde{w}_t \in C([0,T];H)$ (for details see [4]). Since for $\widetilde{g}(\cdot) = g(y(\cdot)) \in L^2(0,T)$ the unique solution to the linear problem $\widetilde{w}$ coincides with the solution w to the nonlinear problem (1.1)-(1.2), the result follows.

The uniqueness of solutions can be easily established. Indeed, assume that w_1 and w_2 are two solutions to (1.1)-(1.2) and define $z = w_1 - w_2$. Then by calculations similar to the ones employed above we deduce that for each $t \in (0,T)$,

$$\|z_t(t)\|^2 + \kappa_1 \|z_{xx}(t)\|^2 + \left(2\kappa_2 - 2L\widetilde{C}_1\right) \int_0^t \|z_{txx}(\tau)\|^2 d\tau \leq 0.$$

Thus $z = 0$ and this completes the proof of Theorem 2.2. □

We conclude this section by pointing out that from the uniqueness of solutions it follows that the Galerkin approximations w^m converges to the unique solution w strongly in $C([0,T];V)$. In the next section we present a numerical solution to a model problem using the Galerkin approximation developed here.

3 Numerical Results

In this section we summarize the numerical implementation scheme and present some of our numerical findings. For the purpose of the numerical results we set

$$\beta(x,t)g(y) = \beta_1(t)\left(k_a\chi_{[x_1,x_2]}(x)\epsilon(t)\right)g(y) \text{ and } f(x,t) = [k_a\chi_{[x_1,x_2]}(x)\epsilon(t)]_{xx}.$$

Using a standard Galerkin scheme, we discretize the problem (1.1)-(1.2) in terms of spline expansions. More precisely, we use modified (for essential boundary conditions) cubic splines on the interval $(0,1)$ with respect to the uniform mesh $\{0, \frac{1}{m}, \frac{2}{m}, \ldots, 1\}$ to approximate (1.1)-(1.2). Using the notation of Section 2 we denote the 1-D cubic splines that are used to discretize (1.1) by $\{\psi_i^m\}_{i=1}^{m-1}$ and the approximating subspace by $V^m = \text{span}\{\psi_i^m\}_{i=1}^{m-1}$.

Choosing the approximate beam solution to be

$$w^m(x,t) = \sum_{i=1}^{m-1} C_i^m(t)\psi_i^m(x),$$

and restricting the infinite dimensional system to the space V^m, we arrive at equations (2.3)-(2.4) with the test function ψ_j replaced by $\psi \in V$. When the test function ψ is chosen in V^m, we obtain the finite dimensional system

$$M^m\ddot{C}^m(t) + \kappa_1 K^m C^m(t) + \kappa_2 K^m \dot{C}^m(t) = \beta_1(t)B^m g(y^m)\epsilon(t) + B^m\epsilon(t),$$

$$y^m(t) = D^m w_t^m(t), \tag{3.1}$$

$$K^m C^m(0) = w_0^m, \qquad M^m \dot{C}^m(0) = w_1^m.$$

Here the coordinate vector representation of $w^m(t,x)$ with respect to the basis $\{\psi_i^m\}_{i=1}^{m-1}$ is $C^m(t) = \begin{bmatrix} C_1^m(t) & C_2^m(t) & \ldots & C_{m-1}^m(t) \end{bmatrix}^T$, with $\dot{C}^m(t)$ represented analogously. In (3.1) the *mass* and *stiffness* matrices are given by

$$M_{ij}^m = \int_0^1 \psi_i^m(x)\psi_j^m(x)\,dx, \quad K_{ij}^m = \int_0^1 [\psi_i^m(x)]_{xx}[\psi_j^m(x)]_{xx}\,dx, \quad i,j = 1,2,\ldots,m-1, \tag{3.2}$$

respectively, and the *input* and *output* vectors by

$$B_i^m = \int_{x_1}^{x_2} k_a[\psi_i^m(x)]_{xx}\,dx, \quad D_i^m = \int_{x_1}^{x_2} k_s[\psi_i^m(x)]_{xx}\,dx, \quad i = 1,2,\ldots,m-1. \tag{3.3}$$

The initial conditions w_0^m, w_1^m are given by

$$[w_0^m]_i = \int_0^1 w_{0xx}(x)[\psi_i^m(x)]_{xx}\,dx \qquad [w_1^m]_i = \int_0^1 w_1(x)\psi_i^m(x)\,dx, \quad i = 1,\ldots,m-1.$$

We first tested our numerical scheme for various values of the discretization index $m = 4, 8, 16, 32, 64$ and 128. To do so, we set $\beta = 0$ and choose the forcing function in (1.1) to be

$$f(x,t) = -\sin(t)x^2(x-1)^2 + 24\kappa_1 \sin(t) + 24\kappa_2 \cos(t).$$

Then one can easily verify that $w(x,t) = \sin(t)x^2(x-1)^2$ solves (1.1)-(1.2). A finite dimensional system can be obtained for this choice of forcing function f using the same technique which led to (3.1). The approximation errors and the percentage values of the approximation errors are depicted in Figure 1. It can be observed that the approximate solutions w^m converge to the true solution w with order $1/m^2$ for the L_2 norm and $1/\sqrt{m}$ for the H_0^2 norm.

The system with a nonzero β and $\epsilon(t) = \sin(2t)$ is simulated for $m = 128$ in which the nonlinear function $g(y)$ is chosen as

$$g(y) = \frac{25y \sin(y)}{1 + y^2},$$

and where the plant parameters are given by $\kappa_1 = 1$, $\kappa_2 = 0.001$, $k_a = 3.5 \times 10^{-3}$, $k_s = 3.5 \times 10^{-4}$, $x_1 = 0.45$, $x_2 = 0.55$ and $T_f = 0.1$. The above choice satisfies the

Second Order Nonlinear Nonlocal Hyperbolic Equation

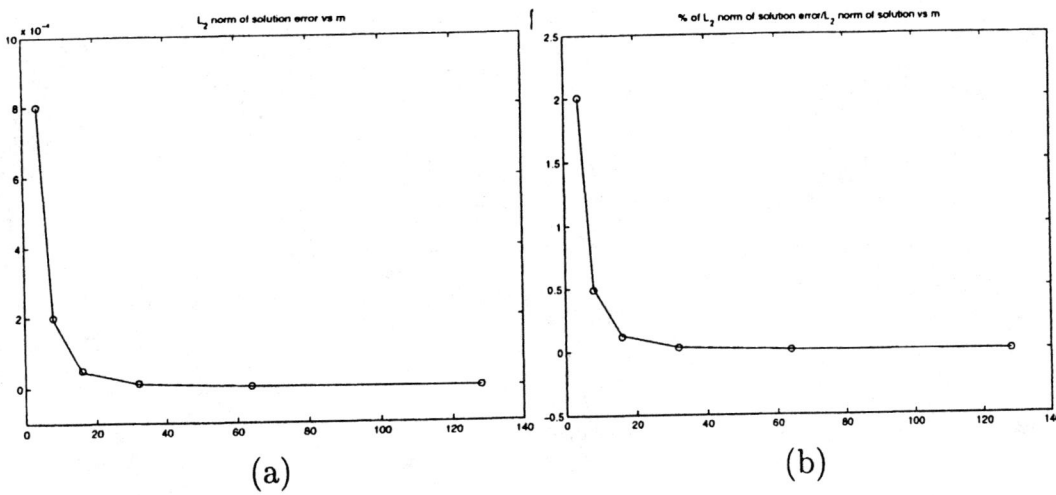

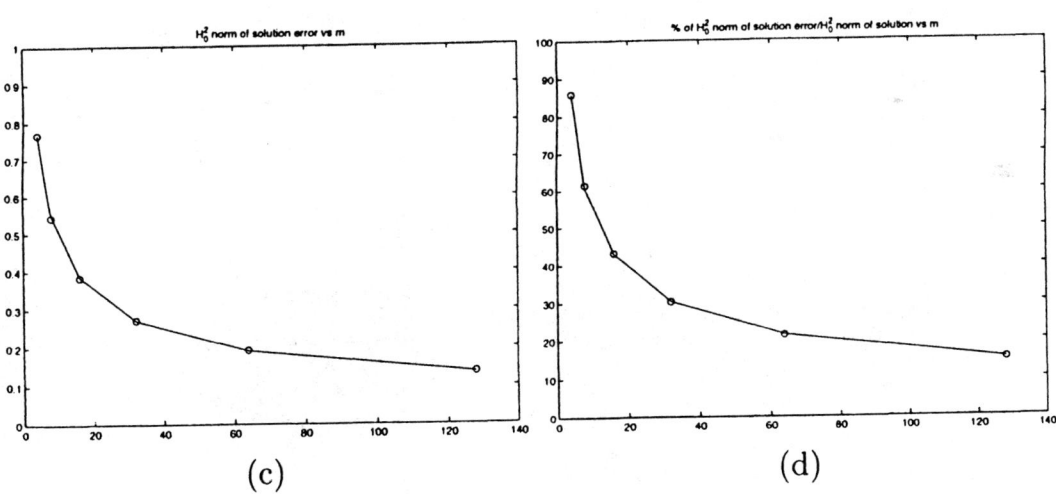

Figure 1: Evolution of (a) L_2 and (c) H_0^2 approximation error norms, and (b) L_2 and (d) H_0^2 percentage approximation error norms vs. m.

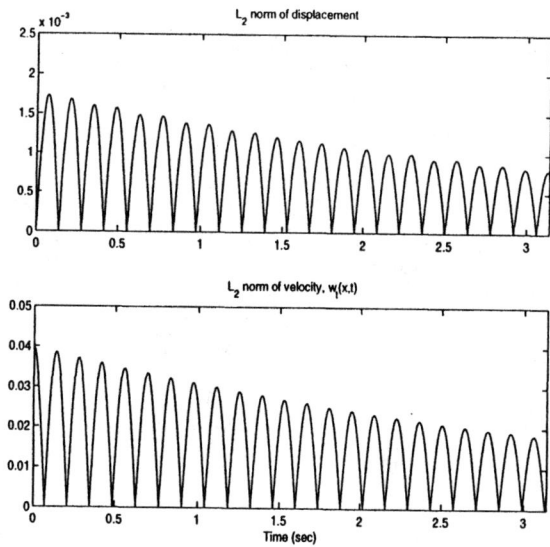

Figure 2: The L^2 norm of $w(t)$ (upper) and $w_t(t)$ (lower).

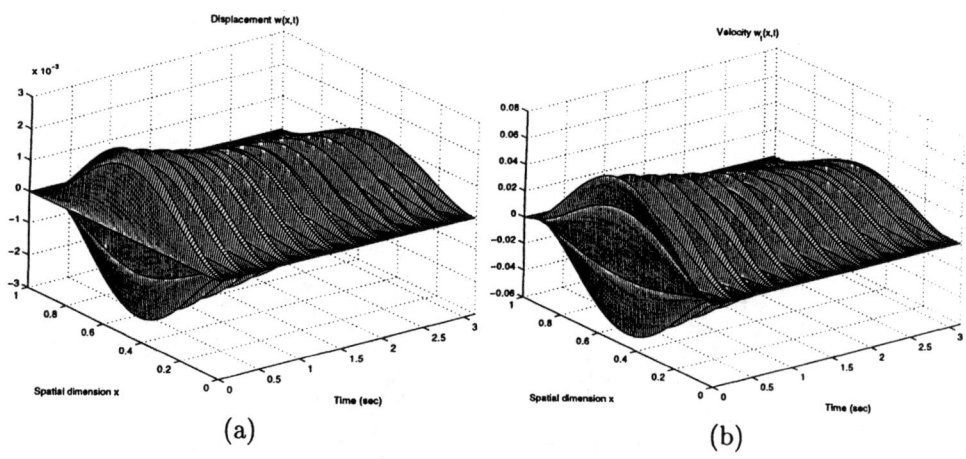

Figure 3: 3-D mesh of (a) $w(x,t)$ and (b) $w_t(x,t)$.

assumptions (A_β) and (A_g). The L_2 norm of the solution $w(t)$ and its derivative $w_t(t)$ are depicted in Figure 2 and the 3-D graphs of w and w_t are presented in Figure 3.

Acknowledgments: The work of S. Reich was partially supported by the Fund for the Promotion of Research at the Technion and by the Technion VPR Fund - E. and M. Mendelson Research Fund.

References

[1] A.S. Ackleh, H.T. Banks and G.A. Pinter, *On a nonlinear beam equation*, Applied Mathematics Letters, to appear.

[2] H. T. Banks, D. S. Gilliam, and V. I. Shubov, *Well-posedness for a one dimensional nonlinear beam*, in **Computation and Control, IV (Bozeman, MT, 1994), Progr. Systems Control Theory, vol. 20,** (Birkhäuser, 1995), pp. 1–21.

[3] H. T. Banks, D. S. Gilliam, and V. I. Shubov, *Global solvability for damped abstract nonlinear hyperbolic systems*, Differential and Integral Equations, 10 (1997), 309–332.

[4] H. T. Banks, K. Ito, and Y. Wang, *Well-posedness for damped second order systems with unbounded input operators*, Differential and Integral Equations, 8 (1995), 587-606.

[5] H. T. Banks, R. C. Smith, and Y. Wang, **Smart Material Structures: Modeling, Estimation and Control** (Wiley-Masson, 1996).

[6] V. Barbu, **Nonlinear Semigroups and Differential Equations in Banach Spaces** (Noordhoff International Publishing, 1976).

[7] R. Dautray and J. L. Lions, **Mathematical Analysis and Numerical Methods for Science and Technology, Volume 5, Evolution Problems I** (Springer, 1993).

[8] M. A. Demetriou, A. S. Ackleh and S. Reich, *Detection and accommodation of second order distributed parameter systems with abrupt changes in the input term: existence and approximation*, Kybernetika, 36 (2000), 117-132.

[9] M. A. Demetriou and M. M. Polycarpou, **Fault accommodation of output-induced actuator failures for a flexible beam with collocated input and output, in Proceedings of the 5th IEEE Mediterranean Conference on Control and Systems**, Phaethon Beach Hotel, Paphos, Cyprus, July 21-23, 1997, CD-ROM publication.

[10] M. A. Demetriou and M. M. Polycarpou, **Fault diagnosis of output-induced actuator failures for a flexible beam with collocated input and output, in Proceedings of the IFAC Symposium on Fault Detection, Supervision and Safety for Technical Processes: SAFEPROCESS'97**, University of Hull, Hull, United Kingdom, August 26-28, 1997.

[11] J. Dosch, D. J. Inman, and E. Garcia, *A self-sensing piezoelectric actuator for collocated control*, Journal of Intelligent Material Systems and Structures, 3 (1992), 166–185.

[12] J. Ha and S. Nakagiri, *Existence and regularity of weak solutions for second order semilinear evolution equations*, Funkcialaj Ekvacioj, 41 (1998), 1-24.

[13] J. L. Lions, **Optimal Control of Systems Governed by Partial Differential Equations** (Springer, 1971).

[14] J. L. Lions and E. Magenes, **Non-Homogeneous Boundary Value Problems and Applications** (Springer, 1972).

[15] M. M. Polycarpou and A. J. Helmicki, *Automated fault detection and accomodation: A learning systems approach*, IEEE Trans. on Systems, Man and Cybernetics, 25 (1995), 1447–1458.

[16] J. Wloka, **Partial Differential Equations** (Cambridge University Press, 1987).

Fully Nonlinear Programming Problems with Closed Range Operators

Sergiu Aizicovici   Ohio University, Athens, Ohio

D. Motreanu   University of Iaşi, Iaşi, Romania

Nicolae H. Pavel   Ohio University, Athens, Ohio

Abstract

The paper provides necessary conditions for the optimality of a pair (y, u) with respect to a locally Lipschitz cost functional $L(y, u)$, subject to $Ay = Cu + B(y, u)$. Here A and C are closed range, densely defined linear operators on some Banach spaces Y and X, while B is a (Gâteaux) differentiable map on $Y \times X$. This extends the result in [1], where the case $B(y, u) = B(u) - F(y)$, with B and F Fréchet differentiable, was studied.

1 Introduction and Main Results

Let X, Y and E be real Banach spaces. Let $A : D(A) \subset Y \to E$ and $C : D(C) \subset X \to E$ be (possibly unbounded) closed linear operators with dense domains $D(A)$ and $D(C)$ in Y and X, respectively. We also consider a (Gâteaux) differentiable map $B : V \times U \to E$, where U and V are open subsets of X and Y, respectively, and a locally Lipschitz function $L : W \to \mathbb{R}$ on an open subset W of $Y \times X$ containing $V \times U$.

This paper is devoted to the following nonlinear programming problem:

(P)
$$\text{(Locally) Minimize } L(y, u)$$
$$\text{subject to } Ay = Cu + B(y, u).$$

The set of constraints for our problem (P) is

$$M = \{(y, u) \in (D(A) \cap V) \times (D(C) \cap U) : Ay = Cu + B(y, u)\}. \quad (1)$$

We assume $M \neq \emptyset$. The tangent cone $T_{(y,u)}M$ of the set M in (1) at a point $(y, u) \in M$ is given by

$$\begin{aligned} T_{(y,u)}M = \{(z, w) \in D(A) \times D(C) : &\exists\, p(t) \to 0 \text{ in } Y \\ &\text{and } q(t) \to 0 \text{ in } X \text{ as } t \to 0^+ \text{ such that} \\ &(y + t(z + p(t)), u + t(w + q(t))) \in M\} \end{aligned} \quad (2)$$

(see Aizicovici, Motreanu and Pavel [1] or Motreanu and Pavel [5], Pavel [6]). In what follows, the superscript $*$ denotes the adjoint of a linear operator. The generalized directional derivative of $L(y, u)$, the generalized gradient of L and the partial generalized gradients of $L(y, u)$ with respect to the variables y and u (in the sense of Clarke [3]) are denoted by $L^0(y, u), \partial L(y, u)$, $\partial_y L(y, u)$ and $\partial_u L(y, u)$, respectively (unless otherwise specified). Finally, $B'(y, u)$, $B_y(y, u)$ and $B_u(y, u)$ denote the Gâteaux derivative and the partial Gâteaux derivatives of $B(y, u)$, respectively.

The basic hypotheses are the following:

(H_1) If $(y, u) \in M$, $z \in D(A)$ and $w \in D(C)$ satisfy

$$Az = Cw + B'(y, u)(z, w), \quad (3)$$

then $(z, w) \in T_{(y,u)}M$.

(H_2) For all $(y, u) \in M$, $R(A - B_y(y, u))$ and $R(C + B_u(y, u))$ are closed in E and either

$$R(A - B_y(y, u)) \subset R(C + B_u(y, u)) \quad (4)$$

or

$$R(C + B_u(y, u)) \subset R(A - B_y(y, u)). \quad (5)$$

Remark 1 *For our main result (Theorem 1) it is sufficient to assume that (H_1) and (H_2) hold for optimal pairs (y, u) of problem (P), only.*

The following simple Lemma will be essentially used in the proof of our main result.

Lemma 1. *Let $F : U \to R$ be a locally Lipschitz function on an open subset of a Banach space X, X_1 a subspace of X and let G be a linear functional on X_1 such that:*
$(1^o)\quad G(v) \leq F^0(x;v), \quad \forall v \in X_1$
where $F^0(x;v)$ is Clarke's generalized directional derivative of F at x in the direction v. Then there is an extension $G_* \in X^*$ of G, i.e. $G_*(v) = G(v)$, $\forall v \in X_1$, with
$(2^o)\quad G_*(v) \leq F^0(x;v), \quad \forall v \in X,$
i.e. $G_* \in \partial F(x)$.

Proof. Recall that $v \to F^0(x;v)$ is positively homogeneous and subadditive on X, with $|F^0(x;v)| \leq l|v|$, $\forall v \in X$, where l is the Lipschitz constant of F near x. By the Hahn-Banach theorem, there is a linear extension G_* of G from X_1 to all of X satisfying Inequality (2°). But $F^0(x;v)$ is continuous at $v = 0$ and we clearly have

$$-F^0(x;-v) \leq G_*(v) \leq F^0(x;v), \quad \forall v \in X$$

which implies the continuity of G_* at zero. □

We now state our main result concerning the necessary conditions of optimality for problem (P).

THEOREM 1. *Under assumptions (H_1) and (H_2), for every (locally) optimal pair $(y,u) \in M$ of problem (P) there exists $\bar{p} \in D(A^*) \cap D(C^*)$ such that*

$$(-(A - B_y(y,u))^*\bar{p}, (C + B_u(y,u))^*\bar{p}) \in \partial L(y,u). \tag{6}$$

If, in addition, L is regular at (y,u) (in the sense of Clarke [3], p. 39), then one has

$$(-(A - B_y(y,u))^*\bar{p} \in \partial_y L(y,u) \text{ and } (C + B_u(y,u))^*\bar{p} \in \partial_u L(y,u). \tag{7}$$

PROOF. First we show that if $(z,w) \in D(A) \times D(C)$ satisfies (3), then

$$L^0(y,u;z,w) \geq 0. \tag{8}$$

Indeed, we know by assumption (H_1) that $(z,w) \in T_{(y,u)}M$. Consequently, the functions $p(t)$ and $q(t)$ exist as stated in (2). The (local) optimality of (y,u) in problem (P) ensures that

$$L(y+t(z+p(t)), u+t(w+q(t))) \geq L(y,u)$$

for all $t > 0$ small enough. Then we derive that

$$0 \leq \limsup_{t \to 0^+} \frac{1}{t}[L(y+t(z+p(t)), u+t(w+q(t))) - L(y,u)]$$

$$\leq \limsup_{t \to 0^+} \frac{1}{t}[L(y+t(z+p(t)), u+t(w+q(t))) - L(y+tp(t), u+tq(t))]$$

$$+ \limsup_{t \to 0^+} \frac{1}{t}[L(y+tp(t)), u+tq(t))) - L(y,u)]$$

$$\leq L^0(y,u;z,w) + l \lim_{t \to 0^+} \|(p(t),q(t))\|_{Y \times X} = L^0(y,u;z,w),$$

for the constant $l > 0$ in the proof of Lemma 1. Here we have used the definition of $L^0(y,u;z,w)$ and the fact that L is locally Lipschitz near (y,u) in $Y \times X$. The claim in (8) is thereby verified.

On the basis of Lemma 1, Inequality (8) implies the existence of $(\xi, \eta) \in Y^* \times X^*$ such that

$$(\xi, \eta) \in \partial L(y,u), \quad \text{and} \quad \langle \xi, z \rangle_{Y^*,Y} + \langle \eta, w \rangle_{X^*,X} = 0 \qquad (9)$$

whenever $(z,w) \in D(A) \times D(C)$ satisfies (3). Indeed, set $X_1 = \{(z,w) \in D(A) \times D(C), \text{ satisfying condition (3)}\}$.
By (H_1), $X_1 \subset T_{(y,u)}M$, so (9) follows from Lemma 1 with G - the null functional on X_1.

If we set $w = 0$ in (9) we find

$$\langle \xi, z \rangle_{Y^*,Y} = 0, \quad \forall z \in D(A) \text{ with } Az = B_y(y,u)z,$$

which can be expressed as

$$\xi \in N(A - B_y(y,u))^\perp. \qquad (10)$$

By (H_2), $R(A - B_y(y,u))$ is closed in E and therefore

$$R((A - B_y(y,u))^*) = N(A - B_y(y,u))^\perp \qquad (11)$$

(see, e.g., Brézis [3], p. 29). Then (10) and (11) imply

$$\xi \in R((A - B_y(y, u))^*). \tag{12}$$

Setting now $z = 0$ in (9), we obviously get

$$\langle \eta, w \rangle_{X^*, X} = 0, \ \forall w \in D(B) \text{ with } Cw + B_u(y, u)w = 0.$$

Reasoning as above we see that

$$\eta \in R((C + B_u(y, u))^*). \tag{13}$$

By (12) and (13) there exist $\tilde{p} \in D(A^*) \subset E^*$ and $\bar{p} \in D(C)^* \subset E^*$ such that

$$(A - B_y(y, u))^* \tilde{p} = \xi, \quad (C + B_u(y, u))^* \bar{p} = \eta, \tag{14}$$

so (9) can be rewritten as

$$\langle \tilde{p}, (A - B_y(y, u))z \rangle_{E^*, E} + \langle \bar{p}, (C + B_u(y, u))w \rangle_{E^*, E} = 0 \tag{15}$$

for all $(z, w) \in D(A) \times D(C)$ satisfying (3).

Assume that Condition (4) in (H_2) holds. Then for each $z \in D(A)$ there is $w \in D(C)$ such that (3) holds. Therefore (15) leads to

$$\langle \tilde{p} + \bar{p}, (A - B_y(y, u))z \rangle_{E^*, E} = 0, \quad \forall z \in D(A). \tag{16}$$

Equivalently, (16) expresses that

$$\tilde{p} + \bar{p} \in R(A - B_y(y, u))^\perp. \tag{17}$$

Making use of the formula

$$R(A - B_y(y, u))^\perp = N((A - B_y(y, u))^*),$$

(see, e.g., Brézis [3], p. 28), from (17) one obtains that

$$\tilde{p} + \bar{p} \in N((A - B_y(y, u))^*), \tag{18}$$

so, by the first equality in (14),

$$(A - B_y(y, u))^* \bar{p} = -(A - B_y(y, u))^* \tilde{p} = -\xi. \tag{19}$$

This, in conjunction with (14) and (9), ensures that

$$(-(A - B_y(y,u))^*\bar{p}, (C + B_u(y,u))^*\bar{p}) = (\xi, \eta) \in \partial L(y,u) \qquad (20)$$

i.e., the conclusion (6) of Theorem 1 is satisfied.

Assume now that (5) is fulfilled. We proceed following the same arguments as in case (4). We note that relations (9-15) remain valid because they are independent of the assumptions (4) and (5). Now, in view of (5), for each $w \in D(C)$ there is $z \in D(A)$ such that (3) holds. Then (15) guarantees that

$$\langle \tilde{p} + \bar{p}, (C + B_u(y,u))w \rangle_{E^*, E} = 0, \quad \forall w \in D(C). \qquad (21)$$

Notice that (21) means that

$$\tilde{p} + \bar{p} \in R(C + B_u(y,u))^\perp. \qquad (22)$$

Combining (22) and the equality

$$R(C + B_u(y,u))^\perp = N((C + B_u(y,u))^*),$$

we find that

$$\tilde{p} + \bar{p} \in N((C + B_u(y,u))^*). \qquad (23)$$

By (23) and (14) it follows that

$$(C + B_u(y,u))^*\tilde{p} = -(C + B_u(y,u))^*\bar{p} = -\eta. \qquad (24)$$

Using (9), (14) and (24) we arrive at

$$(-(A - B_y(y,u))^*(-\tilde{p}), (C + B_u(y,u))^*(-\tilde{p})) = (\xi, \eta) \in \partial L(y,u). \qquad (25)$$

It is clear that (25) proves (6) with $\bar{p} = -\tilde{p}$.

Finally, if the regularity of L at (y, u) is satisfied, then

$$\partial L(y, u) \subset \partial_y L(y, u) \times \partial_u L(y, u)$$

(see Clarke [3], p. 48). Thus property (7) follows from (6) which completes the proof of Theorem 1. □

We present now a sufficient condition to have optimal pairs for problem (P).

THEOREM 2. *Assume that X, Y, E are real Banach spaces, with X and Y reflexive, $A : D(A) \subset Y \to E$ is a closed, densely defined linear operator, $C \in L(X, E)$ and $B : Y \times X \to E$ is weakly-weakly sequentially continuous and (Gâteaux) differentiable. Next suppose that the set $M := \{(y, u) \in D(A) \times X : Ay = Cu + B(y, u)\}$ is nonempty. Let $L : Y \times X \to \mathbb{R}$ be a functional which is weakly lower semicontinuous and bounded from below on M, and for which there exists an $\varepsilon > 0$ such that the set*

$$\{(y, u) \in M : L(y, u) \leq \inf_M L + \varepsilon\} \tag{26}$$

is bounded. (In particular, the last condition is satisfied if L is coercive on M). Then problem (P) admits a solution, that is, there exists $(y, u) \in M$ such that $L(y, u) = \inf_M L$.

PROOF. Let $(y_n, u_n) \in M$ be a minimizing sequence for problem (P), which means that

$$Ay_n = Cu_n + B(y_n, u_n) \tag{27}$$

and

$$L(y_n, u_n) \to \inf_M L. \tag{28}$$

The boundedness of the set introduced in (26) and the convergence in (28) imply that (y_n, u_n) is a bounded sequence. Then the reflexivity of the spaces X and Y ensures that along a subsequence one has

$$y_n \to y \text{ weakly in } Y \text{ and } u_n \to u \text{ weakly in } X, \tag{29}$$

for some element $(y, u) \in Y \times X$. Then, from (29), we derive that $C(u_n) \to C(u)$ weakly in E and $B(y_n, u_n) \to B(y, u)$ weakly in E. We see from (27) that

$$Ay_n \to Cu + B(y, u) \text{ weakly in } E. \tag{30}$$

Since the operator A is linear and closed, it follows that its graph is weakly closed. This enables us to conclude from (29) and (30) that $(y, u) \in M$. The weak lower semicontinuity of L (at (y, u)) and property (28) complete the proof of Theorem 2. □

2 An Example

Let Ω be a bounded domain in $\mathbb{R}^N$, with a smooth boundary $\partial\Omega$. We consider the nonlinear Dirichlet problem

$$\begin{cases} -\Delta y = f(y+u) & \text{in } \Omega \\ y = 0 & \text{on } \partial\Omega, \end{cases}$$

where f is a C^1 diffeomorphism of $\mathbb{R}$ satisfying the conditions
(i) $|f'(t)| \leq K < \lambda_1$, $\forall t \in \mathbb{R}$;
(ii) $|(f^{-1})'(t)| \leq K_0$, $\forall t \in \mathbb{R}$.
Here K and K_1 are positive constants and λ_1 stands for the first eigenvalue of $-\Delta$ on $H_0^1(\Omega)$.

To apply Theorem 1 we take $E = X = Y = U = V = L^2(\Omega)$, $A = -\Delta :$ $H_0^1(\Omega) \cap H^2(\Omega) \subset L^2(\Omega) \to L^2(\Omega)$, $C = 0$ and define $B : L^2(\Omega) \times L^2(\Omega) \to L^2(\Omega)$ by $B(y,u) = f(y+u)$. The mapping B is well defined since by condition (i) above one has

$$|f(t)| \leq K|t| + K_1, \ \forall t \in \mathbb{R}, \tag{31}$$

for some constant $K_1 > 0$.

Let us check that B is Gâteaux differentiable. Indeed, it is seen that

$$\frac{1}{t}\|f(y+u+t(z+w)) - f(y+u) - f'(y+u)(z+w)\|_{L^2(\Omega)}$$

$$= \frac{1}{t}\left(\int_\Omega \left|\int_0^1 \frac{d}{d\tau}f(y+u+t\tau(z+w))d\tau - f'(y+u)(z+w)\right|^2 dx\right)^{\frac{1}{2}}$$

$$= \left(\int_\Omega (z+w)^2 \left|\int_0^1 (f'(y+u+t\tau(z+w)) - f'(y+u))d\tau\right|^2 dx\right)^{\frac{1}{2}}$$

for all $(y,u), (z,w) \in L^2(\Omega) \times L^2(\Omega)$ and $t \in \mathbb{R} \setminus \{0\}$. The Lebesgue Dominated Convergence Theorem and assumption (i) ensure that the right hand side of the above equality converges to 0 as $t \to 0$.

Moreover, the map $(z,w) \in L^2(\Omega) \times L^2(\Omega) \mapsto f'(y+u)(z+w) \in L^2(\Omega)$ is linear and continuous. The continuity follows from the inequalities below

$$\|f'(y+u)(z+w)\|_{L^2(\Omega)} \leq K\|z+w\|_{L^2(\Omega)} \leq K(\|z\|_{L^2(\Omega)} + \|w\|_{L^2(\Omega)}).$$

Therefore the mapping $B : L^2(\Omega) \times L^2(\Omega) \to L^2(\Omega)$ is Gâteaux differentiable, with $B'(y,u)(z,w) = f'(y+u)(z+w)$.

We now show that Condition (H_1) is satisfied. Observe that in our setting, the set

$$M = \{(y,u) \in (H_0^1(\Omega) \cap H^2(\Omega)) \times L^2(\Omega) : -\Delta y = f(y+u)\}$$

is nonempty (cf. [4, Chapter 3]). Hence, to verify (H_1), let $y, z \in H_0^1(\Omega) \cap H^2(\Omega)$ and $u, w \in L^2(\Omega)$ satisfy

$$-\Delta y = f(y+u), \quad -\Delta z = f'(y+u)(z+w) \tag{32}$$

and for any $\varepsilon > 0$ define $\rho : [0, \varepsilon] \to L^2(\Omega)$ by $\rho(0) = 0$ and

$$\rho(t) = \frac{1}{t}(f^{-1}(f(y+u) + tf'(y+u)(z+w)) - y - u - t(z+w)), \; t \in (0, \varepsilon].$$

Then, by the Mean Value Theorem for f^{-1}, one obtains that

$$\|\rho(t)\|_{L^2(\Omega)} =$$

$$\frac{1}{t}(\int_\Omega |f^{-1}(f(y+u) + tf'(y+u)(z+w)) - f^{-1}(f(y+u+t(z+w)))|^2 dx)^{\frac{1}{2}}$$

$$\leq \frac{1}{t} K_0 \left(\int_\Omega |f(y+u) + tf'(y+u)(z+w) - f(y+u+t(z+w))|^2 dx \right)^{\frac{1}{2}}.$$

The Lebesgue Dominated Convergence Theorem yields

$$\rho(t) \to 0 \text{ in } L^2(\Omega) \text{ as } t \to 0^+.$$

This is true because

$$\frac{1}{t}|f(y+u) + tf'(y+u)(z+w) - f(y+u+t(z+w))|^2$$

$$= |f'(y+u)(z+w) - \frac{1}{t}(f(y+u+t(z+w)) - f(y+u))|^2$$

$$\to |f'(y+u)(z+w) - f'(y+u)(z+w)|^2 = 0 \text{ for almost all } x \in \Omega \text{ as } t \to 0^+$$

and by the Mean Value Theorem and assumption (i) we have

$$\frac{1}{t}|f(y+u) + tf'(y+u)(z+w) - f(y+u+t(z+w))|^2$$

$$\leq (|f'(y+u)(z+w)| + K|z+w|)^2 \leq 4K^2|z+w|^2 \in L^1(\Omega).$$

By (32) and the definition of ρ, we have

$$-\Delta(y+tz) = f(y+u)+tf'(y+u)(z+w) = f\left((y+u) + t(z+w+\rho(t))\right), \forall t \in [0,\varepsilon]$$

This is because

$$(y+u) + t(z+w+\rho(t)) = f^{-1}[f(y+u) + tf'(y+u)(z+w)]).$$

Therefore the assertion (H_1) holds true with $p(t) = 0$ and $q(t) = \rho(t)$.

To prove that condition (H_2) holds, we first verify that

$$R(A - B_y(y,u)) = L^2(\Omega).$$

To this end, we show that the problem

$$\begin{cases} -\Delta z - f'(y+u)z = f & \text{in } \Omega \\ z = 0 & \text{on } \partial\Omega \end{cases}$$

has a unique solution $z \in H_0^1(\Omega) \cap H^2(\Omega)$, for every $f \in L^2(\Omega)$. Actually, since

$$\frac{\|\nabla z\|_{L^2(\Omega)}^2}{\|z\|_{L^2(\Omega)}^2} \geq \lambda_1, \forall z \in H_0^1(\Omega),$$

from assumption (i) we have the estimate

$$\int_\Omega (|\nabla z|^2 - f'(y+u)z^2)dx \geq \|\nabla z\|_{L^2(\Omega)}^2 - K\|z\|_{L^2(\Omega)}^2$$

$$\geq \|\nabla z\|_{L^2(\Omega)}^2 - K\lambda_1^{-1}\|\nabla z\|_{L^2(\Omega)}^2 = (1 - K\lambda_1^{-1})\|\nabla z\|_{L^2(\Omega)}^2, \forall z \in H_0^1(\Omega).$$

Consequently, we can apply the Lax-Milgram Theorem to deduce the existence and the uniqueness of the solution z as desired.

To complete the proof of condition (H_2) it remains to justify that $R(C + B_u(y,u)) = R(B_u(y,u))$ is closed in $L^2(\Omega)$. Actually $R(B_u(y,u)) = L^2(\Omega)$. Indeed, for any $g \in L^2(\Omega)$, $\frac{1}{f'(y+u)}g \in L^2(\Omega)$ as well. This is because

$$\left|\frac{1}{f'(y+u)}\right| = |(f^{-1})'(f(y+u))| \leq K_0, \quad \text{a.e. in } \Omega.$$

Thus,
$$g = B_u(y,u)\left(\frac{1}{f'(y+u)}g\right), \quad \text{i.e. } g \in R(B_u(y,u)).$$

As a specific example of a function f satisfying the requirements in our example, we indicate
$$f(t) = kt + \varepsilon g(t), \quad \forall t \in \mathbb{R},$$
where $g \in C^1(\mathbb{R})$ is such that $\max\{|g(t)|, |g'(t)|\} \leq c$, $\forall t \in \mathbb{R}$, with constants $0 < |k| < \lambda_1$ and $c > 0$. Then, if $\varepsilon > 0$ is sufficiently small, the conditions imposed in our example for f are satisfied.

Assuming that the hypotheses in our example are fulfilled, Theorem 1 enables us to find necessary optimality conditions for the problem:

Minimize $\int_\Omega \int_0^{y(x)} g(x,t)dx\, dt + \frac{1}{2}\int_\Omega u^2 dx$

subject to
$$\begin{cases} -\Delta y = f(y+u) & \text{in } \Omega \\ y = 0 & \text{on } \partial\Omega. \end{cases}$$

Here f satisfies (i) and (ii), and $g \in C(\bar\Omega \times R; R)$ satisfies the growth condition
$$|g(x,t)| \leq c_0(1+|t|^{p-1}), \quad \forall (x,t) \in \Omega \times R,$$
with $c_0 > 0$, $2 < p < 2N/(N-2)$, if $N \geq 3$, and any $p > 2$, if $N = 1, 2$. According to Theorem 1, if $(\bar y, \bar u)$ is an optimal pair there exists $\bar p \in H_0^1(\Omega) \cap H^2(\Omega)$ satisfying
$$\begin{cases} -\Delta \bar p - f'(\bar y + \bar u)\bar p = -g(x,\bar y) \\ f'(\bar y + \bar u)\bar p = \bar u \\ -\Delta \bar y = f(\bar y + \bar u). \end{cases}$$

We get that
$$\begin{cases} -\Delta \bar p - \bar u = -g(x,\bar y) \\ f'(\bar y + \bar u)\bar p = \bar u \\ -\Delta \bar y = f(\bar y + \bar u). \end{cases}$$

From this we deduce that
$$\frac{\bar u}{f'(\bar y + \bar u)} \in H_0^1(\Omega) \cap H^2(\Omega)$$
and the optimal pair $(\bar y, \bar u)$ must solve the system
$$\begin{cases} -\Delta \bar y = f(\bar y + \bar u) \\ -\Delta\left(\frac{\bar u}{f'(\bar y + \bar u)}\right) = -g(x,\bar y) + \bar u. \end{cases}$$

References

[1] S. Aizicovici, D. Motreanu and N. H. Pavel (1999). Nonlinear programming problems associated with closed range operators. Appl. Math. Optimiz. 40:211-228.

[2] H. Brézis (1992). Analyse Fonctionnelle. Théorie et Applications. Masson. Paris.

[3] F. H. Clarke (1983). Optimization and Nonsmooth Analysis. John Wiley and Sons, New York.

[4] D.G. De Figueiredo (1989). The Ekeland Variational Principle with Applications and Detours, Springer, Berlin.

[5] D. Motreanu and N. H. Pavel (1999). Tangency, Flow-Invariance for Differential Equations and Optimization Problems, Marcel Dekker, Vol. 219, New York.

[6] N. H. Pavel (1984). Differential Equations, Flow-Invariance and Applications. Pitman Res. Notes Math. 113, London.

Internal Stabilization of the Diffusion Equation

Laura-Iulia Aniţa[1] and Sebastian Aniţa[2]

[1]Faculty of Physics, University "Al.I. Cuza",
Iaşi 6600, Romania
and
[2]Faculty of Mathematics, University "Al.I. Cuza",
Iaşi 6600, Romania

ABSTRACT: In this paper we analyze a stabilizability problem for the diffusion model. We provide results of stabilizability based on spatially localized control, i.e., we show that it is possible to diminish exponentially the density of a diffusive gas, by acting in a nonempty and large enough subset of the spatial domain.

1. INTRODUCTION AND SETTING OF THE PROBLEM

We consider a general mathematical model describing the diffusion of a gas in a bounded domain $\Omega \subset \mathbf{R}^n$ ($n \in \mathbf{N}^*$) with a smooth boundary $\partial\Omega$. Let $y(x,t)$ be the density of the gas in the position $x \in \overline{\Omega}$ at the moment $t \geq 0$. The diffusion is described by the following system:

$$\begin{cases} y_t - \Delta y + a(x)y = m(x)u(x,t), & (x,t) \in Q_T = \Omega \times (0,T) \\ y(x,t) = 0, & (x,t) \in \Sigma_T = \partial\Omega \times (0,T) \\ y(x,0) = y_0(x), & x \in \Omega \end{cases} \quad (1.1)$$

($T \in [0, +\infty]$). Here m is the characteristic function of $\overline{\omega}$, where ω is a nonempty open subset satisfying $\omega \subset\subset \Omega$ and $u(x,t)$ is a control function. So, this is the case when the control acts only on a subset of Ω.

We assume that

$$a \in L^\infty(\Omega), \quad y_0 \in L^\infty(\Omega), \quad y_0(x) > 0 \text{ a.e. } x \in \Omega.$$

It is well known that for any $T \in (0, +\infty)$, there exists $u \in L^2(Q_{\omega,T})$ ($Q_{\omega,T} = \omega \times (0,T)$) such that

$$y^u(x,T) = 0 \quad \text{a.e. } x \in \Omega,$$

where y^u is the solution of (1.1). This means that system (1.1) is exact null controllable. For this we refer to G. Lebeau and L. Robbiano [8].

On the other hand the exact controllability of (1.1) implies the stabilizability (in $L^2(\Omega)$) and that there exists a stabilizing feedback control

$$u(x,t) = -m(x)(Py(t))(x), \tag{1.2}$$

where P is the solution to a certain algebraic Riccati equation (see J.L. Lions [9] and J. Zabczyk [10]). The problem is that the solution of (1.1) corresponding to the feedback control given by (1.2) is not necessary nonnegative (as should be the density $y(x,t)$).

For the exact controllability of the semilinear heat equation we refer to [5] and [7]. The null controllability of a reaction-diffusion system has been investigated in [2].

The stabilization of a certain reaction-diffusion system was studied by L.I. Aniţa in [1].

Our goal is to find necessary and sufficient condtions for "nonnegative" stabilizability (stabilizability with preservation of the nonnegativity of the solution y to (1.1)) and in the affirmative case to indicate a feedback control.

The plan of the paper is the following. In Section 2 we shall present the main result. Section 3 is devoted to some final remarks.

2. THE MAIN RESULT

Denote by λ_1^ω the first eigenvalue corresponding to the following problem:

$$\begin{cases} -\Delta\varphi(x) + a(x)\varphi(x) = \lambda\varphi(x), & x \in \Omega \setminus \overline{\omega} \\ \varphi(x) = 0, & x \in \partial\Omega \cup \partial\omega. \end{cases} \tag{2.1}$$

Internal Stabilization of Diffusion Equation

Theorem 2.1. *System (1.1) is "nonnegative" stabilizable if and only if $\lambda_1^\omega > 0$.*

Proof. First we shall prove that $\lambda_1^\omega \leq 0$ implies that (1.1) is not "nonnegative" stabilizable. Indeed, if there exists a control acting in $\overline{\omega}$ such that

$$\lim_{t \to +\infty} \|y^u(t)\|_{L^2(\Omega)} = 0$$

and

$$y^u(x,t) \geq 0 \quad \text{a.e. in } Q = \Omega \times (0, +\infty),$$

then it is obvious that

$$y^u(x,t) \geq z(x,t)$$

a.e. $(x,t) \in (\Omega \setminus \omega) \times (0, +\infty)$, where z is the solution to

$$\begin{cases} z_t - \Delta z + a(x)z = 0, & (x,t) \in (\Omega \setminus \overline{\omega}) \times (0, +\infty) \\ z(x,t) = 0, & (x,t) \in (\partial\Omega \cup \partial\omega) \times (0, +\infty) \\ z(x,0) = y_0(x), & x \in \Omega. \end{cases} \quad (2.2)$$

The solution z to (2.2) is strictly positive a.e.. On the other hand, using the Fourier development for $z(t)$ in $L^2(\Omega \setminus \overline{\omega})$ we may infer that

$$\|z(t)\|_{L^2(\Omega \setminus \overline{\omega})} \geq M_1 e^{-\lambda_1^\omega t}, \quad \forall t \geq 0,$$

where $M_1 > 0$ is a constant. The conclusion is now obvious.

If $\lambda_1^\omega > 0$, then we shall prove first that there exists a positive $T \in (0, +\infty)$ and a control $u \in L^2(Q_{\omega,T})$ such that

$$y^u(x,t) \geq 0 \quad \text{a.e. in } Q_T$$

and

$$y^u(x,T) = 0 \quad \text{a.e. } x \in \omega.$$

Indeed, if $y_0(x) = 0$ on a subset of ω of positive measure, then we consider the following system:

$$\begin{cases} y_t - \Delta y + a(x)y = -m(x)\rho \cdot \text{sgn}_{L^1(\omega)} y(t), & x \in \Omega, \ t > 0 \\ y(x,t) = 0, & x \in \partial\Omega, \ t > 0 \\ y(x,0) = y_0(x), & x \in \Omega, \end{cases} \quad (2.3)$$

where $\rho > 0$ will be precised later and

$$sgn_{L^1(\omega)} y = \|y\|_{L^1(\omega)}^{-1} y, \quad \text{if } \|y\|_{L^1(\omega)} = 0.$$

In what follows we shall use the next auxiliary result:

Lemma 2.2. *For any $\rho > 0$ large enough, there exists $T \in (0, +\infty)$ such that (2.3) has a unique and nonnegative solution y on the time interval $(0, T)$ and in addition*

$$y(x, T) = 0 \quad a.e. \ x \in \omega.$$

Proof of Lemma 2.2. The operator defined by

$$D(A) = \{y \in W_0^{1,1}(\Omega); \ \Delta y \in L^1(\Omega)\}$$

$$Ay = \Delta y - a(\cdot)y, \quad \forall y \in D(A),$$

is the generator of a compact C_0-semigroup in $L^1(\Omega)$ (see [6]).

For each $\varepsilon > 0$ we consider the approximating system

$$\begin{cases} y_t - \Delta y + a(x)y = -m(x)\rho \cdot \dfrac{y(x,t)}{\|y(t)\|_{L^1(\omega)} + \varepsilon}, & x \in \Omega, \ t > 0 \\ y(x, t) = 0, & x \in \partial\Omega, \ t > 0 \\ y(x, 0) = y_0(x), & x \in \Omega, \end{cases} \quad (2.3)'$$

Since the application

$$y \to \frac{m \cdot y}{\|y\|_{L^1(\omega)} + \varepsilon}$$

(form $L^1(\Omega)$ to $L^1(\Omega)$) is locally Lipschitz we conclude that problem $(2.3)'$ has a unique mild solution $y_\varepsilon \in C(\mathbf{R}^+; L^1(\Omega))$. Using the comparison result for parabolic operators we conclude that if $0 < \varepsilon_1 < \varepsilon_2$, then

$$0 \leq y_{\varepsilon_1}(x, t) \leq y_{\varepsilon_2}(x, t)$$

a.e. On the other hand, since

$$\left\| -m\rho \cdot \frac{y(t)}{\|y(t)\|_{L^1(\omega)} + \varepsilon} \right\|_{L^1(\Omega)} \leq \rho,$$

for almost every $t > 0$ and using the Baras compactness theorem we get that

$$y_\varepsilon \to y \quad \text{in } C([0, T]; L^1(\Omega)),$$

Internal Stabilization of Diffusion Equation

for any $T > 0$. It follows that
$$\|y_\varepsilon\|_{L^1(\omega)} \to \|y\|_{L^1(\omega)}$$
in $C([0,T])$, for any $T > 0$.

Let
$$\tilde{T} = Sup\{T \in [0,+\infty]; \|y(t)\|_{L^1(\omega)} > 0, \forall t \in [0,T]\}.$$

Passing to the limit ($\varepsilon \to 0^+$) in $(2.3)'$ we conclude that y is the solution of (2.3) on $[0,T]$ for any $0 < T < \tilde{T}$. Multiplying (2.3) by $sgn\, y$ and integrating over Ω we get

$$\|y(t)\|_{L^1(\Omega)} - \|y_0\|_{L^1(\Omega)} \leq \|a\|_{L^\infty(\Omega)} \int_0^t \|y(s)\|_{L^1(\Omega)} ds - t\rho.$$

Denote by $w(t) = \|y\|_{L^1(\Omega \times (0,t))}$ and by α a constant satisfying $\alpha > \|a\|_{L^\infty(\Omega)}$. It follows that:
$$z'(t) = \|y(t)\|_{L^1(\Omega)} \leq \|y_0\|_{L^1(\Omega)} + \alpha z(t) - t\rho$$

and so
$$(e^{-\alpha t} z(t))' \leq \|y_0\|_{L^1(\Omega)} e^{-\alpha t} - \rho t e^{-\alpha t}, \quad \forall t \in [0,T].$$

In conclusion
$$e^{-\alpha t} z(t) \leq \|y_0\|_{L^1(\Omega)} \frac{1}{\alpha}(1 - e^{-\alpha t}) - \rho \int_0^t s e^{-\alpha s} ds$$
$$= \|y_0\|_{L^1(\Omega)} \frac{1}{\alpha}(1 - e^{-\alpha t}) - \frac{\rho}{\alpha^2} + \rho e^{-\alpha t}(\frac{t}{\alpha} + \frac{1}{\alpha^2}).$$

Thus
$$\|y(t)\|_{L^1(\Omega)} \leq \|y_0\|_{L^1(\Omega)} - \frac{\rho}{\alpha} e^{\alpha t} + \frac{\rho}{\alpha} + \|y_0\|_{L^1(\Omega)}(e^{\alpha t} - 1).$$

So, for $\rho > 0$ large enough, there exists $T \in (0,+\infty)$ such that (2.3) has a unique and nonnegative solution on the time interval $(0,T)$ and in addition
$$y(x,T) = 0 \quad \text{a.e. } x \in \omega$$

and
$$\|y(t)\|_{L^1(\omega)} > 0 \quad \text{for any } t \in [0,T).$$

Proof of Theorem 2.1 (continued). We conclude that using the feedback control
$$u(t) = -\rho \cdot sgn_{L^1(\omega)} y(t),$$

we get that
$$y^u(x,t) \geq 0 \quad \text{a.e. in } Q_T$$

and
$$y^u(x,T) = 0 \quad \text{a.e. in } \omega.$$
Multiplying (2.3) by y and integrating over Q_T we conclude that
$$y(T) \in L^2(\Omega).$$

For $t > T$ we shall use the following feedback control:
$$u(t) = \mu_y(t),$$
where μ_y is the measure defined by
$$\mu_y(\varphi) = \int_{\partial\omega} \frac{\partial y}{\partial \nu^-}(x)\varphi(x)d\sigma, \quad \forall \varphi \in C(\partial\omega).$$
Here
$$\frac{\partial y}{\partial \nu^-}(x) = \lim_{\varepsilon \to 0^+} \frac{y(x + \varepsilon \nu^-) - y(x)}{\varepsilon}, \quad \forall x \in \partial\omega,$$
where ν^- is the outward normal versor to ω.

So, μ_y is a measure with support in $\partial\omega \subset \overline{\omega}$.

The following system
$$\begin{cases} y_t - \Delta y + a(x)y = \mu_y(t), & (x,t) \in \Omega \times (T, +\infty) \\ y(x,t) = 0, & (x,t) \in \partial\Omega \times (T, +\infty) \\ y(x,T) = h(x) & x \in \Omega \end{cases}$$

($h \in L^2(\Omega)$) has a unique weak solution $y \in C([T,T_1]; L^2(\Omega)) \cap AC([T,T_1]; L^2(\Omega \setminus \overline{\omega})) \cap AC([T,T_1]; L^2(\omega)) \cap L^2(T,T_1; H_0^1(\Omega)) \cap L^2_{loc}(T,T_1; H^2(\Omega\setminus\overline{\omega})) \cap L^2_{loc}(T,T_1; H^2(\omega))$, $\forall T_1 \in (T, +\infty)$ (see [3], [4]) and it satisfies
$$y(x,t) = 0 \quad \text{a.e. } x \in \omega,\ t > 0$$
and the restriction of y to $(\overline{\Omega} \setminus \omega) \times [T, +\infty)$ is obviously the solution to
$$\begin{cases} y_t - \Delta y + a(x)y = 0, & (x,t) \in (\Omega \setminus \overline{\omega}) \times (T, +\infty) \\ y(x,t) = 0, & (x,t) \in (\partial\Omega \cup \partial\omega) \times (T, +\infty) \\ y(x,T) = h(x) & x \in \Omega. \end{cases}$$

Now it is clear that

$$\|y(t)\|_{L^2(\Omega)} = \|y(t)\|_{L^2(\Omega\setminus\omega)}$$
$$\leq M_2 e^{-\lambda_1^\omega t}, \quad \forall t \geq 0$$

and we get the exponential stabilization of the solution y.

3. FINAL REMARKS

The main conclusion of the previous section is that the feedback control

$$u(t) = \begin{cases} -\rho \cdot sgn_{L^1(\omega)} y(t), & t \in [0, T] \\ \mu_y(t), & t > T \end{cases}$$

stabilizes the system (1.1) if and only if $\lambda_1^\omega > 0$.

If we denote by λ_1 the first eigenvalue of the operator

$$D(A) = H_0^1(\Omega) \cap H^2(\Omega),$$
$$Ay = -\Delta y + a(\cdot)y, \quad \forall y \in D(A),$$

then we have

$$\lambda_1 \leq \lambda_1^\omega.$$

If $\lambda_1 > 0$, the system (1.1) can be stabilized by the trivial control $u \equiv 0$.

If $\lambda_1 \leq 0$, then the control $u \equiv 0$ does not stabilize (1.1) (for related results see [1]), but the system can be stabilized if $\lambda_1^\omega > 0$.

In the same manner as in Section 2 can be proved that (1.1) is not exact "non-negative" controllable.

REFERENCES

[1] L.I. Aniţa, Asymptotic behaviour of the solutions of some reaction-diffusion processes, *International J. Appl. Math.*, submitted.

[2] S. Aniţa and V. Barbu, Local exact controllability of a reaction-diffusion system, *Diff. Integral Eqs.*, to appear.

[3] V. Barbu, *Analysis and Control of Nonlinear Infinite Dimensional Systems*, Academic Press, Boston (1993).

[4] V. Barbu, *Partial Differential Equations and Boundary Value Problems*, Kluwer Acad. Publ., Dordrecht (1998).

[5] V. Barbu, Exact controllability of the superlinear heat equation, *Appl. Math. Optim.*, to appear.

[6] H. Brézis and A. Friedman, Nonlinear parabolic equations involving measures as initial conditions, *J. Math. Pures Appl.*, **62** (1983), 73–97.

[7] E. Fernandez-Cara, Null controllability of the semilinear heat equation, *ESAIM: Control, Optim., Calc. Var.*, **2** (1997), 87–107.

[8] G. Lebeau and L. Robbiano, Contrôle exact de l'équation de la chaleur, *Comm. Partial Diff. Eqs.*, **30** (1995), 335–357.

[9] J.L. Lions, *Controlabilité exacte, stabilisation et perturbation de systemes distribués*, RMA 8, Masson, Paris (1988).

[10] J. Zabczyk, *Mathematical Control Theory: An Introsuction*, Birkhäuser, Boston (1992).

Flow-Invariant Sets with Respect to Navier-Stokes Equation

V. BARBU[*] AND N. H. PAVEL[**]

[*]University of Iași, Department of Mathematics, 6600 Iași, Romania

[**]Ohio University, Department of Mathematics, Athens, Ohio 45701, USA

1. Introduction. A new result (Theorem 2.1) on the flow-invariance of a closed subset with respect to a differential equation associated with a nonlinear semigroup generator on Banach spaces is given(the proof will be given in [1]). Applications to the flow-invariance of controlled flux sets (the Enstrophy and Helicity sets) with respect to Navier-Stokes equations are presented.

Recall that K is said to be a flow invariant set with respect to a differential equation $y' = Ay$, if every solution y starting in K (i.e. $y(0) = x \in K$) remains in K as long as it exists (i.e. $y(t) \in K$, for all t in the domain of y). In our cases here, we are using the strong solutions, so actually we deal with the flow invariance of $K \cap D(A)$. We think that the existing results on this topic are not applicable to our cases treated here. Indeed, the general result of R. H. Martin Jr. [8] requires the right hand side A of (2.9) to be continuous and dissipative on K. None of these key conditions (on A) are required here.

Note also that the subsets considered here are closed, but not necessarily convex. A different approach to flow-invariance of such sets with respect to Navier-Stokes equations was given by Barbu and Sritharan [2].

Our general framework is $V \hookrightarrow H \hookrightarrow V'$ algebraically and topologically with V and H real Hilbert spaces, V—densely and compactly embedded in H, V'— the dual of V.

Section 3 is devoted to the structure of the contingent cone $T_K^H(f)$ to K at $f \in K$ in the topology of H. Therefore, applications of Theorem 2.1 are given to the semilinear equations of the form:

$$y' = Cy + Dy, \tag{1.1}$$

where C is a C_0-semigroup generator $S_0(t)$ in H and D is a nonlinear perturbation of C such that $C + D = A$ is a generator of a nonlinear semigroup $S(t)$ as indicated in (2.4).

Recall the general results of Pavel [7] for the flow-invariance of a closed subset K with respect to (1.1). "If D is continuous and dissipative from $K \subset X$ into X", or "D is only continuous on K but $S_0(t)$ is compact for $t > 0$", then K is invariant with respect to (1.1) iff:

$$\frac{1}{h} d(S_0(h)x + hDx; K) \to 0 \text{ as } h \downarrow 0 \tag{1.2}$$

for all $x \in K$, where $d(z; K)$ stands for the distance from z to K.

Again, even this result which is more general than Martin's result above, is not applilcable to our cases here, simply because here D is neither continuous on K nor dissipative.

2. Flow-invariance results in reflexive Banach spaces

Let X be a Banach space of norm $\|\cdot\|$ and let X'. Recall that a (possible multivalued) operator $A : D(A) \subset X \to 2^X$ is said to be ω-dissipative (for some $\omega \in \mathbb{R}$), if for every $\lambda \geq 0$ with $1 - \lambda\omega > 0$, and for every $x_j \in D(A)$, there are $y_j \in Ax_j$, $j = 1, 2$ such that:

$$(1 - \lambda\omega)\|x_1 - x_2\| \leq \|x_1 - \lambda y_1 - (x_1 - \lambda y_2)\| \tag{2.1}$$

or equivalently:

$$\langle y_1 - y_2, J(x_1 - x_2) \rangle \leq \omega \|x_1 - x_2\|^2, \tag{2.2}$$

where $J : X \to X'$ is the duality mapping of X. If J is multivalued, one replaces $J(x_1 - x_2)$ in (2.2), by some $x^* \in J(x_1 - x_2)$.

Flow-Invariant Sets for Navier-Stokes Equation

Suppose in addition to the inequality (2.1) the following range condition holds

$$\overline{D(A)} \subset R(I - \lambda A), \tag{2.3}$$

for all sufficiently small $\lambda > 0$ (precisely, for $\lambda\omega < 1$).

The fundamental result on the generation of nonlinear semigroups $S_A(t) = S(t)$ is the following one (known as the exponential formula of Crandall-Liggett [6]).

Let A be ω-dissipative (i.e. (2.1) holds) satisfying the range condition (2.3). Then

$$\lim_{n \to \infty} (I - \frac{t}{n}A)^{-n}x = S(t) \in \overline{D(A)} \tag{2.4}$$

for all $x \in \overline{D(A)}$ and $t \geq 0$. Moreover $\|S(t)x - S(t)y\| \leq e^{t\omega}\|x - y\|$, $\forall t \geq 0$, $x, y \in \overline{D(A)}$, $S(t+s) = S(t)S(s)$, $S(0) = I$—the identity on X, $\lim_{t \downarrow 0} S(t)x = x$, $\forall x \in \overline{D(A)}$, $\|S(t)x - S(s)x\| \leq |t - s||Ax|\exp(2\omega_0(t+s))$, $t, s \geq 0$, for all $x \in D(A)$, where $\omega_0 = \max\{0, \omega\}$ and $|Ax| = \inf\{\|y\|, y \in Ax\}$. If $s \to S(s)x$ is differentiable at $s = t$, then $u(t) = S(t)x \in D(A)$ and it is the only (strong) solution to the Cauchy problem

$$u'(t) = Au(t), \quad u(0) = x, \quad x \in D(A), \quad t \geq 0 \tag{2.5}$$

In our example applications here, the following hypotheses are fulfilled.

(H1) For every $x \in D(A)$, $t \to S(t)x$ is differentiable at every $t \geq 0$.

So:

$$S(t) : D(A) \to D(A); \quad \lim_{h \downarrow 0} \frac{S(h)x - x}{h} = Ax, \forall x \in D(A) \tag{2.5}'$$

and $u(t) = S(t)x$ is the only strong solution to the problem (2.5). Let K be a closed subset of X. The basic hypothese on the relationship between A and K are given below:

(H2) The projection $P_K(y)$ on K exists for all $y \in D(A)$. Moreover,

$$P_K(D(A)) \subset D(A) \cap K. \tag{2.6}$$

This means, that for every $y \in D(A)$, there is $y_0 = P_K(y) \in D(A) \cap K$ such that the distance $d(y; K)$ from y to K satisfies

$$d(y; K) = \inf\{\|y - z\|, z \in K\} = \|y - y_0\| = d(y; K \cap D(A)) \qquad (2.7)$$

for some $y_0 \in K \cap D(A)$. Recall also the definition of the tangential (contingent) cone $T_K(x)$ to K at $x \in K$ in the sense of Bouligand [4, Ch.1]

$$T_K(x) = \{v \in X, \lim_{h \downarrow 0} \frac{1}{h} d(x + hv; K) = 0\} \qquad (2.8)$$
$$= \{v \in X; \exists r(h) \in X \text{ with } r(h) \to 0 \text{ as } h \downarrow 0 \text{ and } x + h(v + r(h)) \in K\}$$

Note that $T_K(x)$ is a closed cone, even if K is not closed [4, p. 2].

The main result of this section is

Theorem 2.1. *Suppose that A and K satisfy Hypotheses* (H1) *and* (H2). *Then a necessary and sufficient condition for $K \cap D(A)$ to be a flow-invariant set with respect to*

$$y' = Ay, \qquad y(0) = x, \quad x \in K \cap D(A) \qquad (2.9)$$

is

$$\lim_{h \downarrow 0} \frac{1}{h} d(x + hAx; K \cap D(A)) = 0, \quad \forall x \in K \cap D(A). \qquad (2.10)$$

The proof will be given in [1]. Note that Necessity is immediate: Indeed, suppose that the strong solution $y(t) = S(t)x$, of (2.9) with $x \in K \cap D(A)$ remains in K, for all $t \geq 0$. As $S(t)D(A) \subset D(A)$, it follows that actually $S(t)x \in K \cap D(A)$ so $\frac{1}{h}d(x+hAx; K \cap D(A)) \leq \frac{1}{h}\|x+hAx-S(h)x\| = \|\frac{S(h)x-x}{h}-Ax\| \to 0$ as $h \downarrow 0$, so (2.10) is a necessary condition for the flow-invariance of $K \cap D(A)$ with respect to (2.9).

The following simple known lemma will be useful in the next section.

Lemma 2.1. *Let X be a real Hilbert space of inner product $\langle \cdot, \cdot \rangle$ and let $\phi : X \to R$ be a Fréchet differentiable functional. Consider the closed set:*

$$K = \{f \in X; \phi(f) \leq 0\}. \qquad (2.12)$$

Then the tangent cone $T_K(f)$ to K at $f \in K$ is given by:

$$T_K(f) = \begin{cases} X, & \text{if } \phi(f) < 0 \\ \{v \in X; \langle \phi'(f), v \rangle \leq 0\} = M_f, & \text{if } \phi(f) = 0. \end{cases} \quad (2.13)$$

Proof: The inclusion $T_K(f) \subset M_f$ is immediate. Viceversa, the interior M_f^0 of M_f is:

$$M_f^0 = \{v \in X; \langle \phi'(f), v \rangle < 0\} \subset T_K(f).$$

This is because $v \in M_f^0$ implies $f + tv \in M_f^0$ for all $t > 0$ sufficiently small, which is easy to check. Taking into account that the closure $\overline{M_f^0} = M_f$ and that $T_K(f)$ is closed, we get $M_f \subset T_K(f)$, which completes the proof.

3. Flow-invariance of flux sets

We will present several consequences of Theorem 2.1 to the particular case $X = H$—a real Hilbert space of norm $|\cdot|$ and inner product $\langle \cdot, \cdot \rangle$, and $K \subset V$, with K—closed in H. Precisely, we will assume that

$$V \hookrightarrow H \hookrightarrow V' \quad (3.1)$$

algebraically and topologically, and the embedding of V in H is dense and compact (i.e. the bounded subsets in V are relatively compact in H). V' is the dual of V, and H' is identified with H. Denote by $P_K^H(y)$ the projection of y on K in (the norm of) H, and by $T_K^V(x)$ and $T_K^H(x)$ the tangential cones of K at $x \in K$ in V and H, repectively. Denote by $\ll \cdot, \cdot \gg$ and $\|\cdot\|$ the inner product and the norm of V, respectively. Let also $A : D(A) \subset H \to H$ satisfy (2.1)+(2.3). Denote by $S(t) = e^{tA}$ the semigroup generated by A (see (2.4)). The basic hypotheses of this section are:

(C3.1) K is a subset of V, which is closed in H. Every subset of K which is bounded in H is bounded in V, too.

(C3.2) $P_K^H(D(A)) \subset D(A)$.

(C3.3) $Ax \in T_K^H(x), \quad \forall x \in K \cap D(A)$, i.e. Ax is tangent (in H) to K at x, in the sense of (2.8).

We first note that under (C1), the projection of y on K exists, i.e. there is $y_0 \in K$, such that $d(y; K) = |y - y_0|$. This is because the minimizing sequence $z_n \in K$; $\lim_{n \to \infty} |y - z_n| = d(y; K)$ is bounded in H so by (C1) is also bounded in V, so it is relatively compact in H, i.e. z_n contains convergent subsequences z_{n_k} in H. Say $z_{n_k} \to y_0$. Then $P_K^H(y) = y_0$.

A direct consequence of Theorem 2.1 is

Corollary 3.1. *Under Hypotheses (C1)–(C3), $K \cap D(A)$ is a flow invariant set with respect to $y' = Ay$ (i.e., $e^{tA}x \in K \cap D(A)$, $\forall x \in K \cap D(A)$).*

Hypotheses (C3.1) and (C3.2) are satisfied in the case of subsets of the form

$$K = \{f \in V; \|f\|^2 \leq \varphi(|f|^2) + \rho\} = \{f \in V; \phi(f) \leq 0\} \quad (3.2)$$

where

$$\phi(f) = \|f\|^2 - \varphi(|f|^2) - \rho, \quad \rho \geq 0 \quad (3.3)$$

and

$$\varphi \in C^1(\mathbb{R}_+), \quad \varphi(0) = 0, \quad \varphi'(r) \geq 0, \quad \forall r \in \mathbb{R}_+. \quad (3.4)$$

Clearly, K is not necessarily convex.

In applications to heat propagation, we will use

$$V = H_0^1(\Omega) \quad H = L^2(\Omega), \quad V' = H^{-1}(\Omega) \quad (3.5)$$

Here $\Omega \subset \mathbb{R}^n$, is a bounded open subset with smooth boundary $\Gamma = \partial \Omega$. In this case

$$\|f\| = |\nabla f| = |\mathrm{grad} f|_{L^2(\Omega)} \quad (3.6)$$

so K represents a constraint on the flux $q = \nabla f$, i.e.,

$$\int_\Omega q^2(x)\, dx \leq \varphi\left(\int_\Omega e^2(x)\, dx\right) + \rho \quad (3.7)$$

where $e(x) = f(x)$ is regarded as the energy of the flow $f = f(x)$, in Ω. In order to verify the tangential condition (C3) we need to find the tangential cone $T_K^H(f)$ to K as in (3.2) at $f \in K$, in the norm of H. Precisely, this is given by:

Lemma 3.1.

$$T_K^H(f) = \begin{cases} H, & \text{if } \phi(f) < 0 \\ H, & \text{if } \phi(f) = 0 \text{ and } f \bar{\in} D(J) \\ \{v \in H; \langle -J(f) + \varphi'(|f|^2)f, v \rangle \geq 0\} \equiv M_f^H, \\ & \text{if } \phi(f) = 0 \text{ and } f \in D(J). \end{cases} \quad (3.8)$$

Here J is the duality mapping of V (the cannonical isomorphism from V into V').

Finally

$$D(J) = \{f \in V; J(f) \in H\}. \quad (3.9)$$

Clearly, in the case of $V = H_0^1(\Omega)$ and $H = L^2(\Omega)$, Lemma 3.1 becomes:

Corollary 3.2. Let

$$K = \{f \in H_0^1(\Omega), |\nabla f|^2 \leq \varphi(|f|^2) + \rho\} \quad (3.10)$$

with φ as in (3.4) and $\phi(f) = |\nabla f|^2 - \varphi(|f|^2) - \rho$. Then

$$T_K^{L^2(\Omega)}(f) = \begin{cases} L^2(\Omega), & \text{if } \phi(f) < 0 \\ L^2(\Omega), & \text{if } \phi(f) = 0 \text{ and } f \bar{\in} H^2(\Omega) \\ \{v \in L^2(\Omega), \langle \nabla f + \varphi'(|f|^2)f, v \rangle \geq 0\}. \end{cases} \quad (3.11)$$

Proof: Indeed, in this case $J(f) = -\Delta f$ with $D(\Delta) = H_0^1(\Omega) \cap H^2(\Omega)$. Clearly, in this case, $f \in K$ and $f \bar{\in} H^2(\Omega)$ means $f \bar{\in} D(\Delta)$.

Remark 3.1. In the proof of Lemma 3.1 it is essential to observe that:

$$T_K^V V(f) \subset T_K^H(f), \quad \forall f \in K. \quad (3.12)$$

This is because (with the notations in (2.8)) $r(h) \to 0$ in V implies $r(h) \to 0$ in H, as $h \downarrow 0$. Now, with K as in (3.2), it follows from Lemma 2.1 that:

$$T_K^V(f) = \begin{cases} V, & \text{if } \phi(f) < 0 \\ \{v \in V; \ll f, v \gg -\varphi'(|f|^2)\langle f, v \rangle \leq 0\} = M_f^V & \text{if } \phi(f) = 0. \end{cases} \quad (3.13)$$

Indeed, ϕ is continuous from V into $\mathbb{R}$ and

$$\ll \phi'(f), v \gg = \ll f, v \gg -\varphi'(|f|^2)\langle f, v \rangle \quad (3.14)$$

so (3.13) follows from (2.17).

Proof of Lemma 3.1: If $\phi(f) < 0$, then by (3.12) and (3.13) we have

$$V = T_K^V(f) \subset T_K^H(f) \subset H. \tag{3.15}$$

This, the density of V in H and the fact that $T_K^H(f)$ is closed in H, yield $T_K^H(f) = H$.

If $\phi(f) = 0$ and $f \bar{\in} D(J)$, then

$$\begin{aligned} M_f^0 &= \{v \in V; \ll f, v \gg -\varphi'(|f|^2)\langle f, v\rangle = 0\} \\ &= \{v \in V; \ll f - \varphi'(|f|^2)J^{-1}(f), v \gg = 0\} \end{aligned} \tag{3.16}$$

is dense in H. Indeed, if by contradiction, this were not the case, then the closure $\overline{M}_f^0$ would be strictly included in H. This implies the existence of an $\eta \in H$, $\eta \neq 0$ with η orthogonal on $\overline{M}_f^0$, i.e. $\langle \eta, v\rangle = 0$, for all $v \in M_f^0$, so $\ll J^{-1}(\eta), v \gg = 0, \forall v \in M_f^0$. This and (3.16) (the orthogonal space of M_f^0 is one dimensional) imply that

$$f - \varphi'(|f|^2)J^{-1}(f) = tJ^{-1}(\eta), \quad \text{for some } t \in \mathbb{R}$$

so $f \in D(J)$, which is absurd. On the other hand, in view of (3.13)

$$M_f^0 \subset M_f^V = T_K^V(f) \subset T_K^H(f) \subset H$$

which implies $T_K^H(f) = H$ for $\phi(f) = 0$ and $f \bar{\in} D(J)$. Finally, if $f \in D(J)$, than we can replace $\ll f, v \gg = \langle J(f), v\rangle$ in (3.13) so for $\phi(f) = 0$ and $f \in D(J)$,

$$T_K^V(f) = \{v \in V; \ll -J(f) + \varphi'(|f|^2)f, v\rangle \geq 0\} \subset T_K^H(f)$$

which yields $M_f^H \subset T_K^H(f)$.

For proving the converse inclusion let $v \in T_K^H(f)$, i.e. there is $r(t) \to 0$ in H as $t \downarrow 0$ such that $f + t(v + r(t)) = f + tv_t \in K$, with $v_t = v + r(t) \to v$ in H. We derive (for $\phi(f) = 0$) $\ll f, v_t \gg \leq \varphi'(|f|^2)\langle f, v_t\rangle + \alpha(t)$ with $\alpha(t) \to 0$

as $t \downarrow 0$ and $\ll f, v_t \gg = \langle J(f), v_t \rangle$ which implies $\langle J(f), v \rangle \leq \varphi'(|f|^2)\langle f, v \rangle$ i.e. $T_K^H(f) \subset M_f^H$ and the proof is complete.

4. Flow invariant sets for Navier-Stokes equations.

We will discuss the invariance of Enstrophy and Helicity sets with respect to the Navier-Stokes system:

$$
\begin{aligned}
& y_t - \nu \Delta y + (y \cdot \nabla) y = \nabla p, \quad \text{in } (0, T) \times \Omega \\
& \nabla \cdot y = \operatorname{div} y(t, x) = 0, \quad \text{in } (0, T) \times \Omega \\
& y(0, x) = y_0(x), \quad \text{in } \Omega \\
& y(t, x) = 0, \quad \text{on } (0, T) \times \partial\Omega,
\end{aligned}
\tag{4.1}
$$

where Ω is a bounded domain of $\mathbb{R}^n$, with smooth boundary $\partial\Omega$, $y_0 \in (L^2(\Omega))^n$. $\nabla = (\frac{\partial}{\partial x_1}, \ldots, \frac{\partial}{\partial x_n})$ is the gradient operator, and $\eta(x) = (\eta_1, \ldots, \eta_n)$ is the outward normal to $\partial\Omega$ at $x \in \partial\Omega$, ν is positive and $p = p(t, x)$ is the pressure [2].

The basic functional spaces are

$$V = \{y \in (H_0^1(\Omega))^n, \ \operatorname{div} y(x) = 0, \text{ in } \Omega\} \tag{4.2}$$

$$H = \{y \in (L^2(\Omega))^n, \ \operatorname{div} y(x) = 0, \text{ in } \Omega, y(x) \cdot \eta(x) = 0, \text{ on } \partial\Omega\}. \tag{4.3}$$

For $y \in (H^1(\Omega))^n$, $y = (y_1, \ldots, y_n)$, by definition

$$\nabla y = (\nabla y_1, \ldots, \nabla y_n) \in (L^2(\Omega))^{n^2},$$

$$\Delta y = (\Delta y_1, \ldots, \Delta y_n). \tag{4.4}$$

Denote by

$$\ll \cdot, \cdot \gg, \quad \|\cdot\|, \quad \langle \cdot, \cdot \rangle, \quad |\cdot| \tag{4.5}$$

the inner products and norms in V and H, respectively.

Finally, $(y \cdot \nabla) y$ is the following n-component function of $(L^2(\Omega))^n$

$$(y \cdot \nabla) y = (y \cdot \nabla y_1, \ldots, y \cdot \nabla y_n) \tag{4.6}$$

with
$$y \cdot \nabla y_k = \sum_{j=1}^{n} y_j \cdot y_{k,j}, \quad y_{k,j} = \frac{\partial y_k}{\partial x_j}, \quad k = 1, \ldots, n. \tag{4.7}$$

Recall also the formula of "integration by parts:
$$\int_{\Omega} \nabla f \cdot y dx = \int_{\partial \Omega} fy \cdot \eta d\sigma - \int_{\Omega} f \mathrm{div}\, y dx, f \in H^1(\Omega), y \in (H^1(\Omega))^n \tag{4.8}$$

In view of (4.8), the gradient $\nabla p(t,x)$ (with respect to x) is orthogonal on H ($\int_{\Omega} \nabla p(t,x) \cdot y = 0$ for all $y \in H$, with $p(t) \in H^1(\Omega)$, $(p(t))(x) = p(t,x)$). Therefore, multiplying formally (4.1) by $w \in V$, in the sense of the inner product of H, we also have $\langle \nabla p(t), w \rangle = 0$, so (4.1) can be represented in H as:
$$\begin{aligned} y' + \nu Ay + By &= 0, \quad t \in (0,T) \\ y(0) &= y_0, \quad \nu > 0 \end{aligned} \tag{4.9}$$

where
$$(By, w) = b(y, y, w), \quad \forall y, w \in V \tag{4.10}$$

with
$$b(y, z, w) = \sum_{i,j=1}^{n} \int_{\Omega} y_i D_i z_j w_j dx = \sum \int_{\Omega} y_i z_{j,i} w_j dx \tag{4.11}$$

and $A \in L(V, V')$, with
$$\langle Au, v \rangle = \sum_{i=1}^{n} \int_{\Omega} \nabla u_i \cdot \nabla v_i dx, \quad \forall u, v \in V \tag{4.12}$$

In order to show that $\nu A + B$ is a semigroup generator, one proceeds by (trunchiation) of B, i.e. the quatization
$$B_N(y) = \begin{cases} By, & \text{if } \|y\| \leq N, y \in D(A) \\ \frac{N^2}{\|y\|^2} By, & \text{if } \|y\| > N, y \in D(A). \end{cases} \tag{4.13}$$

It follows that for $\alpha = \alpha_N$ sufficiently large, $\nu A + B_N + \alpha_N I$ is m-accretive in H (see [2]). Therefore, for $y_0 \in D(A)$, the Cauchy problem
$$\begin{aligned} y'_N + \nu Ay_N + B_N(y_N) &= 0, \quad t \in (0,T) \\ y_N(0) &= y_0 \end{aligned} \tag{4.14}$$

Flow-Invariant Sets for Navier-Stokes Equation 49

has a unique solution

$$y_N \in W^{1,\infty}([0,T]; H) \cap L^\infty(0,T; D(A)).$$

Moreover, by standard estimates([2],[10],[11], there is an interval T^* with $T^* = T^*(\|y_0\|) < T$ if $n = 3$, and $T^* = T$ if $n = 2$, such that

$$\|y_N(t)\| \leq N, \quad \forall t \in [0, T^*]$$

for N sufficiently large. This means that $y = y_N$ satisfies $B_N(y_N) = By$, so $y = y_N$ is the strong solution to (4.9). In other words it suffices to check the condition of Theorem 2.2, with $-\nu A - B_N$ in place of A. In what follows we shal assume that $n = 2, 3$.

Flow invariance of the Enstrophy sets

Let K be so called the Enstrophy set [2].

$$K = \{f \in V, |\nabla \times f|^2 \leq \varphi(|f|^2) + \rho\}, \quad \rho > 0 \tag{4.15}$$

where $\nabla \times f = \mathrm{curl} f(x)$. It is true (although not immediat!) that the $(L^2(\Omega))^n$ norm of $\mathrm{curl} f$, satisfies

$$|\nabla \times f| = |\nabla f| = \|f\|. \tag{5.16}$$

The tangent cone $T_K^H(f)$ is given by (3.8) with $J(f) = Af$. Clearly, K is closed in H. Therefore, the invariance of $K \cap D(A)$ is equivalent to the tangency of $v = -\nu A - B_N f$ to K to $K \cap D(A)$, i.e. to the inequality in (3.8), i.e.

$$E(f) = \langle Af - \varphi'(|f|^2)f, \nu Af + B_N f \rangle \geq 0 \tag{4.17}$$

with $f \in D(A) \cap K$, $\|f\|^2 = |\nabla f|^2$, $\|f\|^2 - \varphi(|f|^2) - \rho = 0$. Indeed, $E(f)$ can be estimated as follows:

$$E(f) = \nu |Af|^2 + \langle Af, B_N f \rangle - \nu \varphi'(|f|^2)\langle Af, f \rangle \tag{4.18}$$

as $\langle B_N f, f \rangle = 0$ due to $\langle f, f \rangle = b(f, f, f) = 0$ (as $b(f, g, w) = -b(f, w, g)$). Similarly to inequalities in (4.6),

$$\|f\|^2 \leq \frac{1}{\lambda_1}|Af|^2, \quad \langle Af, f \rangle = \|f\|^2 \tag{4.19}$$

where λ_1 is the first eigenvalue of A. Finally

$$\langle B_N f, Af \rangle \leq |\langle Bf, Af \rangle| = |b(f, f, Af)| \leq c\|f\||Af|^2 \tag{4.20}$$

for some $c > 0$. Therefore, with $r = |f|^2$, we have

$$E(f) \geq (\nu - c\|f\| - \frac{\nu}{\lambda_1}\varphi'(r))|Af|^2 \geq 0 \tag{4.21}$$

if

$$\nu - \frac{\nu}{\lambda_1}\phi'(r) - c(\phi(r) + \rho)^{1/2} \geq 0 \tag{4.22}$$

as $\|f\| = (\varphi(r) + \rho)^{1/2}$.

To conclude, we have proved the following result on the invariance of K in (4.15):

Theorem 4.1. *Let $\varphi : \mathbb{R}_+ \to \mathbb{R}_+$ be of class $C^1(\mathbb{R}_+)$, and*

$$\lambda_1^{-1}\varphi'(r) + c\nu^{-1}\sqrt{\varphi(r) + \rho} \leq 1 \tag{4.23}$$

where λ_1 is the first eigenvalue of A, and $c > 0$ is the best constant in (4.20). Then the subset $K \cap D(A)$ is flow invariant with respect to the Navier-Stokes equation (4.9).

Flow invariance of Helicity sets.

With the same technique as in the previous subsection, let us study the invariance of the "Helicity Set" (see also [2],[9])

$$K = \{f \in V; |\int_\Omega f(x) \cdot \nabla \times f(x) dx|^2 + \lambda^2 \int_\Omega |\nabla f(x)|^2 dx \leq \rho\} \tag{4.24}$$

where $\nabla \times f = \text{curl} f$. The helicity set has an important role in fluid mechanics and in particular it is an invariant set of Euler's equation ([9]). With the previous notations, K can be written as:

$$K = \{f \in V; \phi(f) = \langle f, \text{curl} f \rangle^2 + \lambda^2 \|f\|^2 - \rho^2 \leq 0\} \tag{4.25}$$

Flow-Invariant Sets for Navier-Stokes Equation

where λ and k are positive constants. Clearly, K is closed in H. Set $\varphi(f) = \langle f, \operatorname{curl} f \rangle$. Taking into account that

$$\langle y, \operatorname{curl} v \rangle = \langle \operatorname{curl} y, v \rangle, \quad y, v \in V, \tag{4.26}$$

$J(f) = Af$, and

$$\langle \phi'(f), v \rangle = \langle 4\varphi(f) \operatorname{curl} f, v \rangle + 2\lambda^2 \ll f, v \gg \tag{4.27}$$

for $v \in V$, we derive as in the case of (3.8) (as $\ll f, v \gg = \langle Af, v \rangle$ for $f \in D(A)$)

$$T_K^H(f) = \begin{cases} H, & \text{if } \phi(f) < 0 \\ H, & \text{if } \phi(f) = 0, f \bar{\in} D(A) \\ \{v \in V, \langle -\lambda^2 Af - 2\varphi(f) \operatorname{curl} f, v \rangle \geq 0\} & \text{if } \phi(f) = 0, \text{ and } f \in D(A). \end{cases} \tag{4.28}$$

Therefore, the only fact we have to check for the invariance of K, is the last inequality in (4.28) with $v = -\nu Af - B_N f$, i.e.

$$E(f) = \langle \lambda^2 Af + 2\varphi(f) \operatorname{curl} f, \nu Af + B_N f \rangle \geq 0 \tag{4.29}$$

for $\phi(f) = 0$ and $f \in D(A)$.

First let's note that (4.25) (i.e. the definition of K) implies

$$|\varphi(f)| = |\langle f, \operatorname{curl} f \rangle| \leq \rho.$$

We also have (see (4.20))

$$\lambda^2 |\langle Af, B_N(f) \rangle| \leq \lambda^2 c \|f\| |Af|^2 \leq c\lambda\rho |Af|^2$$
$$\langle B_N f, \operatorname{curl} f \rangle \leq |b(f, f, \operatorname{curl} f)| = 0. \tag{4.30}$$
$$|2\varphi(f) \langle \operatorname{curl} f, \nu Af \rangle| \leq 2\rho\nu |Af| \|f\| \leq 2\rho\nu \lambda_1^{-1/2} |Af|^2$$

due to (4.19) and $|\operatorname{curl} f| = \|f\| \leq \lambda_1^{1/2} |Af|$.

Combining these inequalities, we clearly get

$$E(f) \geq (\lambda^2 \nu - \lambda c \rho - 2\rho\nu \lambda_1^{-1/2}) |Af|^2. \tag{4.31}$$

Thus, if

$$\lambda^2 \nu \geq \lambda c \rho + 2\rho\nu \lambda_1^{-1/2}, \tag{4.32}$$

we have $E(f) \geq 0$, i.e. $-\nu Af - B_N f \in T_K^H(f)$. Therefore we have proved:

Theorem 4.2. *Suppose that* (4.32) *holds. Then the helicity set defined in* (4.24) *is a flow invariant set with respect to the abstract Navier-Stokes equation* (4.9). *More precisely, the solutions starting in* $K \cap D(A)$ *remain in* $K \cap D(A)$ *as long as they exist in the future.*

The proof of formula $b(f, f, \operatorname{curl} f) = 0$ in (4.30) is given in [1].

References

[1] V. Barbu and N. H. Pavel, *Flow-Invariant Closed Sets with Respect to Nonlinear Semigroup Flows*, to appear.

[2] V. Barbu and S. Sritharan, *Flow-invariance preserving feedback controllers for the Navier-Stokes equation*, J. Math. Anal. Appl.(to appear).

[3] N. H. Pavel, *Invariant sets for a class of semilinear equations of evolutions*, Nonlinear Anal. TMA, 1(1977), pp. 187-196.

[4] H. Brezis, *On a characterization of flow-invariant sets*, Comm. Pure Appl. Math., 23(1970), pp. 261-263.

[5] D. Motreanu and N. H. Pavel, *Tangency, flow invariance and optimization problems*, Pure and Appl. Math., A series of monographs and textbooks, Marcel, Vol. 219, Marcel Dekker, New York, Basel, 1999.

[6] V. Barbu, *Nonlinear Semigroups and Differential Equations in Banach Spaces*, Noordhoff, Leyden 1975.

[7] N. H. Pavel, *Semilinear equations with dissipative time-dependent domain perturbations*, Isreal J. Math., 46(1983), pp. 103-122.

[8] R. H. Martin, *Differential equations on closed subsets of a Banach space*, Trans. Amer. Math. Soc., 179(1973), pp. 399-414.

[9] V. I. Arnold and B. A. Khesin, Topological Methods in Hydrodynamics, Springer-Verlag, 1998.

[10] P. Constantin and C. Foias, Navier-Stokes Equations, The University of Chicago Press, Chicago, 1988.

[11] R. Temam, *Navier-Stokes Equations; Theory and Numerical Analysis*, North Holland, Amsterdam, 1984.

Numerical Approximation of the Riccati Equation via Fractional Steps Method

Tudor BARBU[1] and Costică MOROŞANU[2]

[1]Institute of Mathematics of Romanian Academy,
6600 Iaşi, Romania

[2]Department of Mathematics, University "Al.I.Cuza",
6600 Iaşi, Romania

Abstract. *In this paper we prove the convergence of an iterative scheme of fractional steps type for the nonlinear Ricatti equation. The use of this method simplifies the numerical algorithms due to its decoupling feature. A numerical algorithm and numerical results are presented.*

1. INTRODUCTION

Consider the Ricatti equation

$$(1.1) \quad \begin{cases} P'(t) + A^*P(t) + P(t)A + P(t)BN^{-1}B^*P(t) = Q, & t \in (0,T] \\ P(0) = P_0, \end{cases}$$

where $A, Q \in L(\mathbb{R}^n, \mathbb{R}^n)$, $B \in L(\mathbb{R}^n, \mathbb{R}^m)$, $N \in L(\mathbb{R}^n, \mathbb{R}^n)$, $N = N^*$,

$Q = Q^* \geq 0$.

For any integer $M \geq 1$, let us associate to the time-interval $[0,T]$ the equidistant partition $0 < \varepsilon < \cdots < M\varepsilon = T$ of length $\varepsilon = T/M$ and, corresponding to this, the following approximating scheme (the Lie-Trotter scheme)

(1.2) $\begin{cases} P'_\varepsilon(t) + P_\varepsilon(t)BN^{-1}B^*P_\varepsilon(t) = 0 & t \in [i\varepsilon, (i+1)\varepsilon] \\ P_\varepsilon(i\varepsilon) = Z_\varepsilon((i+1)\varepsilon) + \varepsilon e^{-A^*\varepsilon}Qe^{-A\varepsilon} \end{cases}$

(1.3) $\begin{cases} Z'_\varepsilon(t) + A^*Z_\varepsilon(t) + Z_\varepsilon(t)A = 0 & t \in [i\varepsilon, (i+1)\varepsilon] \\ Z_\varepsilon(i\varepsilon) = P_\varepsilon^-(i\varepsilon), \quad P_\varepsilon^-(0) = P_0 \end{cases}$

for $i = 0, 1, \cdots, M-1$, where $P_\varepsilon^-(i\varepsilon) = \lim_{t \nearrow i\varepsilon} P_\varepsilon(t)$.

Taking into account Theorem 3 in [1], we may view $P_\varepsilon(t)$, a.e. $t \in [0,T]$, as an approximate solution for the unknown $P(t)$ in (1.1). So, in the sequel, we will deal with the computation of the approximate solution for P_ε in (1.2)-(1.3). The advantage of this approach consists in simplify the numeric calculus of the approximation to $P(t)$.

2. THE MAIN RESULT

In this section we will concern with the nonlinear algebraic equation (1.2). Firstly, we will introduce the discrete state equation. So, using a standard implicit method, (1.2) becomes

(2.1) $\quad P_\varepsilon^{i+1} + \varepsilon P_\varepsilon^{i+1} BN^{-1}B^* P_\varepsilon^{i+1} = Z_\varepsilon^{i+1}, \quad i = 0, 1, \cdots, M-1,$

where we have denoted $P_\varepsilon(i\varepsilon) = P_\varepsilon^i$, $Z_\varepsilon^{i+1} = Z_\varepsilon((i+1)\varepsilon) + \varepsilon e^{-A^*\varepsilon}Qe^{-A\varepsilon}$. Because B and N are self-adjoint then $BN^{-1}B^*$ can be equivalently written as $\bar{B}\bar{B}^*$.

At any time level $i+1$, $i = 0, 1, \cdots, M - 1$, we associate to equation (2.1) the following iterative linear process

$$(2.2) \qquad P_{\varepsilon,k+1} + \varepsilon P_{\varepsilon,k} \bar{B}\bar{B}^* P_{\varepsilon,k+1} = Z_{\varepsilon,k}.$$

The main result, Theorem 1 below, is concerned with the convergence of the scheme (2.2) for $k \longrightarrow \infty$.

Theorem 1. *If $Z_\varepsilon^i \geq 0$, $i = 1, 2, \cdots, M$, and $P_{\varepsilon,k} \geq 0$, for all $k \in \mathbb{N}$, then the sequence $\{P_{\varepsilon,k}\}$ is convergent to P_ε^{i+1} in $L(\mathbb{R}^n, \mathbb{R}^n)$, $i = 0, 1, \cdots, M - 1$.*

Proof. From (2.2) we have

$$\|\bar{B}^* P_{\varepsilon,k+1} X\|^2 \leq \left(Z_\varepsilon^{i+1} X, \bar{B}\bar{B}^* P_{\varepsilon,k+1} X\right)$$

$$\leq \|\bar{B}^*\| \|Z_\varepsilon^{i+1}\| \|\bar{B}^* P_{\varepsilon,k+1} X\|, \quad \forall X = 0.$$

Then

$$(2.3) \qquad \|\bar{B}^* P_{\varepsilon,k+1} X\|^2 \leq C, \quad \forall X = 0.$$

We have also

$$(2.4) \qquad P_{\varepsilon,k+1} - P_{\varepsilon,k} + \varepsilon P_{\varepsilon,k} \bar{B}\bar{B}^* (P_{\varepsilon,k+1} - P_{\varepsilon,k})$$

$$+ \varepsilon (P_{\varepsilon,k} - P_{\varepsilon,k-1}) \bar{B}\bar{B}^* P_{\varepsilon,k} = 0.$$

So

$$\|\bar{B}^*(P_{\varepsilon,k+1} - P_{\varepsilon,k})\| \leq \varepsilon \|\bar{B}\| \|\bar{B}^*(P_{\varepsilon,k} - P_{\varepsilon,k-1})\| \|\bar{B}\| \|P_{\varepsilon,k}\|$$

i.e.,

$$(2.5) \qquad \|\bar{B}^*(P_{\varepsilon,k+1} - P_{\varepsilon,k})\| \leq (C\varepsilon)^k, \quad \forall k \in \mathbb{N}.$$

From (2.5) we may conclude that

$$\{\bar{B}^*(P_{\varepsilon,k})\} \longrightarrow 0 \quad \text{as } k \longrightarrow \infty, \tag{2.6}$$

for $\varepsilon > 0$ small enough.

From (2.2) it follows

$$P_{\varepsilon,k+1} + \varepsilon P_{\varepsilon,k} T = Z_\varepsilon^{i+1} + \varepsilon R_{\varepsilon,k}, \tag{2.7}$$

where $R_{\varepsilon,k} \longrightarrow 0$ as $k \longrightarrow \infty$ and $T = \lim_{k \to \infty} \bar{B}\bar{B}^* P_{\varepsilon,k}$. Then, we have the inequality

$$\|P_{\varepsilon,k+1} - P_{\varepsilon,k}\| \leq \varepsilon \|P_{\varepsilon,k} - P_{\varepsilon,k-1}\| \|T\| + \varepsilon \|R_{\varepsilon,k}\|$$

and taking into account (2.6), we can derive that $P_{\varepsilon,k}$ is convergent as claimed.

3. NUMERICAL APPROACH

This section is devoted to numerical approximation of the solution corresponding to problem (1.2)-(1.3).

As regards the equation (1.3), we know that the solution is given by

$$Z_\varepsilon^{i+1} = e^{-\varepsilon A^*} P_\varepsilon^-(i\varepsilon) e^{-\varepsilon A} + e^{-\varepsilon A^*} Q e^{-\varepsilon A}, \quad i = \overline{0, M-1}. \tag{3.1}$$

Next, for convenience we re-write the iterative linear process (2.2) associated to the nonlinear equation (1.2), namely

$$P_{\varepsilon,k+1} + \varepsilon P_{\varepsilon,k} B N^{-1} B^* P_{\varepsilon,k+1} = Z_\varepsilon^{i+1}. \tag{3.2}$$

Algorithm FRAC_k

1. Initialize parameters
   $\varepsilon, M, kmax, A, B, Q, N, P_0$;
   $P_\varepsilon^-(0) = P_0$;
   for $i = 0$ to $M - 1$ do
   2. Compute Z_ε^{i+1} from (3.1);
   3. Compute P_ε^{i+1} solving (3.2);
      $P_{\varepsilon,0} = Z_\varepsilon^{i+1}$;
      for $k = 0$ to $kmax$ do
         $P_{\varepsilon,k+1} = (I + \varepsilon P_{\varepsilon,k} B N^{-1} B^*)^{-1} Z_\varepsilon^{i+1}$;
      End-for
      $P_\varepsilon^{i+1} = P_{\varepsilon,kmax}$;
   End-for;
End.

The MATLAB code listed below can be used in computing the approximate solution $P_\varepsilon(T)$.

```
A=[1.4 -.208 6.175 -5.676; -.581 -4.29 0 .675;
    1.067 4.273 -6.654 5.893; -.048 4.273 1.343 -2.104];
Q=[1 0 1 -1; 0 1 0 0; 1 0 1 -1; -1 0 -1 1];
B=[0 0; 5.679 0; 1.136 -3.146; 1.136 0];
P0=100*eye(4);
M=11;
e=0.5/M;
N=700*eye(2);
kmax=1;
PM=P0;
```

The graph of Z is

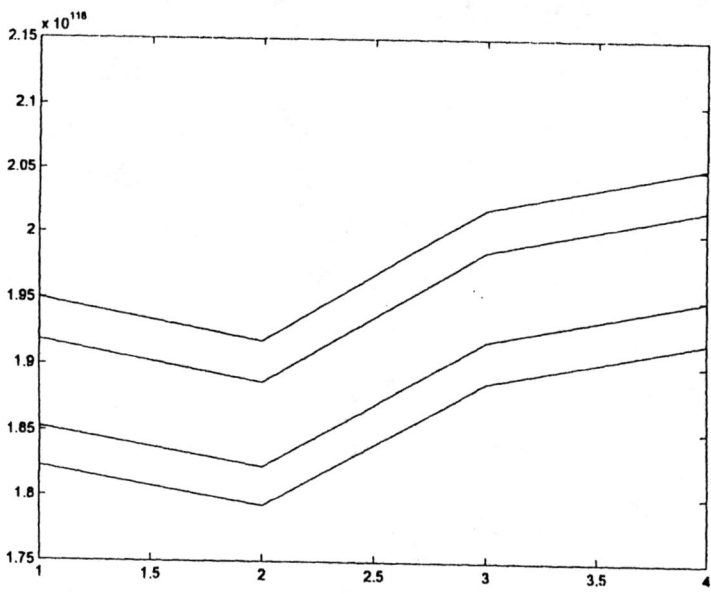

The graph of PK is

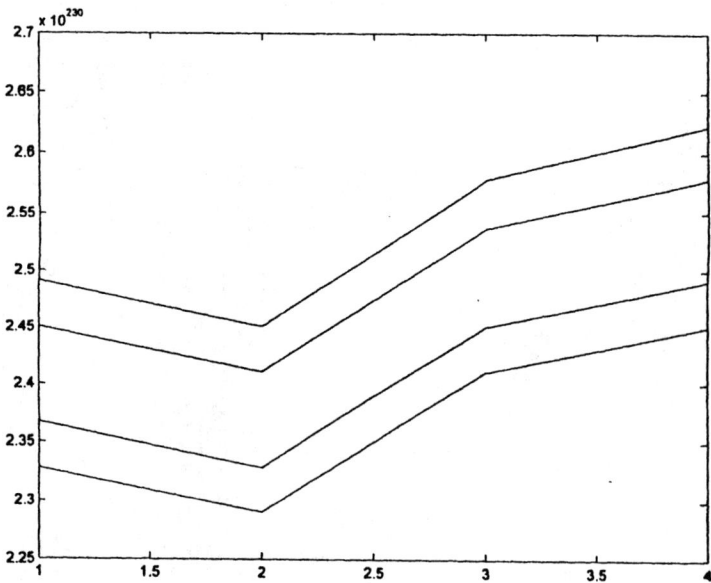

```
for i=0:M
    Z=exp(-e*A')*PM*exp(-e*A)+e*exp(-e*A')*Q*exp(-e*A);
    PK=PM;
    for k=1:kmax
        PK=(eye(4)+e*PK*B*inv(N)*B')'*Z;
    end
    PM=PK;
end
disp(Z);
disp(PK);
plot(Z);
plot(PK);
```

We have obtained the following numerical results for Z and P_K. They are presented exactly as they appear in the MATLAB output screen

$$\text{The matrix } Z$$

$$Z = 1.0e+118 * \begin{pmatrix} 1.6134 & 1.5870 & 1.6703 & 1.6979 \\ 1.4973 & 1.4728 & 1.5501 & 1.5758 \\ 1.8670 & 1.8365 & 1.9328 & 1.9648 \\ 1.5041 & 1.4795 & 1.5571 & 1.5829 \end{pmatrix}$$

$$\text{The matrix } PK$$

$$PK = 1.0e+230 * \begin{pmatrix} 0.0 & 0.0 & 0.0 & 0.0 \\ 1.6972 & 1.6695 & 1.7570 & 1.7861 \\ 3.2879 & 3.2341 & 3.4038 & 3.4601 \\ 2.3082 & 2.2705 & 2.3895 & 2.4291 \end{pmatrix}$$

REFERENCES

[1] V. Barbu, Approximation of the Hamilton-Jacobi equations via Lie- Trotter product formula, *Control-Theory and Advanced Technology*, vol. **4** (1988), No. 2, pp. 189-208.

[2] V. Barbu and G. Da Prato, Hamilton-Jacobi equations in Hilbert spaces, Research Notes in Mathematics, **86**, Pitman, London.

Asymptotic Analysis of the Telegraph System with Nonlinear Boundary Conditions

L. Barbu*, E. Cosma*, Gh. Moroşanu**, and W.L. Wendland**

*Ovidius University, Math. Dept., Bd. Mamaia 124, 8700 Constanţa, Romania

**Stuttgart University, Math. Inst. A, Phaffenwaldring 57, 70569 Stuttgart, Germany

Abstract. The boundary layer function method of Vishik and Lyusternik is used to investigate the behavior of the solution of the problem (1.1), (1.2), (1.3) below as the small parameter ε tends to zero.

1. Introduction

In the domain $D_T := \{(x,t); 0 < x < 1, 0 < t < T\}$ we consider the telegraph system

$$\begin{cases} \varepsilon u_t + v_x + ru = f_1, \\ v_t + u_x + gv = f_2, \end{cases} \quad (1.1)$$

with the following initial and boundary conditions:

$$u(x,0) = u_0(x), \quad v(x,0) = v_0(x), \quad 0 \leq x \leq 1, \quad (1.2)$$

$$\begin{cases} u(0,t) = -\beta_1(v(0,t)), \\ u(1,t) = \beta_2(v(1,t)), \quad 0 \leq t \leq T, \end{cases} \quad (1.3)$$

where $g \geq 0$, $r > 0$ are constants, $f_1, f_2 : \overline{D_T} \to \mathbb{R}$, $\beta_1, \beta_2 : \mathbb{R} \to \mathbb{R}$ are given functions, and ε is a positive parameter.

---

[1] The third author's work has been supported by the German Research Foundation DFG (Deutsche Forschungsgemeinschaft), under the project **DFG We 659/35-2**.

This problem (which will be called P_ε) is a model for an electrical circuit with nonlinear resistors at the ends $x = 0$ and $x = 1$, where u represents the current flowing in the line, and v is the voltage across the line. The parameter ε represents the specific inductance, while the specific capacitance, which usually multiplies v_t in the telegraph system, is assumed to be equal to 1. This does not restrict the generality of the problem. For more details concerning the physical model, we refer the reader, e.g., to [Br1], [Br2], [BrM], [CK2], [Mo].

In the case of integrated circuits, we have a system of $2n$ equations instead of (1.1), but the treatement of the corresponding problem P_ε is essentially similar.

We will assume that $\varepsilon > 0$ is a small parameter. If we put $\varepsilon = 0$, then our problem P_ε becomes a parabolic boundary value problem, say P_0. (Notice that such a parabolic model, extended to the case of n telegraph systems which are connected by means of some appropriate boundary conditions, is applicable to on-chip and interchip interconnections for many digital systems, where the distributed inductances of the conductive layers are negligible [MaN]).

However, from a physical point of view, the parameter ε appearing in our problem P_ε is just small, not zero. So, we ask ourselves if, by putting $\varepsilon = 0$, the resulting parabolic model P_0 still describes the physical phenomenon properly. In other words, the question is whether the solution of the reduced model, P_0, is "sufficiently" close to the solution of the hyperbolic model P_ε. Using the Vishik-Lyusternik method, an asymptotic expansion of the solution of P_ε will be constructed. The presence of some corrector in that asymptotic expansion shows that the reduced parabolic model P_0 is not valid in $\overline{D}_T$, with respect to the C-norm (i.e., the norm of uniform convergence in $\overline{D}_T$).

In Section 2 we will indeed derive a formal zeroth order asymptotic expansion for the solution $(u_\varepsilon, v_\varepsilon)$ of problem P_ε. The segment $\{(x,0); 0 \leq x \leq 1\}$ is a boundary layer and the expansion of u_ε contains a corresponding corrector (boundary layer function), with respect to the C-norm.

In Section 3 we will state some sufficient conditions for the existence, uniqueness, and higher regularity of the solutions to problems P_ε and P_0 (see Theorem 3.1). These conditions are independent of ε.

Finally, in Section 4, we obtain some estimates for the remainder terms of the expansion established in Section 2, with respect to the C-norm. This justifies our asymptotic expansion.

Notice that, if we consider the L^p-norm ($1 \leq p < \infty$) instead of the C-norm, then our corrector (see (2.2), (2.4), and (2.9) below) is not necessary any more (actually, it can be included in the remainder), and this means that our problem is a regularly perturbed one with respect to the L^p-norm (cf., e.g., [E, Chapter 1, §1.1]). On the other hand, taking into account the form of the corrector, we can see that the solution of the reduced problem P_0 approximates uniformly the solution of P_ϵ in any rectangle $[0,1] \times [\delta, T]$ with $\delta > 0$.

Many similar, singularly perturbed boundary value problems can be studied by the same technique, in particular, the case of a nonlinear partial differential system instead of (1.1) as well as the case when the boundary conditions contain time derivatives of the unknowns (see [BaM1], [BaM2], [BaM3], and [MoP]).

The theory of singular perturbations is a huge field, with many different subdivisions, and many considerable efforts have been done in this area. For general or more advanced information we refer the reader to the books: [CK1], [D], [E], [I], [JF], [L], [MaNP], [VBK], and the list may continue. In particular, a large amount of work has been dedicated to hyperbolic-parabolic singularly perturbed problems. See, for instance, [HW1], [HW2], [HW3], and the references therein. However, our problem P_ϵ with the nonlinear boundary conditions (1.3) is considered here for the first time. One of the main difficulties we encounter here is to prove results of higher regularity for the solutions to problems P_ϵ and P_0. This is necessary for the existence of our asymptotic expansion as well as for proving uniform estimates for the remainder components.

2. A formal asymptotic expansion for the solution of P_ε

Let $U_\varepsilon = (u_\varepsilon, v_\varepsilon)$ be a smooth solution of problem P_ε. The existence of such a solution will be discussed in the next section.

Our problem P_ε is a singularly perturbed one, with respect to the C-norm. To this end, let us remark that if the solution U_ε of P_ε converges uniformly in $\overline{D}_T$ to the solution of P_0 then we necessarily have

$$v_0'(x) + ru_0(x) = f_1(x, 0), \quad 0 \leq x \leq 1. \tag{2.1}$$

If (2.1) is not satisfied then that uniform convergence cannot be true and U_ε has a singular behavior in the neighbourhood of the segment $\{(x,0), 0 \leq x \leq 1\}$. This segment is our boundary layer.

According to the Vishik-Lyusternik method (see, e.g., [VBK]), wee seek a solution $U_\varepsilon = (u_\varepsilon, v_\varepsilon)$ of P_ε of the form

$$U_\varepsilon(x,t) = U_0(x,t) + V_0(x,\tau) + R(x,t,\varepsilon), \tag{2.2}$$

where
$\tau = t/\varepsilon$ is the boundary layer variable (or rapid variable);
$U_0(x,t) = (X(x,t), Y(x,t))$ is the first term of the regular series;
$V_0(x,\tau) = (X_1(x,\tau), Y_1(x,\tau))$ is the corresponding boundary layer function;
$R(x,t,\varepsilon) = (R_1(x,t,\varepsilon), R_2(x,t,\varepsilon))$ is the remainder (of order zero).

In order to find V_0 we will take into account the fact that V_0 should be "small enough" far from the boundary layer; more precisely $V_0 \longrightarrow 0$ as $\tau \longrightarrow \infty$, for every x. Now, we will formally substitute (2.2) into (1.1) and then equate the coefficients of the like powers of $\varepsilon^k (k = -1, 0)$, separating those depending on t from those depending on τ. First, we can see that Y_1 satisfies the problem

$$\begin{cases} Y_{1\tau} = 0, \\ Y_1 \to 0, \quad \text{for } \tau \to \infty \text{ and for every } x, 0 \leq x \leq 1. \end{cases}$$

Therefore,

$$Y_1 \equiv 0. \tag{2.3}$$

Also, X_1 satisfies the equation

$$X_{1\tau} + rX_1 = 0,$$

and, hence,

$$X_1 = \alpha(x)e^{-r\tau}, \tag{2.4}$$

where the function α will be found later. Furthermore, assuming that the remainder R is "small enough", we can see that X and Y satisfy the so-called unperturbed system

$$\begin{cases} Y_x + rX = f_1, \\ Y_t + X_x + gY = f_2, \quad \text{in } D_T. \end{cases} \tag{2.5}$$

Clearly, (2.5) can be written in the form

$$X = (1/r)(f_1 - Y_x), \tag{2.6}$$

Telegraph System with Nonlinear Boundary Conditions

$$Y_t - (1/r)Y_{xx} + gY = f_2 - (1/r)f_{1x}. \tag{2.7}$$

Finally, the remainder R satisfies the system

$$\begin{cases} \varepsilon R_{1t} + R_{2x} + rR_1 = -\varepsilon X_t, \\ R_{2t} + R_{1x} + gR_2 = -X_{1x}. \end{cases} \tag{2.8}$$

Now, let us formally substitute (2.2) into (1.2). We remark that X_1 has to compensate for the initial condition for X which is not necessary any more because X satisfies an algebraic equation. So we get

$$\alpha(x) = u_0(x) + (1/r)(v_0'(x) - f_1(x,0)), \ 0 \le x \le 1, \tag{2.9}$$

and the following initial conditions

$$Y(x,0) + Y_1(x,0) = v_0(x) \Leftrightarrow Y(x,0) = v_0(x), \ 0 \le x \le 1, \tag{2.10}$$

$$R_1(x,0,\varepsilon) = 0, \ R_2(x,0,\varepsilon) = 0. \tag{2.11}$$

Analogously, by substituting (2.2) into (1.3), we obtain

$$X(0,t) + \beta_1(Y(0,t)) = 0, \ X(1,t) - \beta_2(Y(1,t)) = 0, \ 0 \le t \le T, \tag{2.12}$$

$$\begin{cases} R_1(0,t,\varepsilon) + \beta_1(R_2(0,t,\varepsilon) + Y(0,t)) - \beta_1(Y(0,t)) = 0, \\ R_1(1,t,\varepsilon) - \beta_2(R_2(1,t,\varepsilon) + Y(1,t)) + \beta_2(Y(1,t)) = 0, \ 0 \le t \le T. \end{cases} \tag{2.13}$$

From (2.6) and (2.12) we get

$$\begin{cases} -(1/r)Y_x(0,t) + \beta_1(Y(0,t)) = -(1/r)f_1(0,t), \\ (1/r)Y_x(1,t) + \beta_2(Y(1,t)) = (1/r)f_1(1,t), \ 0 \le t \le T, \end{cases} \tag{2.14}$$

$$X_1(0,\tau) = X_1(1,\tau) = 0 \Leftrightarrow \alpha(0) = \alpha(1) = 0. \tag{2.15}$$

Summarizing, the reduced problem, (P_0), is defined by (2.7) – (2.10) – (2.14), while the problem satisfied by the remainder terms is given by (2.8) – (2.11) – (2.13).

3. Existence and higher regularity for the perturbed and for the reduced problem

The aim of this section is to state some sufficient conditions for the existence and higher order regularity of the solutions to problems P_ε and P_0. This requires higher regularity of u_0, v_0, f_1, f_2, β_1, β_2 as well as higher order compatibility conditions. We need higher regularity not only for the existence of our asymptotic expansion (i.e., the existence of all functions involved in this expansion), but also for proving estimates for the remainder components.

The main difficulties in the study of these problems are due to the nonlinearity of the boundary conditions (1.3) and (2.14).

Theorem 3.1. *Assume that*

$$r > 0, \ g \geq 0, \ \beta_1, \ \beta_2 \in W^{2,\infty}_{loc}(\mathbb{R}), \ \beta_1' \geq 0, \ \beta_2' \geq 0; \tag{3.1}$$

$$\begin{cases} f_1 \in W^{2,1}(0,T; H^1(0,1)), \\ f_2 \in W^{2,1}(0,T; L^2(0,1)) \cap C^1([0,T]; C[0,1]), \\ f_1(\cdot,0) \in H^3(0,1), \ f_2(\cdot,0) \in H^2(0,1) \ f_1(0,\cdot), \ f_2(1,\cdot) \in W^{3,1}(0,T); \end{cases} \tag{3.2}$$

$$u_0 \in C^1([0,1]), \ v_0 \in H^4(0,1); \tag{3.3}$$

and that the following compatibility conditions are satisfied,

$$\begin{cases} u_0(0) + \beta_1(v_0(0)) = 0, \\ u_0(1) - \beta_2(v_0(1)) = 0, \end{cases} \tag{3.4}$$

$$\begin{cases} f_1(0,0) - ru_0(0) - v_0'(0) = 0, \\ \beta_1'(v_0(0))\big(f_2(0,0) - gv_0(0) - u_0'(0)\big) = 0, \\ f_1(1,0) - ru_0(1) - v_0'(1) = 0, \\ \beta_2'(v_0(1))\big(f_2(1,0) - gv_0(1) - u_0'(1)\big) = 0, \end{cases} \tag{3.5}$$

$$\begin{cases} r^{-1}v_0^{(3)}(0) - gv_0'(0) + f_{2x}(0,0) - r^{-1}f_{1xx}(0,0) - r\beta_1'(v_0(0)) \times \\ \quad \times \big(r^{-1}v_0''(0) - gv_0(0) + f_2(0,0) - r^{-1}f_{1x}(0,0)\big) = f_{1t}(0,0), \\ r^{-1}v_0^{(3)}(1) - gv_0'(1) + f_{2x}(1,0) - r^{-1}f_{1xx}(1,0) + r\beta_2'(v_0(1)) \times \\ \quad \times \big(r^{-1}v_0''(1) - gv_0(1) + f_2(1,0) - r^{-1}f_{1x}(1,0)\big) = f_{1t}(1,0). \end{cases} \tag{3.6}$$

Then problem P_ε has a unique solution $U_\varepsilon = (u_\varepsilon, v_\varepsilon) \in C^1(\overline{D}_T)^2$, and also problem P_0 has a unique solution $Y \in H^2(0,T; H^1(0,1))$.

For the proof of Theorem 3.1 and for other related details the reader is refered to [AMH], [BaM3], and [BaM4].

Remark. *From the conditions (3.4), (3.5), and (3.6) there follows (2.15).*

4. Estimates for the remainder terms of the asymptotic expansion

The main result of this section is formulated in the following theorem.

Theorem 4.1. *Assume that (3.1) – (3.6) are satisfied. Then, for every $\varepsilon > 0$, problem P_ε has a unique solution U_ε of the form (2.2), where U_0 is determined from (2.6) and (2.7) – (2.10) – (2.14), $V_0(x,\tau) = (\alpha(x)e^{-r\tau}, 0)$, with $\alpha(x)$ given by (2.9). The remainder R satisfies (2.8) – (2.11) – (2.13).*

Moreover, we have the following estimate

$$\|R(\cdot,\cdot,\varepsilon)\|_{C(\overline{D}_T)^2} \leq M(\varepsilon^{1/8}),$$

where M depends on r, g, β_1, β_2, u_0, v_0, f_1, f_2, and T.

Proof. We can basically use the technique of [BaM1], [BaM3], and [MoP]. However, for completeness, we will give here the full proof. By Theorem 3.1, the problem (2.8) – (2.11) – (2.13) has a unique, smooth solution,

$$R(x,t,\varepsilon) := (R_1(x,t,\varepsilon), R_2(x,t,\varepsilon)) \text{ for } (x,t) \in \overline{D}_T.$$

Let $H := (L^2(0,1))^2$ which is a Hilbert space, with the scalar product

$$\langle p, q \rangle := \varepsilon \int_0^1 p_1(x)p_2(x)dx + \int_0^1 q_1(x)q_2(x)dx \text{ for any } p,q \in H,$$

$$p := (p_1, p_2), \quad q := (q_1, q_2).$$

Denote by $\|\cdot\|$ the corresponding Hilbert norm. We define the operator

$$B_\varepsilon(t) : D(B_\varepsilon(t)) \subset H \to H.$$

$$D(\mathcal{B}_\varepsilon(t)) := \{(p,q) \in (H^1(0,1))^2, \ p(0) + \beta_1(q(0) + Y(0,t)) = \beta_1(Y(0,t)),$$
$$p(1) - \beta_2(q(1) + Y(1,t)) = -\beta_2(Y(1,t))\},$$
$$\mathcal{B}_\varepsilon(t)(\mathrm{col}(p,q)) := (\varepsilon^{-1}q' + r\varepsilon^{-1}p, \ p' + gq), \quad 0 \le t \le T.$$

It is known that $\mathcal{B}_\varepsilon(t)$ is a maximal monotone operator for $0 \le t \le T$ [Mo, Chapter III, §4], and the problem (2.8) − (2.11) − (2.13) may be written as the following Cauchy problem in H.

$$\begin{cases} R'_\varepsilon(t) + \mathcal{B}_\varepsilon(t)(R_\varepsilon(t)) = F_\varepsilon(t), & 0 \le t \le T, \\ R_\varepsilon(0) = 0, \end{cases} \quad (4.1)$$

where $R_\varepsilon(t) := R(\cdot, \varepsilon)$, $F_\varepsilon(t) := (-X_t(\cdot, t), -X_{1x}(\cdot, \tau))$, $0 < t < T$.

Because $\mathcal{B}_\varepsilon(t)$ is a maximal monotone operator and $(0,0) \in D(\mathcal{B}_\varepsilon(t))$, we obtain that R_ε is the unique strong solution of problem (4.1).

Let us multiply $(4.1)_1$ by $R_\varepsilon(t)$ and then integrate over $[0,t]$:

$$\|R_\varepsilon(t)\|^2 + 2\int_0^t \langle \mathcal{B}_\varepsilon(s)R_\varepsilon(s), R_\varepsilon(s)\rangle ds = 2\int_0^t \langle F_\varepsilon(s), R_\varepsilon(s)\rangle ds, \quad (4.2)$$

where $\langle \cdot, \cdot \rangle$ denotes the scalar product of $L^2(0,1)$. Because $\langle \mathcal{B}_\varepsilon(s)R_\varepsilon(s), R_\varepsilon(s)\rangle \ge 0$ for $0 \le s \le T$ (notice that β_1, β_2 are both nondecreasing), we get

$$\|R_\varepsilon(t)\|^2 \le 2\left(\int_0^t \|F_\varepsilon(s)\|^2 ds\right)^{1/2} \left(\int_0^t \|R_\varepsilon(s)\|^2 ds\right)^{1/2}. \quad (4.3)$$

On the other hand,

$$\|F_\varepsilon(s)\|^2 = \varepsilon \int_0^1 X_s(x,s)^2 dx + \left(\int_0^1 \alpha'(x)^2 dx\right) e^{-2rs/\varepsilon},$$

and therefore

$$\int_0^T \|F_\varepsilon(s)\|^2 ds \le M_1 \varepsilon. \quad (4.4)$$

Denote $h_\varepsilon(t) := \int_0^t \|R_\varepsilon(s)\|^2 ds$. From (4.3), we can deduce that $h'_\varepsilon(s) \le 2\sqrt{M_1 \varepsilon}\sqrt{h_\varepsilon(s)}$, and integrating with respect to s over $[0,t]$, we get

$$h_\varepsilon(t) \le M_2 \varepsilon \text{ and } h'_\varepsilon(t) \le 2\sqrt{M_1\varepsilon}\sqrt{M_2\varepsilon} = M_3\varepsilon. \quad (4.5)$$

From $(4.5)_2$, we can deduce that

$$\|R_\varepsilon(t)\|^2 = \varepsilon\|R_1(\cdot,t,\varepsilon)\|_0^2 + \|R_2(\cdot,t,\varepsilon)\|_0^2 \le M_3\varepsilon. \quad (4.6)$$

where $\|\cdot\|_0$ denotes the norm of $L^2(0,1)$.

As β_1, β_2 are nondecreasing, we have

$$\int_0^T \langle \mathcal{B}_\varepsilon(s) R_\varepsilon(s), R_\varepsilon(s)\rangle ds \geq g\int_0^T \|R_2^2(\cdot, s, \varepsilon)\|_0^2 ds + r\int_0^T \|R_1(\cdot, s, \varepsilon)\|_0^2 ds. \quad (4.7)$$

From (4.2), (4.4), (4.6), and (4.7) we can deduce that

$$\int_0^t (r\|R_1(\cdot, s, \varepsilon)\|_0^2 + g\|R_2(\cdot, s, \varepsilon)\|_0^2) ds \leq M_4 \varepsilon, \quad 0 \leq t \leq T. \quad (4.8)$$

Because $Y \in H^2(0, T; H^1(0,1))$ (cf. Theorem 3.1 above), we obtain from (2.6) that $X \in H^2([0,T]; L^2(0,1))$, so

$$Q_\varepsilon(t) := R'_\varepsilon(t) = \left(R_{1t}(\cdot, t, \varepsilon), R_{2t}(\cdot, t, \varepsilon)\right)$$

is the unique solution of the Cauchy problem in H:

$$\begin{cases} Q'_\varepsilon(t) + \mathcal{E}_\varepsilon(t)(Q_\varepsilon(t)) = F'_\varepsilon(t), & 0 < t < T, \\ Q_\varepsilon(0) = F_\varepsilon(0), \end{cases} \quad (4.9)$$

where $\mathcal{E}_\varepsilon(t): D(\mathcal{E}_\varepsilon(t)) \subset H \to H$,

$$D(\mathcal{E}_\varepsilon(t)) = \big\{(p, q) \in H^1(0,1), \ p(0) + \beta'_1(R_2(0, t, \varepsilon) + Y(0, t))(q(0)+$$

$$+Y_t(0, t)) = \beta'_1(Y(0, t))Y_t(0, t), \ p(1) - \beta'_2(R_2(1, t, \varepsilon) + Y(1, t))[q(1) + Y_t(1, t)] =$$

$$= -\beta'_2(Y(1, t))Y_t(1, t)\big\},$$

$$\mathcal{E}_\varepsilon(t)((p, q)) = \mathcal{B}_\varepsilon(t)((p, q)) \ \text{ for } \ 0 \leq t \leq T.$$

Now, we multiply $(4.9)_1$ by $Q_\varepsilon(t)$ and then integrate over $[0, t]$:

$$\|Q_\varepsilon(t)\|^2 + 2r\|R_{1t}\|^2_{L^2(D_T)} + 2g\|R_{2t}\|^2_{L^2(D_T)}+$$

$$+2\int_0^t \left(\beta'_1(R_2(0, s, \varepsilon) + Y(0, s)) - \beta'_1(Y(0, s))\right) Y_s(0, s) R_{2s}(0, s, \varepsilon) ds+$$

$$+2\int_0^t \left(f'_0(R_2(1, s, \varepsilon) + Y(1, s)) - f'_0(Y(1, s))\right) Y_s(1, s) R_{2s}(1, s, \varepsilon) ds \leq$$

$$\leq \|Q_\varepsilon(0)\|^2 + 2\int_0^t \|F'_\varepsilon(s)\| \|Q_\varepsilon(s)\| ds. \quad (4.10)$$

Obviously, $\|Q_\varepsilon(0)\| \leq M_5$.

In order to continue the proof, we need the following auxiliary result:

Lemma 4.2. *If the assumptions of Theorem 4.1 hold, then $\|v(\cdot,\cdot,\varepsilon)\|_{C(\overline{D}_T)} \leq C$, where C is independent of ε.*

Proof. We define the operator $\mathcal{B}_{1\varepsilon}: D(\mathcal{B}_{1\varepsilon}) \subset H \to H$ by

$$D(\mathcal{B}_{1\varepsilon}) := \{col(p,q) \in H^1(0,1),\ p(0) + \beta_1(q(0)) = 0,\ p(1) - \beta_2(q(1)) = 0\},$$

$$\mathcal{B}_{1\varepsilon}(col(p,q)) = (\varepsilon^{-1}q' + r\varepsilon^{-1}p,\ p' + gq).$$

It is obvious that problem P_ε may be written as the following Cauchy problem in H,

$$\begin{cases} U'_\varepsilon(t) + \mathcal{B}_{1\varepsilon}(U_\varepsilon(t)) = G_\varepsilon(t), & 0 < t < T \\ U_\varepsilon(0) = (u_0, v_0), \end{cases} \tag{4.11}$$

where $U_\varepsilon(t) := (u(\cdot,t,\varepsilon), v(\cdot,t,\varepsilon))$, $G_\varepsilon(t) := (\varepsilon^{-1}f_1(\cdot,t), f_2(\cdot,t))$.

Now, it is easily seen that

$$\|U'_\varepsilon(t)\| \leq \|G_\varepsilon(0) - \mathcal{B}_{1\varepsilon}((u_0,v_0))\| + + \int_0^t \|G'_\varepsilon(s)\| ds \leq C_1 \varepsilon^{-1/2} \text{ for } 0 \leq t \leq T.$$

Therefore,

$$\varepsilon\|u_t(\cdot,t,\varepsilon)\|_0^2 + \|v_t(\cdot,t,\varepsilon)\|_0^2 \leq C_1^2 \varepsilon^{-1}. \tag{4.12}$$

From (4.6) we derive

$$\|v(\cdot,t,\varepsilon)\|_0 \leq \|Y(\cdot,t)\|_0 + \|R_2(\cdot,t,\varepsilon)\|_0 \leq C_2, \tag{4.13}$$

$$\|u(\cdot,t,\varepsilon)\|_0 \leq \|X(\cdot,t)\|_0 + \|X_1(\cdot,\tau)\|_0 +$$

$$+ \|R_1(\cdot,t,\varepsilon)\|_0 \leq C_3,\ 0 \leq t \leq T,\ \varepsilon > 0. \tag{4.14}$$

On the other hand, from (1.1a), $v_x = f_1 - \varepsilon u_t - ru$, in D_T, and, further, according to (4.12) and (4.14), we can see that

$$\|v_x(\cdot,t,\varepsilon)\|_0 \leq C_4 \text{ for } 0 \leq t \leq T \text{ and } \varepsilon > 0. \tag{4.15}$$

Using now (4.13) and (4.15) it follows that $\|v(\cdot,\cdot,\varepsilon)\|_{C(\overline{D}_T)} \leq C$, as asserted (notice that $H^1(0,1) \hookrightarrow C[0,1]$), which completes the proof of the lemma.

Telegraph System with Nonlinear Boundary Conditions

We now continue the proof of Theorem 4.1. By Lemma 4.2 it follows that

$$\int_0^t \left(\beta_1'(v(0,s,\varepsilon)) - \beta_1'(Y(0,s))\right) Y_s(0,s)(v_s(0,s,\varepsilon) - Y_s(0,s))ds \geq$$

$$\geq \int_0^t \left(\beta_1'(v(0,s,\varepsilon))v_s(0,s,\varepsilon)\right) Y_s(0,s)ds - K_1^2 -$$

$$- \int_0^t \beta_1'(Y(0,s))Y_s(0,s)v_s(0,s,\varepsilon)ds \geq$$

$$\geq \beta_1(v(0,s,\varepsilon))Y_s(0,s)\big|_0^t - \beta_1'(Y(0,s))Y_s(0,s)v(0,s,\varepsilon)\big|_0^t -$$

$$- \int_0^t \beta_1(v(0,s,\varepsilon))Y_{ss}(0,s)ds +$$

$$+ \int_0^t \left(\beta_1'(Y(0,s))Y_s(0,s)\right)_s v(0,s,\varepsilon)ds - K_1^2 \geq -K_2^2 \quad \text{for} \ 0 \leq t \leq T. \tag{4.16}$$

Analogously,

$$\int_0^t \left(f_0'(v(1,s,\varepsilon)) - f_0'(Y(1,s))\right) Y_s(1,s) R_{2s}(1,s)ds \geq -K_3^2. \tag{4.17}$$

From (4.16), (4.17) and (4.10) we can deduce that

$$\|Q_\varepsilon(t)\|^2 + 2\left(r\|R_{1t}\|^2_{L^2(D_T)} + g\|R_{2t}\|^2_{L^2(D_T)}\right) \leq$$

$$\leq M_6 + 2\int_0^t \|F_\varepsilon'(s)\| \|Q_\varepsilon(s)\| ds \quad \text{for} \ 0 \leq t \leq T. \tag{4.18}$$

On the other hand,

$$2\int_0^t \|F_\varepsilon'(s)\| \|Q_\varepsilon(s)\| ds \leq 2\varepsilon \int_0^t \|X_{ss}(\cdot,s)\|_0 \|R_{1s}(\cdot,s,\varepsilon)\|_0 ds +$$

$$+ 2(r/\varepsilon)\|\alpha'\|_0 \int_0^t e^{-rs/\varepsilon} \|R_{2s}(\cdot,s,\varepsilon)\|_0 ds \leq$$

$$\leq M_7 + \varepsilon \int_0^t \|R_{1s}(\cdot,s,\varepsilon)\|_0^2 ds + + (r/\varepsilon)\int_0^t e^{-rs/\varepsilon}\left(\|R_{2s}(\cdot,s,\varepsilon)\|_0^2 ds \tag{4.19}$$

(because

$$2(r/\varepsilon)K \int_0^t e^{-rs/\varepsilon} |f(s)| ds \leq$$

$$\leq 2(r/\varepsilon) \int_0^t (1/2)(K^2 e^{-rs/\varepsilon} + e^{-rs/\varepsilon} |f(s)|^2)ds =$$

$$= K^2(1 - e^{-rt/\varepsilon}) + (r/\varepsilon)\int_0^t e^{-rs/\varepsilon}|f(s)|^2)ds.$$

where $K = \|\alpha'\|_0, |f(s)| = \|R_{2s}(\cdot, s, \varepsilon)\|_0)$.

From (4.18) and (4.19) we obtain that

$$\|Q_\varepsilon(t)\|^2 \leq M_8 + \int_0^t (1 + r/\varepsilon)e^{-rs/\varepsilon}\|Q_\varepsilon(s)\|^2 ds \text{ for } 0 \leq t \leq T$$

and so, by Gronwall's lemma, $\|Q_\varepsilon(t)\| \leq M_9$.

Therefore,

$$\varepsilon\|R_{1t}(\cdot, t, \varepsilon)\|_0^2 + \|R_{2t}(\cdot, t, \varepsilon)\|_0^2 \leq M_{10}, \tag{4.20}$$

$$\|R_{1t}\|_{L^2(D_T)}^2 \leq M_{11}, \|R_{1t}\|_{L^2(D_T)}^2 \leq M_{12}. \tag{4.21}$$

But

$$\|R_i(\cdot, t, \varepsilon)\|_0^2 \leq 2\|R_{it}\|_{L^2(D_T)}\|R_i\|_{L^2(D_T)}, \; i = 1, 2,$$

so, using (4.8) and (4.21), we get

$$\|R_i(\cdot, t, \varepsilon)\|_0^2 \leq M_{13}\varepsilon^{1/2}, \; i = 1, 2. \tag{4.22}$$

From (2.8), (4.6) and (4.21) we can obtain

$$\|R_{kx}(\cdot, t, \varepsilon)\|_0 \leq M_{14}, \; k = 1, 2. \tag{4.23}$$

Let $t \in [0, T]$. From (4.22) we obtain that there exists $x_{t\varepsilon}^i \in [0, 1]$ such that $R_i^2(x_{t\varepsilon}^i, t, \varepsilon) \leq M_{16}\varepsilon^{1/2}$. Because

$$R_i^2(x, t, \varepsilon) = R_i^2(x_{t\varepsilon}^i, t, \varepsilon) + 2\int_{x_{t\varepsilon}^i}^x R_{i\xi}(\xi, t, \varepsilon)R_i(\xi, t, \varepsilon)d\xi \leq$$

$$\leq R_i^2(x_{t\varepsilon}^i, t, \varepsilon) + 2\|R_i(\cdot, t, \varepsilon)\|_0 \|R_{ix}(\cdot, t, \varepsilon)\|_0,$$

we have that $R_i^2(x, t, \varepsilon) \leq M_{17}\varepsilon^{1/4}, \; i = 1, 2$, so

$$\|R(\cdot, \cdot, \varepsilon)\|_{C(\overline{D}_T)^2} \leq M\varepsilon^{1/8}.$$

Q.E.D.

References

[AHM] S. Aizicovici, V. M. Hokkanen, and Gh. Moroşanu, Existence and regularity for a class of nonlinear hyperbolic boundary value problems, *J. Math. Anal. Appl.*, submitted.

[BaM1] L. Barbu and Gh. Moroşanu, Asymptotic analysis of the telegraph equations with non-local boundary value conditions, *PanAmer. Math. J.* **8** (1998), no. 4, 13-22.

[BaM2] L. Barbu and Gh. Moroşanu, A first order asymptotic expansion of the solution of a singularly perturbed problem for the telegraph equations, *Applicable Analysis* **72**(1-2) (1999), 111-125.

[BaM3] L. Barbu and Gh. Moroşanu, *Asymptotic Analysis of Some Boundary Value Problems with Singular Perturbations*, Publ. House of the Romanian Academy, Bucharest, 2000 (in Romanian).

[BaM4] L. Barbu and Gh. Moroşanu, High regularity for the solution of a parabolic boundary value problem, in preparation.

[Br1] R.K. Brayton, Bifurcation of periodic solutions in a nonlinear difference-differential equation of neutral type, *Quart. Appl. Math.* **24** (1966/67), 215-224.

[Br2] R.K. Brayton, Nonlinear oscillations in a distributed network, *Quart. Appl. Math.* **24** (1966/67), 289-301.

[BrM] R.K. Brayton and W.L. Miranker, A stability theory for nonlinear mixed initial boundary value problems, *Arch. Rational Mech. Anal.* **17** (1964), 358-376.

[CK1] J.D. Cole and J. Kevorkian, *Multiple Scale and Singular Perturbation Methods*, Appl. Math. Sciences **114**, Springer, New York, 1996.

[CK2] K. L. Cooke and D. W. Krumme, Differential-difference equations and nonlinear initial boundary value problems, *J. Math. Anal. Appl.* **24** (1968), 372-387.

[D] M. van Dyke, *Perturbation Methods in Fluid Mechanics*, Annotated Edition, The Parabolic Press, Stanford, California, 1975.

[E] W. Eckhaus, *Mathched Asymptotic Expansions and Singular Perturbations*, Mathematics Studies **6**, North Holland/American Elsevier, 1973.

[HW1] G.C. Hsiao and R.J. Weinacht, A singularly perturbed Cauchy problem, *J. Math. Anal. Appl.* **71** (1979), 242-250.

[HW2] G.C. Hsiao and R.J. Weinacht, Singular perturbations for a weakly nonlinear hyperbolic equation *Appl. Anal.* **10** (1980), 221-229.

[HW3] G.C. Hsiao and R.J. Weinacht, Singular perturbations for a semi-linear hyperbolic equation. *SIAM J. Math. Anal.* **14** (1983), 1168-1179.

[I] M. Il'in, *Matching of Asymptotic Expansions of Solutions of Boundary Value Problems*, Translations of Math. Monographs **102**, Amer. Math. Society, Providence, Rhode Island, 1992 (translated from the Russian by V.V. Minakhin).

[JF] E.M. de Jager and J. Furu, *The Theory of Singular Perturbations*, Appl. Math. and Mechanics, Vol.42, North Holland/Elsevier, Amsterdam, 1996.

[L] J.L. Lions, *Perturbations Singulieres dans les Problemes aux Limites et en Controle Optimal*, Lecture Notes in Math. **323**, Springer, Berlin-Heidelberg-New York, 1973.

[MaN] C.A. Marinov and P. Neittaanmaki, *Mathematical Models in Electrical Circuits: Theory and Applications*, Kluwer, Dordrecht, 1991.

[MaNP] V. Maz'ya, S. Nazarov and A. Plamenevskii, *Asymptotic Theory of Elliptic Boundary Value Problems in Singularly Perturbed Domains*, Volumes I and II, Birkhauser, Basel, 2000.

[Mo] Gh. Moroşanu, *Nonlinear Evolution Equations and Applications*, Reidel, Dordrecht, 1988.

[MoP] Gh. Moroşanu and A. Perjan, The singular limit of telegraph equations, *Comm. Appl. Nonlinear Anal.* **5** (1998), no. 1, 91-106.

[VBK] A.B. Vasilieva, V.F. Butuzov and I.V. Kalashev, *The Boundary Function Method for Singular Perturbation Problems*, SIAM, Philadelphia, 1995.

Global Existence for a Class of Dispersive Equations

Radu C. Cascaval
University of Missouri
Columbia, MO 65211

Abstract

We study the existence of solutions for a class of dispersive equations which includes the generalized Korteweg-de Vries equation and the Benjamin-Ono equation. An explicit method for generating solutions is provided by a semilinear Hille-Yosida type semigroup theory, allowing to obtain - at the same time - estimates for different norms of the solution. The semilinear Hille-Yosida theory has been developed in a general setting by S. Oharu and T. Takahashi and extended by J. Goldstein to cover equations such as the generalized KdV equation, for the real line case [23] [25]. One of our purposes here is to show that this semigroup approach extends to a larger class of dispersive equations, including the generalized Benjamin-Ono equation, and applies to the real line as well as to the periodic case.

1 Introduction

In this paper we study the global existence of smooth solutions and the well-posedness of the following Cauchy problem (written in general form):

$$\frac{du}{dt} - Mu_x + F(u)_x = 0, \qquad \text{(NDE)}$$
$$u(0) = u_0,$$

with the linear dispersive term $-Mu_x$, M being the linear operator given by a multiplication operation in the Fourier space: $\widehat{Mv}(\xi) = |\xi|^{2\beta}\widehat{v}(\xi), \beta \geq \frac{1}{2}$. The nonlinearity is of the form $F(u)_x = F'(u)u_x$, with $F: \mathbb{R} \to \mathbb{R}$ a (sufficiently) smooth function satisfying the following growth condition:

$$\limsup_{|r| \to \infty} \frac{F'(r)}{|r|^p} < \infty \text{ for some } p < 4\beta. \qquad (\mathcal{C})$$

In many situations it is instructive to consider the particular nonlinearity $F(u)_x = u^p u_x$, for some power $p \geq 1$, (with the convention $u^p = |u|^{p-1}u$ for non-integer powers p). For $p = 1$, (NDE) generalizes the KdV equation ($\beta = 1$), the Benjamin-Ono ($\beta = \frac{1}{2}$) and the fifth-order KdV ($\beta = 2$). For other integer powers $p = k$, one obtains the (KdV$_k$) equations (see below), which exhibit many interesting features. We address the question of global existence of solutions and well-posedness in the Sobolev spaces $H^s(\mathbb{T})$. Our results are also valid in the context of $H^s(\mathbb{R})$, but we shall not bother to focus on this case since it is similar to and slightly easier than the $H^s(\mathbb{T})$ case.

We begin by mentioning some of the previous work done in the literature. Many authors have studied the initial-value problem for the generalized KdV equation:

$$u_t + u_{xxx} + F(u)_x = 0, \qquad \text{(GKdV)}$$
$$u(0) = u_0,$$

and, in particular,

$$u_t + u_{xxx} + u^k u_x = 0, \qquad \text{(KdV}_k\text{)}$$
$$u(0) = u_0.$$

Here $x \in \mathbb{R}$ (real line case) or $x \in \mathbb{T}$ (periodic case). The first result on global well-posedness for (GKdV) was proved by Bona and Smith [8] in 1978, for the classical KdV equation, i.e. $F(u) = \frac{u^2}{2}$. Later, Kato [29], in 1983, developed a theory for quasi-linear evolution equations, and applied it to equation (GKdV). This was the first place where it was realized that global existence of solutions depends in a precise way on the nonlinearity. In a series of papers, e.g. [1], [7], Bona et al have studied the more general forms of this equation, with generalized dispersion relations. See also Dix [16], Ginibre-Velo [20], [21], Saut [42].

It is worth noting that the signs of the dispersion term and of the nonlinearity play an important role in the global existence of the solution $u(t)$. As it will be seen in Chapter 4, global existence in time is guaranteed by the uniform boundedness of the $H^\beta(\mathbb{T})$ norm of the solutions. If condition $(\mathcal{C})$ is assumed, this uniform bound can be automatically derived from the conserved quantities. The condition $p < 4\beta$ imposed in $(\mathcal{C})$ is essential to guarantee the boundedness of the norm mentioned above. For $\beta = 1$ and $p \geq 4$, Bona et al, [5], obtained numerical evidence of blow-up of u_x in finite time. It is widely believed that the Cauchy problem is not globally well posed for $\beta = 1$ and $p \geq 4$, but no complete proof is available. In general, for $p \geq 4\beta$, one can still get uniform bounds for the H^β-norm of the solution if the initial data are small enough (in a sense to be made precise later).

One interesting problem that has been recently addressed in the literature is that of obtaining optimal results with respect to the smoothness of the solution. That is, one is interested in solving the Cauchy problem for "rough" data. This direction has been pursued in the literature through the leading work of Kenig, Ponce, Vega (see [32], [33], [34], [35]), Bourgain ([10]), Staffilani ([44]) and many others. Most of the methods used are applied also to other dispersive equations, such as the nonlinear Schrödinger equation and the Zakharov system. A related question is the growth of the Sobolev norms of solutions of such systems. Results have been obtained in this direction, for example, in [44].

Despite the legitimate interest in these directions of research, in the present work we do not concentrate on (optimal) small values of s. Instead we plan to study the relationship between the dispersion effects and the strength of the nonlinearity in the usual functional setting of the Sobolev spaces.

The main result in this paper is:

Theorem 1. *The Cauchy problem (NDE) is globally well-posed in $H^s(\mathbb{T})$ with $s = \max\{2\beta, \frac{3}{2} + \varepsilon\}$ for some $\varepsilon > 0$, provided that the growth condition $(\mathcal{C})$ is satisfied and we are in any of the following cases:*

- $\beta = 1$ and $s = 2$;

- $\beta = \frac{1}{2}$ and $s \in (\frac{3}{2}, 2]$ and $F'(u) = u$;

- $\beta > \frac{3}{2}$ and $s = 2\beta$.

For $\frac{1}{2} < \beta \leq \frac{3}{2}$, $\beta \neq 1$, (NDE) is locally well-posed in $H^s(\mathbb{T})$, with $s = 2\beta$ for $\beta > \frac{3}{4}$ and $s \in (\frac{3}{2}, 2\beta + 1]$ for $\frac{1}{2} < \beta \leq \frac{3}{4}$.

In particular, we obtain well-posedness results for (GKdV), when $\beta = 1$.

Theorem 2. *The Cauchy problem for the periodic (GKdV) equation is globally well-posed in $H^2(\mathbb{T})$.*

Theorems 1 and 2 also hold in the real-line case, i.e. on the spaces $H^s(\mathbb{R})$, where $s = \max\{2\beta, \frac{3}{2} + \varepsilon\}$ and ε chosen as in Theorem 1. In both cases, the growth condition $(\mathcal{C})$ on the nonlinearity is assumed.

The plan of the paper is as follows. Section 2 describes the abstract semilinear Hille-Yosida theory used in the proof of Theorem 1. The proof of the abstract theorem is contained in Section 3. Finally, in section 4 we prove that the operators and functionals assiciated with (NDE), satisfy all the assumptions that make the abstract semilinear theory work. We conclude with a list of open problems.

2 Semilinear Hille-Yosida theory

The abstract semilinear Hille-Yosida theory has been developed in [39], [40], and adapted in [25] for some situations not covered by the standard Crandall-Liggett theory. Here is the setting of the abstract theorem.

Let X be a real Hilbert space with norm $\|\cdot\|$ and inner product $(\cdot, \cdot)$. In the applications we may consider a complex Hilbert space $(H, \langle \cdot, \cdot \rangle)$ to be a real one $(H, (\cdot, \cdot))$ by taking $(x, y) = Re \langle x, y \rangle$.

C is a subset of X, $u_0 \in C$. $\varphi = (\varphi_1, \ldots, \varphi_N)$ are N lower semi-continuous functionals on X, with $D(\varphi_k) = C$, for $k = 1, \ldots, N$. For $\alpha = (\alpha_1, \ldots, \alpha_N) \in \mathbb{R}^N$, define $C_\alpha = \{u \in C | \varphi_k(u) \leq \alpha_k \text{ for } k = 1, \ldots, N\}$.

Of concern is the abstract Cauchy problem in X

$$\frac{du}{dt} = Au + B(u), \qquad (1)$$
$$u(0) = u_0.$$

Throughout the rest of this chapter we make the following assumptions.
(A1) $A : D(A) \subset X \to X$ is the linear generator of a C_0-semigroup

$\{T(t)\}_{t\geq 0}$ with

$$\|T(t)\| \leq e^{\omega_0 t} \text{ for some } \omega_0 \in \mathbb{R}.$$

(A2) In addition, assume that $\overline{D(A) \cap C_\alpha} \subset C$, for all α.
$B : C \to X$ is a nonlinear operator satisfying:
(B1) B is weakly locally sequentially continuous, i.e.

$$\{u_n\} \subset C_\alpha \text{ for some } \alpha > 0 \text{ and } u_n \to u \text{ implies } Bu_n \rightharpoonup Bu. \quad (2)$$

(Here $\to$ [resp. $\rightharpoonup$] means norm [resp. weak] convergence in X.)
(B2) B is locally quasi-dissipative in the sense that:
For all $\alpha > 0$, there exists ω_α such that

$$(Bu - Bv, u - v) \leq \omega_\alpha |u - v|^2, \text{ for all } u, v \in C_\alpha \quad (3)$$

Then the following theorem (due to Goldstein, Oharu, Takahashi [25]) holds.

Theorem 3. *Let $a_k, b_k \geq 0$ be fixed numbers, $k \in 1, \ldots N$. Assume the following condition holds true:*
For $v \in C, \varepsilon > 0$, there exists $0 < \lambda_0 = \lambda_0(v, \varepsilon)$ such that for all $0 < \lambda < \lambda_0$, there exists $u = u_\lambda \in D(A) \cap C$ satisfying

$$|u - \lambda Au - \lambda B(u) - v| < \lambda \varepsilon, \quad (4)$$

$$\varphi_k(u) \leq \frac{\varphi_k(v) + (b_k + \varepsilon)\lambda}{1 - a_k \lambda}, \quad k = 1, \ldots, N. \quad (5)$$

Then there exists a nonlinear semigroup $\{S(t)\}_{t \geq 0}$ of continuous mappings from C to C with the following properties:

$$S(t+s) = S(t)S(s), S(0) = I, \quad (6)$$

$$S(t)u_0 = T(t)u_0 + \int_0^t T(t-s)B(S(s)u_0)ds, \quad (7)$$

$$\varphi_k(S(t)u_0) \leq e^{a_k t}(\varphi_k(u_0) + b_k t), \, t \geq 0, \quad k = 0 \ldots N-1, \quad (8)$$

and the mapping $u_0 \mapsto u(\cdot) = S(\cdot)u_0$ is continuous from C into $C(\mathbb{R}_+, X)$ and locally Lipschitzian for $u_0 \in D(A)$, i.e.

$$|u(t) - v(t)| \leq e^{\Omega t} |u_0 - v_0|, \text{ for some } \Omega = \Omega(T, \alpha)$$

whenever $u_0, v_0 \in D(A) \cap C_\alpha, t \in [0, T]$.

Note that the growth condition (8) has as special cases linear growth (when $a_k = 0$) as well as exponential growth (when $b_k = 0$). The proof is given in a slightly more general setting in the next section. The theorem actually holds if X is a general Banach space. In this case, (3) should be replaced by:

$$B - \omega_\alpha I \text{ is dissipative on } C_\alpha.$$

Condition (3) is correctly stated if $(\cdot, \cdot)$ is interpreted as a semi-inner product (in the sense of Lumer) on X, i.e., for $x, y \in X$:

$$(x, y) = Re\phi(x), \text{ where } \phi \in X^* \text{ is such that } \phi(y) = \|y\|_X^2 = \|\phi\|_{X^*}^2. \tag{9}$$

This fact is basically contained (but not proved) in [23], [25]. For a formulation of our result in general Banach spaces with N functionals, see Remark 2 at the end of Section 3. In the next section, we give the proof of our theorem, generalizing the estimates on the functional φ_j.

3 Abstract theorem. Existence of solutions

Of concern is the abstract Cauchy problem in X

$$\frac{du}{dt} = Au + B(u), \tag{10}$$
$$u(0) = x_0 \in X.$$

We assume all the hypothesis on A, B, C and φ from the preceding section, i.e. (A1),(A2), (B1) and (B2), hold throughout this chapter.

Let $g = g(r) > 0$ be a C^1 increasing function and $m = m(t, \alpha)$ the maximal solution of the initial value problem

$$m'(t) = g(m(t)), \, t \geq 0, \tag{11}$$
$$m(0) = \alpha,$$

where $\alpha \in \mathbb{R}$. For the sake of simplicity, we will assume that the maximal solution is defined for all $t \geq 0$. Otherwise, when we say "for $t \geq 0$" we mean "for all $t \in [0, T_{max})$". When necessary, we will write $m(t, \alpha) = m_g(t, \alpha)$.

Here is the main abstract theorem, formulated for a single functional φ ($N = 1$). The general case is very similar and will be explained later.

Theorem 4. *Assume that, for each $x_0 \in C$, the following hypothesis holds:*

$$(\mathcal{H}) \begin{cases} \text{For every } \varepsilon > 0, \text{ there exists } \delta = \delta(x_0, \varepsilon) \text{ such that} \\ \text{for all } h \in (0, \delta), \text{ there exists } x_h \in D(A) \cap C \text{ satisfying} \\ \quad |x_h - x_0 - h(Ax_h + B(x_h))| \leq \varepsilon h, \\ \quad \frac{1}{h}[\varphi(x_h) - \varphi(x_0)] \leq g(\varphi(x_h)) + \varepsilon. \end{cases}$$

Then there exists a nonlinear semigroup $\{S(t) : C \to C\}_{t \geq 0}$ of continuous operators on C with the following properties:

$$S(t)x_0 = T(t)x_0 + \int_0^t T(t-s)BS(s)ds, \qquad (12)$$

$$\varphi(S(t)x_0) \leq m(t, \varphi(x_0)), \qquad (13)$$

for all $x_0 \in C$ and all $t \geq 0$.

Most of Theorem 3 is obtained from Theorem 4 when we use N functions $g_k(r) = a_k r + b_k$, for $k = 1, \ldots, N$. Extra arguments are needed for the conclusion involving local Lipschitz conditions.

The idea of the proof follows the work of Kobayashi, [36], and consist of constructing, for each $x_0 \in C$, the map $t \to S(t)x_0$ as the limit of a sequence of approximations $\{u_n(t)\}_n$, where $u_n(t)$ are step functions constructed by means of an implicit difference scheme. The only addition here is the estimates using the functionals φ_j.

Let $x_0 \in C_{r_0}$ (for some fixed r_0) and $T > 0$ be fixed. Choose also $\eta > 1$ (close to 1) and $R = R(r_0, T)$ such that

$$R > m_{\eta g}(r_0, T) = m_g(\eta r_0, \eta T).$$

Denote $\omega = \omega_R$ the dissipativity constant for the operator $A + B_R$ on C_R. Throughout this section, B_R is the restriction of B to C_R. Let l be a lower bound for φ on C_R. Let $K = \max\{\omega, \sup_{l \leq m \leq R} g'(m)\}$ and denote $\widehat{\varepsilon} = \frac{\eta - 1}{\eta K}$.

Lemma 5. *Assume the hypothesis $(\mathcal{H})$ holds. Then, for $x_0 \in C_{r_0}$ and $0 < \varepsilon < \widehat{\varepsilon}$, the family*

$$\Gamma_{x_0, \varepsilon} := \{(x_k, h_k)_{k=1}^n \mid (x_k)_{k=1}^n \subset D(A) \cap C_R, 0 < h_k < \varepsilon,$$

$$\sum_{k=1}^{n-1} h_k < T, \sum_{k=1}^{n} h_k \geq T,$$

$$|x_k - x_{k-1} - h_k(Ax_k + B(x_k))| \leq \varepsilon h_k, \qquad (14)$$

$$\frac{1}{h_k}[\varphi(x_k) - \varphi(x_{k-1})] \leq g(\varphi(x_k)) + \varepsilon, \text{ for } k = 1, \ldots, n\} \qquad (15)$$

is nonempty.

Proof. The existence of a sequence (x_k, h_k) satisfying the requirements above follows inductively, using the hypothesis $(\mathcal{H})$, which can be conveniently written as a "tangential condition":

For all $\varepsilon > 0, x \in C_r$ such that there exists $\delta > 0$, such that

$$h \in (0, \delta) \text{ implies } dist(x, Range(I - h(A + B_R))) < \varepsilon h, \qquad (16)$$

or,

$$\lim_{h \to 0} \frac{1}{h} dist(x, Range(I - h(A + B_R))) = 0, \text{ for all } x \in C_r, \qquad (17)$$

where $r < R$ must satisfy $m_{\eta g_\varepsilon}(r, T) < R$ (see Proposition 6 below).

Let $\varepsilon < \widehat{\varepsilon}$ be fixed and denote $\overline{\delta} = \overline{\delta}(x, \varepsilon)$ the supremum of $\delta, \delta < \varepsilon$, satisfying (16).

The induction argument is as follows. Assuming we have generated $x_i \in C_R, i = 1, ..., k-1$, $\sum_{i=1}^{k-1} h_i < T$, we can find $\overline{\delta} = \overline{\delta}(x_{k-1}, \varepsilon)$, $h_k \in (0, \overline{\delta})$ and $x_k = x_{h_k}$ as in $(\mathcal{H})$. We choose h_k, in each step, such that

$$\frac{1}{2}\overline{\delta}(x_{k-1}, \varepsilon) < h_k < \varepsilon. \qquad (18)$$

Our first claim is that $x_k \in C_R$ for as long as $\sum_{i=1}^{k-1} h_i < T$. Indeed, we can prove that

$$\varphi(x_i) \le m_{\eta g_\varepsilon}(\varphi(x_{i-1}), h_i), \text{ for all } i = 1, ..., k; \qquad (19)$$

here $g_\varepsilon := g + \varepsilon$. This follows from the following

Proposition 6.

$$(I - hg)^{-1}z \le m_{\eta g}(z, h), \text{ for any } z < R, h < \widehat{\varepsilon}. \qquad (20)$$

To prove the above inequality, it is enough to rewrite it as

$$z \le m_{\eta g}(z, h) - hg(m_{\eta g}(z, h)) =: \Phi(h).$$

Recall that g is increasing function. Then $\Phi'(h) = \eta g(m) - g(m) - hg'(m)\eta g(m) \ge (\eta - 1 - hK\eta)g(m) \ge 0$, for $h < \widehat{\varepsilon}$, which implies $\Phi(h) \ge \Phi(0) = z$, if $0 < h < \widehat{\varepsilon}$. This proves the claim of Proposition 6.

Global Existence for a Class of Dispersive Equations

Using (20) with $g_\varepsilon = g + \varepsilon$ instead of g and the fact that $\varphi(x_i) \leq (I - h_i g_\varepsilon)^{-1} \varphi(x_{i-1})$ (by construction, see (15)) we get

$$\begin{aligned} \varphi(x_i) &\leq (I - h_i g_\varepsilon)^{-1} \varphi(x_{i-1}) \\ &\leq m_{\eta g_\varepsilon}(\varphi(x_{i-1}), h_i) \end{aligned}$$

and therefore (19) holds. Iterating (19), we obtain

$$\begin{aligned} \varphi(x_k) &\leq m_{\eta g_\varepsilon}(\varphi(x_{k-1}), h_k) \\ &\leq m_{\eta g_\varepsilon}(m_{\eta g_\varepsilon}(\varphi(x_{k-2}), h_{k-1}), h_k) = m_{\eta g_\varepsilon}(\varphi(x_{k-2}), h_{k-1} + h_k) \\ &\cdots \\ &\leq m_{\eta g_\varepsilon}(\varphi(x_0), \sum_{i=1}^{k} h_i) \\ &\leq m_{\eta g_\varepsilon}(r_0, T) \leq m_{\eta g_{\bar\varepsilon}}(r_0, T) < R. \end{aligned}$$

As mentioned above, this guarantees the induction can continue until $\sum_{k=1}^{n} h_k \geq T$ for some n. We claim that this is indeed reached in a finite number of steps, which is the conclusion of Lemma 5. We argue by contradiction. Suppose that we can indefinitely generate (x_i, h_i), with $\sum_{i=1}^{k} h_i < T$, for every integer $k \geq 1$. Then $s_0 := \sum_{k=1}^{\infty} h_k \leq T < \infty$. Without loss of generality, we can assume $x_0 \in \mathcal{D}(A)$. One reaches a contradiction by showing that the sequence $(x_k) \subset C_R$ has a limit point $x^* \in C_R$ which fails to satisfy the hypothesis $(\mathcal{H})$.

In a first step, one proves by double induction (see [36]) that for all $i \geq j \geq 0$,

$$|x_i - x_j| \leq e^{\eta\omega(t_i + t_j)}[(t_i - t_j)|\mathcal{B}x_0| + \varepsilon(t_i + t_j)]. \tag{21}$$

or, more generally,

$$|x_i - x_j| \leq e^{\eta\omega(t_i + t_j - 2t_p)}[(t_i - t_j)|\mathcal{B}x_0| + \varepsilon(t_i + t_j - 2t_p)] \tag{22}$$

for every $i \geq j \geq p$. As we take $i, j \to \infty$, $t_i, t_j \to s_0$ in (22) and we conclude

$$\limsup_{i,j \to \infty} |x_i - x_j| \leq 2\varepsilon(s_0 - t_p) e^{2\eta\omega(s_0 - t_p)}$$

for all p. Now take $p \to \infty$, $t_p \to s_0$, which leads us to the conclusion

$$\lim_{i,j \to \infty} |x_i - x_j| = 0$$

i.e., the sequence $\{x_k\}_{k \geq 1}$ is Cauchy, therefore convergent to some $x^* \in C_R$.

By the tangential condition, there exists $\delta^* = \delta(x^*, \frac{\varepsilon}{2})$ such that for all $0 < \lambda < \delta^*$, there exists $\xi_\lambda \in D(A) \cap C$ such that

$$|\xi_\lambda - x^* - \lambda(A\xi_\lambda + B(\xi_\lambda))| < \lambda \frac{\varepsilon}{2}. \tag{23}$$

Because $h_k \to 0$ (as $\sum_{k=1}^{\infty} h_k \leq T < \infty$), and $\frac{1}{2}\overline{\delta}(x_k, \varepsilon) \leq h_k$ (by our choice of h_k, see (18)), it follows that $\overline{\delta}(x_k, \varepsilon) \to 0$ as $k \to \infty$. Thus there exists k^* such that $\overline{\delta}(x_k, \varepsilon) < \frac{\delta^*}{2}$, for all $k > k^*$. This guarantees the existence of an infinite sequence $(\lambda_k)_k \in [\frac{\delta^*}{2}, \delta^*]$ and $(\xi_{\lambda_k})_k \in \mathcal{D}(A) \cap C_R$ satisfying, for all $\lambda < \delta^*$,

$$|\xi_\lambda - x_k - \lambda_k(A\xi_\lambda + B(\xi_\lambda))| \geq \varepsilon \lambda_k, \text{ for all } k \geq k^*.$$

If $\lambda^* \in [\frac{\delta^*}{2}, \delta^*]$ is an accumulation point for $(\lambda_k)_k$, then we have

$$|\xi_{\lambda^*} - x_k - \lambda_k(A\xi_{\lambda^*} + B(\xi_{\lambda^*}))| \geq \varepsilon \lambda_k, \text{ for all } k \geq k^*,$$

and, in the limit $k \to \infty$,

$$|\xi_{\lambda^*} - x^* - \lambda^*(A\xi_{\lambda^*} + B(\xi_{\lambda^*}))| \geq \varepsilon \lambda^*,$$

which contradicts (23).

The contradiction we reached shows that our assumption made earlier, that we can generate an infinite sequence $(x_k, h_k)_k$, is false. Therefore the set $\Gamma_{x_0, \varepsilon}$ is indeed nonempty. This concludes the proof of Lemma 5. □

Lemma 5 allows us to construct an approximate solution to the initial value problem

$$\begin{aligned} u'(t) &= Au(t) + B(u(t)), t \geq 0, \\ u(0) &= x_0 \in C_{r_0}. \end{aligned} \tag{24}$$

More precisely, for a finite sequence $(x_k, h_k)_{k=1}^n \in \Gamma_{x_0,\varepsilon}$, we define the step function $u_\varepsilon : [0,T] \to C_R \cap \mathcal{D}(A)$, such that

$$u_\varepsilon(t) = x_k \text{ if } t \in [t_{k-1}, t_k) \tag{25}$$

for $k = 1, ..., n$. What remains to be proved is that for $\varepsilon = \varepsilon_n \to 0$, the sequence $\{u_\varepsilon(t)\}$ converges uniformly to a continuous function $u(t)$, which turns out to be the desired $S(t)x_0$, satisfying

$$u(t) = T(t)x_0 + \int_0^t T(t-s)B(u(s))ds$$

for all $t \in [0,T]$, i.e. u is the mild solution for the initial value problem (24).

The convergence of the approximate solutions constructed in the previous section follows from the following estimate (see [36]).

Lemma 7. *Let $0 < \widetilde{\varepsilon} < \varepsilon$ and $\{(x_k, h_k)\}_k \in \Gamma_{x_0,\varepsilon}$ and $\{(\widetilde{x}_j, \widetilde{h}_j)\}_j \in \Gamma_{x_0,\widetilde{\varepsilon}}$ be finite sequences constructed as in the proof of Lemma 5. Then, for all $u \in D(\mathcal{B})$,*

$$|x_k - \widetilde{x}_j| \leq e^{\eta\omega(t_i+\widetilde{t}_j)}\{|x_0 - u| + [(t_k - \widetilde{t}_j)^2 + \varepsilon t_k + \widetilde{\varepsilon}\widetilde{t}_j]^{\frac{1}{2}}|\mathcal{B}u| + \varepsilon t_k + \widetilde{\varepsilon}\widetilde{t}_j\} \tag{26}$$

for all k, j.

Based on this estimate and given $\varepsilon = \varepsilon_n \to 0$, the sequence of approximate solutions $u_n(t) = u_{\varepsilon_n}(t)$ constructed in (25) converges to a continuous function $u : [0,T] \to C_R$, which turns out to be a mild solution for the initial value problem (24).

Indeed, for each n, we choose $\left(x_k^{(n)}, h_k^{(n)}\right)_k \in \Gamma_{x_0,\varepsilon_n}$ and denote $t_k^{(n)} = \sum_{i=1}^k h_i^{(n)}$. Let $u \in \mathcal{D}(\mathcal{B})$ arbitrary but fixed. Fix also $t \in [0,T]$. Let $m, n \in \mathbb{N}$. One can find k and j such that $t \in [t_{k-1}^{(n)}, t_k^{(n)}) \cap [t_{j-1}^{(m)}, t_j^{(m)})$. We then apply Lemma 7 with $\varepsilon = \varepsilon_n, \widetilde{\varepsilon} = \varepsilon_m$:

$$|u_n(t) - u_m(t)| \leq e^{\eta\omega(t_k^{(n)}+t_j^{(m)})}\{|x_0 - u| + \tag{27}$$
$$[(t_k^{(n)} - t_j^{(m)})^2 + \varepsilon_n t_k^{(n)} + \varepsilon_m t_j^{(m)}]^{\frac{1}{2}}|\mathcal{B}u| + \varepsilon_n t_k^{(n)} + \varepsilon_m t_j^{(m)}\}.$$

Letting $n, m \to \infty$, we get

$$\limsup_{n,m\to\infty} |u_n(t) - u_m(t)| \le e^{2\eta\omega t}|x_0 - u|, \text{ for all } u \in \mathcal{D}(\mathcal{B}).$$

Because $\mathcal{D}(\mathcal{B}) = \mathcal{D}(A) \cap C_R$ is dense in C_R we obtain

$$\lim_{n,m\to\infty} |u_n(t) - u_m(t)| = 0, \text{ uniformly for } t \in [0, T].$$

This implies that $(u_n)_n$ is convergent to a function $u : [0, T] \to C_R$:

$$\lim_{n\to\infty} u_n(t) = u(t), \text{ uniformly for } t \in [0, T].$$

Reasoning in the same way, but for $t \neq s$, we get

$$|u(t) - u(s)| \le e^{2\omega T}[|x_0 - v| + |t - s||\mathcal{B}v|], \text{ for all } v \in \mathcal{D}(\mathcal{B}). \qquad (28)$$

Because $\mathcal{D}(\mathcal{B})$ is dense in C_R, it follows from the above inequality that $u : [0, T] \to C_R$ is continuous.

Define $S(t)x_0 := u(t)$. It is important to note at this point that, when $x_0 \in \mathcal{D}(\mathcal{B}) = \mathcal{D}(A) \cap C_R$, we have more than continuity. Choosing $v = x_0$ in the inequality (28),

$$|u(t) - u(s)| \le e^{2\omega T}|t - s||Ax_0|, \qquad (29)$$

for $t, s \in [0, T]$, i.e. u is Lipschitz continuous on $[0, T]$, for $x_0 \in \mathcal{D}(\mathcal{B})$.
What is left to prove is that, for any $x_0 \in C_R$, $u(t)$ satisfies

$$u(t) = T(t)x_0 + \int_0^t T(t-s)B(u(s))ds, \qquad (30)$$

$$\varphi(u(t)) \le m(t, \varphi(x_0)), \text{ for } t \in [0, T]. \qquad (31)$$

Let $u_n(t)$ be the approximate solution constructed above and let $f \in \mathcal{D}(A^*)$. If we denote $\epsilon_k^{(n)} := x_k^{(n)} - x_{k-1}^{(n)} - h_k^{(n)}(Ax_k^{(n)} + B(x_k^{(n)}))$, then we know (see $(\mathcal{H})$) that $\left|\epsilon_k^{(n)}\right| \le \varepsilon_n h_k^{(n)}$. Then

$$\begin{aligned}
\left\langle x_k^{(n)}, f \right\rangle &= \left\langle x_{k-1}^{(n)}, f \right\rangle + h_k^{(n)}\left\langle x_k^{(n)}, A^*f \right\rangle + h_k^{(n)}\left\langle B(x_k^{(n)}), f \right\rangle + \left\langle \epsilon_k^{(n)}, f \right\rangle \\
&= \left\langle x_{k-1}^{(n)}, f \right\rangle + \int_{t_{k-1}^{(n)}}^{t_k^{(n)}} [\langle u_n(s), A^*f \rangle + \langle B(u_n(s)), f \rangle + \langle \epsilon_n(s), f \rangle] ds,
\end{aligned}$$

Global Existence for a Class of Dispersive Equations

which implies

$$\langle u_n(t), f\rangle = \langle x_0, f\rangle + \int_0^{t_k^{(n)}} [\langle u_n(s), A^*f\rangle + \langle B(u_n(s)), f\rangle + \langle \epsilon_n(s), f\rangle]ds$$

for $t \in [t_{k-1}^{(n)}, t_k^{(n)}]$. Letting $n \to \infty$, we deduce

$$\langle u(t), f\rangle = \langle x_0, f\rangle + \int_0^t [\langle u(s), A^*f\rangle + \langle B(u(s)), f\rangle]ds. \qquad (32)$$

In the formulas above, $\epsilon_n(t) = \epsilon_k^{(n)}$ for $t \in [t_{k-1}^{(n)}, t_k^{(n)}]$ and $\left|\int_0^{t_k^{(n)}} \epsilon_n(s)ds\right| \leq \varepsilon_n t_k^{(n)} \to 0$ as $n \to \infty$. The integrand in (32) is continuous in s, because u is continuous in X and, consequently, $B(u)$ weakly continuous in X (see (2)). Therefore $\langle u(t), f\rangle$ is differentiable in t and

$$\frac{d}{dt}\langle u(t), f\rangle = \langle u(t), A^*f\rangle + \langle B(u(t)), f\rangle. \qquad (33)$$

Let $\widetilde{u}(t) := T(t)x_0 + \int_0^t T(t-s)B(u(s))ds$. We want to prove that $u(t) \equiv \widetilde{u}(t)$.

We have, for $f \in \mathcal{D}(A^*)$,

$$\langle \widetilde{u}(t), f\rangle = \langle T(t)x_0, f\rangle + \int_0^t \langle T(t-s)B(u(s)), f\rangle ds;$$

thus

$$\begin{aligned}\frac{d}{dt}\langle \widetilde{u}(t), f\rangle &= \left\langle \frac{d}{dt}T(t)x_0, f\right\rangle + \int_0^t \left\langle \frac{d}{dt}T(t-s)B(u(s)), f\right\rangle ds + \langle B(u(t)), f\rangle \\ &= \langle u(t), A^*f\rangle + \langle B(u(t)), f\rangle \\ &= \frac{d}{dt}\langle u(t), f\rangle.\end{aligned}$$

This shows that $\langle \widetilde{u}(t) - u(t), f\rangle = \langle \widetilde{u}(0) - u(0), f\rangle = 0$ for all $f \in D(A^*)$, and consequently (30) holds, i.e.

$$u(t) = T(t)x_0 + \int_0^t T(t-s)B(u(s))ds. \qquad (34)$$

To prove inequality (31), let $u_n = u_{\varepsilon_n}$ be given the approximate solutions. We know from the estimate (19) that

$$\varphi(u_n(t)) \leq m_{\eta g_{\varepsilon_n}}(\varphi(x_0), t) = m_{g_{\varepsilon_n}}(\eta\varphi(x_0), \eta t)$$

whenever $\varepsilon_n < \frac{\eta-1}{\eta M}$ for some constant M, independent of n. Thus, for sequences $\eta \to 1$ and $n \to \infty$, we infer that

$$\varphi(u(t)) \leq \liminf_{n\to\infty} \varphi(u_n(t)) \leq m_g(\varphi(x_0),t). \qquad (35)$$

i.e. (13) holds. Here we used the lower semicontinuity of φ.

Note that all the above results hold true for all $x_0 \in C = \bigcup_{r>0} C_r$. To complete the proof of Theorem 4, it remains to prove the uniqueness of the solution $u(t) = S(t)x_0$, with $x_0 \in C$. This is the claim of the following

Proposition 8. *The mild solution of the Cauchy problem*

$$\begin{aligned} u'(t) &= Au(t) + B(u(t)), t \geq 0, \\ u(0) &= x_0 \in C, \end{aligned} \qquad (36)$$

that is, the strongly continuous function u satisfying (34) and (35), is unique.

Proof. Let $x_0, y_0 \in C_r$ and $T > 0$. Denote $R := m(r,T)$.

Assuming there are two functions, $u(t)$ and $v(t)$, both satisfying (34), we have

$$u(t), v(t) \in C_R, \text{ for all } t \in [0,T].$$

As $B - \Omega I$ is dissipative operator on C_R with some constant Ω, we can apply the standard techniques to obtain that

$$\|u(t) - v(t)\| \leq e^{\Omega t}\|x_0 - y_0\|, \text{ for } t \in [0,T].$$

This clearly implies the uniqueness of the Cauchy problem (36). The proof of Theorem 4 is now complete.

Remark 1. In Theorem 4, only the lower-semicontinuity of the functional(s) φ was needed. The converse of Theorem 4 holds in the special case when the set C and the functional(s) $\varphi : X \to \overline{\mathbb{R}}$ is (are) convex. This was proved in [39] for the case of a single functional.

Remark 2. Theorem 4 can be generalized by considering X a general Banach space and N functionals $\varphi_1, \varphi_2, \ldots, \varphi_N : X \to \overline{\mathbb{R}}$, instead of just one functional. In this case $C = \bigcap_i \mathcal{D}(\varphi_i)$. The proof follows the same lines as in the Hilbert space case, with the inner product replaced by the semi-inner product (in the sense of Lumer) that was defined in (9). We omit the details, which are tedious but routine.

Remark 3. Our proof of the abstract theorem is related to the ideas of Kobayashi [36], who was inspired by the historic Crandall-Liggett paper [15]. L.C. Evans ([17], [18]) extended Kobayashi's construction from the context of $du/dt = Au$ to the time-dependent operator context of $du/dt = A(t)u$. This enables us to generalize our abstract result to $du/dt = A(t)u + B(t)u$, and to, for instance,

$$u_t - \alpha(t)Mu_x + F(t,u)_x = 0,$$

a nonstationary version of (NDE). We have no specific applications in mind for this equation so we do not give a precise formulation here.

4 Proof of the Main Theorem 1

We will fit our concrete nonlinear dispersive equation (NDE) in the general framework of the semilinear Hille-Yosida theory presented in the preceding section.

The linear operator $A = D^{2\beta}\partial_x$, with domain $\mathcal{D}(A) = H^{2\beta+1}(\mathbb{T})$, is skew-adjoint on any of the spaces $H^s(\mathbb{T})$, and it generates a group of isometries $\{T(t) | t \in \mathbb{R}\}$ on each $H^s(\mathbb{T})$. This is easy to see since $\widehat{Au}(\xi) = i\xi|\xi|^{2\beta}\hat{u}(\xi)$ for $u \in H^s(\mathbb{T})$. The same argument applies to the restriction of A to $\dot{H}^{2\beta+1}(\mathbb{T}) = V_3$, and $\{T(t) | t \in \mathbb{R}\}$ is also a group of isometries on $\dot{H}^s(\mathbb{T})$. In particular,

$$|T(t)v|_k = |v|_k, \text{ for } v \in V_k, k = 0,1,2,3.$$

Define a nonlinear operator B on $\dot{H}^1(\mathbb{T})$ by

$$Bu = -\partial F(u) = -F'(u)\partial u.$$

Clearly, $B : \dot{H}^1(\mathbb{T}) \to L^2(\mathbb{T})$ and, more generally, $B : \dot{H}^k(\mathbb{T}) \to \dot{H}^{k-1}(\mathbb{T})$. Note that the only thing needed so far is for F to be sufficiently smooth.

Then the following properties hold (for a proof, we refer to [12]:

Lemma 9. *(i) If $w_n \to w$ in $L^2(\mathbb{T})$ and $\sup_n |w_n|_{H^{2\beta}} \leq M < +\infty$, then*
$(I - \lambda A)^{-1}B(w_n) \to (I - \lambda A)^{-1}B(w)$ *in $H^{2\beta}$, for all real $\lambda \neq 0$.*

(ii) If $w_n \to w$ in $L^2(\mathbb{T})$ and $\sup_n |w_n|_{H^{2\beta+1}} \leq M < +\infty$, then $B(w_n) \to B(w)$ in $L^2(\mathbb{T})$.

Lemma 10. *(i) For each real $\lambda \neq 0$, the operators $\pm \mathcal{A}_\lambda = \pm (I - \lambda A)^{-1} B$ are locally quasi-dissipative on $H^{2\beta}(\mathbb{T})$, i.e. for all $r > 0$, there exists $\omega = \omega(r) \in \mathbb{R}$ such that*

$$\left| ((I - \lambda A)^{-1} B(v) - (I - \lambda A)^{-1} B(w), v - w)_{H^{2\beta}} \right| \leq \frac{\omega}{\lambda} |v - w|^2_{H^{2\beta}}$$

whenever $|v|_{H^{2\beta}}, |w|_{H^{2\beta}} \leq r$.

(ii) $\pm B$ are locally quasi-dissipative in $L^2(\mathbb{T})$ on bounded sets in $H^{2\beta+1}(\mathbb{T})$, i.e. for all $r > 0$, there exists $\omega = \omega(r) \in \mathbb{R}$ such that

$$|(B(v) - B(w), v - w)_{L^2}| \leq \omega |v - w|^2_{L^2}$$

whenever $|v|_{H^{2\beta+1}}, |w|_{H^{2\beta+1}} \leq r$.

Let us introduce the following functionals:

$$\varphi_0(u) = \frac{1}{2} \int_\mathbb{T} |u|^2, \quad \text{on } V_0 \tag{37}$$

$$\varphi_1(u) = \frac{1}{2} \int_\mathbb{T} |D^\beta u|^2 - \int_\mathbb{T} G(u), \quad \text{on } V_1 \tag{38}$$

$$\varphi_2(u) = \frac{1}{2} \int_\mathbb{T} |D^{2\beta} u|^2 - \frac{4\beta+1}{4\beta+2} \int_\mathbb{T} F(u) D^{2\beta} u + \frac{4\beta+1}{4\beta+2} \int_\mathbb{T} I(u), \quad \text{on } V_2 \tag{39}$$

$$\varphi_3(u) = \frac{1}{2} \int_\mathbb{T} |D^{2\beta+1} \partial u + \partial F(u)|^2, \quad \text{on } V_3. \tag{40}$$

Here $G(u) = \int_0^u F(r) dr$ and $I(\cdot)$ satisfies $I''(u) = (F'(u))^2$, $I(0) = I'(0) = 0$. In the integrable cases, Korteweg-de Vries ($\beta = 1$) and Benjamin-Ono ($\beta = \frac{1}{2}$), with $F'(u) = u$, these are just three of an infinite list of invariant functionals.

These functionals are lower semi-continuous on the corresponding spaces and, in particular, on $H^{2\beta+1}$. They are also bounded on bounded sets in V_j, more precisely:

B is a bounded set in V_j, for $0 \leq j \leq k$ and $k = 0, 1, 2, 3$, implies $\sup\{\varphi_j(f) : f \in B\} < \infty$ for all $0 \leq j \leq k$.

Let $g = g(r) > 0$ be a C^1 increasing function and $m = m(t, \alpha)$ the maximal solution of the initial value problem

$$m'(t) = g(m(t)), \ t \geq 0, \tag{41}$$
$$m(0) = \alpha,$$

Global Existence for a Class of Dispersive Equations

where $\alpha \in \mathbb{R}$. When necessary, we will write $m(t, \alpha) = m_g(t, \alpha)$.

Our main result concerning the existence of solutions for the nonlinear dispersive equation (NDE) is stated here:

Theorem 11. *Let $\beta \geq \frac{1}{2}$. Under assumption (C) on the growth of the nonlinearity, the Cauchy problem (NDE) is well-posed in $H^s(\mathbb{T})$, with $s = \max\{2\beta, \frac{3}{2} + \varepsilon\}$, ($\varepsilon > 0$ is as in Theorem 1), for arbitrary initial data u_0. Moreover, there exists a C^1 function g such that the following estimates hold when $u_0 \in H^s(\mathbb{T})$:*

$$\varphi_0(u(t)) = \varphi_0(u_0), \tag{42}$$
$$\varphi_1(u(t)) = \varphi_1(u_0), \tag{43}$$
$$\varphi_2(u(t)) \leq m_g(t, \varphi_2(u_0)), \text{ for } t \geq 0, \tag{44}$$

where m_g is the solution of the initial value problem (41). If $u_0 \in H^{2\beta+1}(\mathbb{T})$, then there exists ω_0 depending only on a bound for $|u|_{H^s}$ on $[0, T]$ so that

$$\varphi_3(u(t)) \leq e^{\omega_0 t} \varphi_3(u_0), \text{ for } t \in [0, T],$$

for any $T > 0$.

The choice of the function g mentioned in the theorem depends on β and on the nonlinearity F. Our well-posedness theorem is global in time, provided that the solution m of $dm/dt = g(m)$ exists for all $t \geq 0$. We have shown this to be true whenever $\beta > \frac{3}{2}$ and when $\beta = 1$. It also holds for $\beta = \frac{1}{2}$ and F any quadratic polynomial. For any other case of a pair (β, F) for which one can show that m exists globally in time, then (NDE) is globally well-posed. Thus, local well-posedness assertions in our theorems are actually global in all such cases where m exists globally. Since (β, F) determine many possible choices of g, it seems likely that our result are of a global nature in cases where we do not assert this.

From now on it is sufficient to assume $F(0) = 0$ and to restrict all our calculations to initial data belonging to $\mathring{H}^s$, i.e. with zero mean, instead of the whole H^s. This is possible (without loss of generality) due to the fact that the mean of the solution $u(t)$ is constant in time, so $F(u)$ in the equation can be replaced by $\tilde{F}(u) = F(u + c) - F(c)$, $c = \int u$.

The proof of Theorem 11 will be a consequence of the abstract theorem. If $p \geq 4\beta$ in the assumption (C) on the growth of F, global existence of solutions is still guaranteed, provided that the H^β norm of the solution is uniformly bounded. This can be accomplished, for example, by requiring the smallness of the initial data. We refer to [12] for details.

The next lemma asserts that, under assumption $(\mathcal{C})$, boundedness of the functionals φ_j implies boundedness of solutions in the appropriate Sobolev norms.

Lemma 12. *(i) For $\alpha_0, \alpha_1 > 0$, there exists $\theta_1 = \theta_1(\alpha_0, \alpha_1)$ such that $w \in V_1, \varphi_0(w) \leq \alpha_0, \varphi_1(w) \leq \alpha_1$ implies $|w|_1 \leq \theta_1$.*

(ii) For $\alpha_0, \alpha_1, \alpha_2 > 0$, there exists $\theta_2 = \theta_2(\alpha_0, \alpha_1)$ such that $w \in V_2, \varphi_0(w) \leq \alpha_0, \varphi_1(w) \leq \alpha_1, \varphi_2(w) \leq \alpha_2$ implies $|w|_2 \leq \theta_2$.

(iii) For $\alpha_0, \alpha_1, \alpha_2, \alpha_3 > 0$, there exists $\theta_3 = \theta_3(\alpha_0, \alpha_1)$ such that $w \in V_3, \varphi_j(w) \leq \alpha_j$, $j = 0, 1, 2, 3$, implies $|w|_3 \leq \theta_3$.

Proof. Recall the growth condition $(\mathcal{C})$ imposed on F. We assumed that there exists $p < 4\beta$ such that

$$\limsup_{|r| \to \infty} \frac{F'(r)}{|r|^p} < \infty. \tag{45}$$

Hence, there are constants C, C' such that

$$F'(r) \leq C |r|^p + C'. \tag{46}$$

Because $F(0) = 0$ and $G(0) = 0$, where $G(w) = \int_0^w F(\xi)d\xi$, it follows that (with different constants K, K')

$$G(r) \leq K |r|^{p+2} + K' |r|^2 \text{ for all } r \in \mathbb{R}.$$

Thus,

$$\begin{aligned}
\int_{\mathbb{T}} G(w) dx &\leq K |w|_{L^{p+2}}^{p+2} + K' |w|^2 \\
&\leq K |w|_\infty^p |w|^2 + K' |w|^2 \\
&\leq C |w|^{p(1-\frac{1}{2\beta})+2} \left|D^\beta w\right|^{\frac{p}{2\beta}} + K' |w|^2 \\
&\leq \frac{4\beta - p}{4\beta p} \left(C |w|^{p+2-\frac{p}{2\beta}}\right)^{\frac{4\beta}{4\beta-p}} + \frac{1}{4} \left|D^\beta w\right|^2 + K' |w|^2.
\end{aligned}$$

Here we used the Gagliardo-Nirenberg inequatity and Young's inequality

$$ab \leq \frac{(a)^r}{r} + \frac{b^s}{s}, \quad r = \frac{4\beta}{4\beta - p}, \quad s = \frac{4\beta}{p} \quad (r, s > 1, \frac{1}{r} + \frac{1}{s} = 1).$$

Global Existence for a Class of Dispersive Equations

The assumption $p < 4\beta$ was essential in the previous calculation (to guarantee that $r, s > 1$). As a result of the previous estimates we get

$$\frac{1}{4}|D^\beta w|^2 \leq \overline{K}|w|^{\frac{2(p(2\beta-1)+4\beta)}{4\beta-p}} + K'|w|^2 + \varphi_1(w),$$

which implies (i).

To prove (ii), note that

$$\begin{aligned}
-(F(w), D^{2\beta}w) &= (\partial_x F(w), \partial_x^{-1} D^{2\beta} w) \\
&= (F'(w), \partial_x^{-1} D^{2\beta} w \partial_x w) = \int F'(w) \partial_x^{-1} D^{2\beta} w \partial_x w \\
&\leq C \int_\mathbb{T} |w|^p \left|\partial_x^{-1} D^{2\beta} w \partial_x w\right| + C' \int_\mathbb{T} \left|\partial_x^{-1} D^{2\beta} w \partial_x w\right| \\
&\leq C |w|_\infty^p \left|D^\beta w\right|^2 + C' \left|D^\beta w\right|^2 \\
&\leq C \left|D^\beta w\right|^{p+2} + C' \left|D^\beta w\right|^2,
\end{aligned}$$

and also

$$\begin{aligned}
(I(u), 1) &= \int_\mathbb{T} I(u) \leq C_1 |u|^2 + C_2 |u|_{L^{2p+2}}^{2p+2} \\
&\leq C |D^\beta u|^{2p+2}.
\end{aligned}$$

Thus, from the definition of φ_2,

$$\begin{aligned}
\frac{1}{2}|D^{2\beta} w|^2 &= \varphi_2(w) - \frac{4\beta+1}{4\beta+2}(F(w), D^{2\beta}w) - \frac{4\beta+1}{4\beta+2} \int_\mathbb{T} I(u) \\
&\leq \varphi_2(w) + C_1 |D^\beta w|^{p+2} + C' |D^\beta w|^2,
\end{aligned}$$

i.e. (ii) holds.

The proof of (iii) is straightforward:

$$\begin{aligned}
|D^{2\beta+1} w| &\leq \varphi_3(w) + |\partial F(w)| \leq \varphi_3(w) + |F'(w)|_\infty |\partial u| \\
&\leq \varphi_3(w) + C |w|_{2\beta}.
\end{aligned}$$

$\square$

We need one more result, which will be used in the proof of Theorem 14.

Lemma 13. *Consider the following expression.*
$$\Psi_\beta(z) := \left(D^{2\beta}\left(F'(z)\partial z\right) - F'(z)D^{2\beta}\partial z - (2\beta+1)\partial F'(z)D^{2\beta}z, D^{2\beta}z\right).$$
(i) For $\beta > \frac{1}{2}$ we have that, for all $z \in H^{2\beta+1}$,
$$\Psi_\beta(z) \leq C|Dz|_\infty |D^{2\beta}z|^2. \tag{47}$$
where C depends only on a bound for $|D^\beta z|$.
(ii) For $\beta = \frac{1}{2}$ and $\beta = 1$, one can improve the estimate above, in the sense that $\Psi_1(z) \leq C_1|Dz|^4$ when $\beta = \frac{1}{2}$, with C_1 depending only on $|D^{1/2}z|$, and $\Psi_2(z) \leq C|D^2z|^{3/2}$ in the case $\beta = 1$, for some C_2 depending only on $|Dz|$.

Remark. This result can be reformulated as follows. One can construct a function $g = g(r)$ (depending on β), which, in general, has superquadratic growth as $r \to \infty$, such that
$$\Psi_\beta(z) \leq C g(\varphi_2(z)), \tag{48}$$
where C depends only on $|D^\beta z|$. For $\beta > \frac{3}{2}$, g can be chosen to be linear. For $\beta = 1$, g can be chosen to be sublinear and for $\beta = \frac{1}{2}$, g can be chosen to be quadratic. What is remarkable is that in the special case $F'(u) = u$ and $\beta = 1$ or $\beta = \frac{1}{2}$ we get $\Psi_\beta(z) = 0$, for all z.

The proof of (i) relies on estimates obatined from the product rule and chain rule for fractional order derivatives, for which we refer to [21], [28], [35]. The improved estimates mentioned in (ii) can be found in [12], so we will not reproduce them here.

The last step in the proof of the main theorem is to solve the resolvent equation. Here is the result needed, which makes the hypothesis of the abstract theorem.

Theorem 14. *Let $v \in V_3$ satisfy $\|v\|_{H^{2\beta+1}} \leq r, \varepsilon > 0$. There exists $\lambda_0 = \lambda_0(r, \varepsilon)$ such that, for all real $|\lambda| < \lambda_0$, there exists an unique $u = u_\lambda \in V_3$ satisfying*
$$u - \lambda D^{2\beta}\partial u + \lambda \partial F(u) = v, \tag{49}$$
$$\varphi_0(u) \leq \varphi_0(v) + |\lambda|\varepsilon, \tag{50}$$
$$\varphi_1(u) \leq \varphi_1(v) + |\lambda|\varepsilon, \tag{51}$$
$$\varphi_2(u) \leq \varphi_2(v) + |\lambda|(g(\varphi_2(u)) + \varepsilon), \tag{52}$$
$$\varphi_3(u) \leq \frac{\varphi_3(v)}{1 - \omega_0|\lambda|}. \tag{53}$$

Here g is the function satisfying (48), depending on β and the nonlinearity F, and ω_0 can be chosen to depend only on a bound for $|Dv|_{L^\infty}$.

Proof. Fix $r > 0$ and $\varepsilon > 0$ arbitrarily. Let $v \in V_3$, with $|v| < r$, and choose α_0, α_1 be positive numbers such that $\varphi_0(v) + \varepsilon < \alpha_0, \varphi_1(v) + \varepsilon < \alpha_1$. By Lemma 12, there exists $\theta_1 = \theta_1(\alpha_0, \alpha_1)$ such that for all $z \in H^\beta$, with $\varphi_0(z) \leq \alpha_0, \varphi_1(z) \leq \alpha_1$ implies $|z|_1 \leq \theta_1$.

For the function g defined by (48), let $m_g(t, \alpha)$ be the maximal solution of the initial value problem (11). Choose

$$\alpha_2 \geq m_{g_\varepsilon}(\tau, \varphi_2(v)). \tag{54}$$

Here τ is sufficiently small (depending only on $\varphi_2(v)$) such that the right hand side of (54) is finite.

From Lemma 12 we conclude that there exists $\theta_2 = \theta_2(\alpha_0, \alpha_1, \alpha_2)$ such that, for all $z \in H^{2\beta}$, with $\varphi_0(z) \leq \alpha_0, \varphi_1(z) \leq \alpha_1, \varphi_2(z) \leq \alpha_2$ implies $|z|_2 \leq \theta_2$.

For later purposes, let

$$\rho = \sup\{|F''(w)\partial w|, w \in V_2, |w|_2 \leq \theta_2\} \tag{55}$$

and

$$\sigma_1 = \sup\{|F'(w)\partial w|, w \in V_2, |w|_2 \leq \theta_2\}, \tag{56}$$
$$\sigma_2 = \sup\{|D^{2\beta}F(w)|, w \in V_2, |w|_2 \leq \theta_2\}, \tag{57}$$
$$\sigma_3 = \sup\{|D^{2\beta+1}F(w)|, w \in V_3, |w|_3 \leq \theta_3\}, \tag{58}$$

where we choose $\theta_3 = |v|_3 + 2\sigma_1$.

Because F is assumed to be smooth enough (at least in $C^{2\beta+1}$), there exists $\delta = \delta(|v|_3, \varepsilon) > 0$ such that, for all $w \in V_3, |v - w| \leq \delta$ and $|w|_j \leq max\{\theta_j, |v|_j + \sigma_{j+1}\}$, $j = 0, 1, 2$, the following hold true:

$$|F(v) - F(w)|\theta_3 < \frac{\varepsilon}{2}, \tag{59}$$
$$|D^{2\beta}F(v) - D^{2\beta}F(w)|\theta_3 < \varepsilon_1, \tag{60}$$
$$|F'(v) - F'(w)|_\infty \theta_3(|v|_2 + \sigma_3) < \varepsilon_2, \tag{61}$$
$$|I'(v) - I'(w)|\theta_3 < \varepsilon_3, \tag{62}$$

where ε_j are chosen such that $3\varepsilon_1 + 2\varepsilon_2 + \epsilon_3 < \frac{\varepsilon}{2}$. Let $K = sup\{|g'(r)|, r < \theta_3\}$ and denote $\lambda_0 = \min\{\tau, \frac{\delta}{\theta_3}, \frac{1}{\rho}, \frac{1}{K}\}$. Consider $\lambda \in (-\lambda_0, \lambda_0), \lambda \neq 0$ be arbitrary but fixed.

Consider the set

$$K = \{w \in V_3 | |w - v| \leq |\lambda|\theta_3, |w|_j \leq \theta_j \text{ for all } j = 0, 1, 2, 3\}. \tag{63}$$

Note that K is a compact convex set in $L^2(\mathbb{T})$.

We seek fixed points for the operator $\Gamma : K \to X$ defined by

$$\Gamma w := (I - \lambda A)^{-1}(v + \lambda B(w)).$$

By linearity of A,

$$\Gamma w = (I - \lambda A)^{-1} v + \lambda (I - \lambda A)^{-1} B(w).$$

As we saw in Lemma 9, the operator $\lambda(I - \lambda A)^{-1}B$ is L^2-continuous on bounded sets in $H^{2\beta}$, thus Γ is a continuous operator on K. In order to apply the Schauder-Tichonov fixed point principle, we have to ensure that Γ leaves K invariant, i.e.

$$\Gamma(K) \subset K. \tag{64}$$

We now prove (64). With $v \in H^{2\beta+1}(\mathbb{T})$ fixed, let $w \in K$ be arbitrary. Denote $z = \Gamma w$. We will show that $z \in K$. Since

$$z - \lambda D^{2\beta} \partial z = v - \lambda \partial F(w), \tag{65}$$

we obtain

$$\begin{aligned}
|z - v|^2 &= \left(\lambda D^{2\beta} \partial z, z - v\right) - (\lambda \partial F(w), z - v) \\
&= \lambda \left(D^{2\beta} \partial v, z - v\right) - \lambda \left(\partial F(w), z - v\right) \\
&\leq |\lambda| \left(\left|D^{2\beta} \partial v\right| + |\partial F(w)|\right) |z - v| \\
&\leq |\lambda| \left(|v|_3 + \sigma_1\right) |z - v|,
\end{aligned}$$

or

$$|z - v| \leq |\lambda| (|v|_3 + \sigma_1) \leq |\lambda| \theta_3.$$

From (65) we also obtain

$$\begin{aligned}
|z|_3 &= \left|D^{2\beta} \partial z\right| \leq \frac{1}{|\lambda|} |z - v| + |\partial F(w)| \\
&\leq |v|_3 + \sigma_1 + \sigma_1 \leq \theta_3.
\end{aligned}$$

In order to conclude that $z \in K$ we only need to show

$$|z| \leq \theta_0, \left|D^\beta z\right| \leq \theta_1, \left|D^{2\beta} z\right| \leq \theta_2,$$

Global Existence for a Class of Dispersive Equations

or, in view of Lemma 12, it is enough to prove

$$\varphi_0(z) \leq \alpha_0, \qquad (66)$$
$$\varphi_1(z) \leq \alpha_1, \qquad (67)$$
$$\varphi_2(z) \leq \alpha_2. \qquad (68)$$

From (65), we obtain

$$\begin{aligned}
|z|^2 &\leq (v,z) - (\lambda \partial F(w), z) \\
&= (v,z) - \lambda \left(\partial F(w) - \partial F(z), z\right) \\
&\leq |v||z| + |\lambda| |\partial F(w) - \partial F(z)| |z| \\
&\leq (|v| + |\lambda|\varepsilon) |z|,
\end{aligned}$$

i.e. $|z| \leq |v| + |\lambda|\varepsilon$. We conclude that $\varphi_0(z) \leq \varphi_0(v) + |\lambda|\varepsilon < \varphi_0(v) + \varepsilon < \alpha_0$.

Next, multiply (65) by $D^{2\beta}z$ to obtain

$$\begin{aligned}
|D^\beta z|^2 &= (D^{2\beta}z, z) = (D^{2\beta}z, v) - \lambda\left(D^{2\beta}z, \partial F(w)\right) \\
&= (D^\beta z, D^\beta v) + \lambda\left(D^{2\beta}\partial z, F(w)\right) \\
&\leq \frac{1}{2}|D^\beta z|^2 + \frac{1}{2}|D^\beta v|^2 + \lambda\left(D^{2\beta}\partial z, F(w)\right).
\end{aligned}$$

Now we can estimate

$$\begin{aligned}
\frac{1}{\lambda}[\varphi_1(z) - \varphi_1(v)] &= \frac{1}{\lambda}\left[\frac{1}{2}|D^\beta z|^2 - \frac{1}{2}|D^\beta v|^2 - \int_{\mathbb{T}}(G(z) - G(v))\,dx\right] \\
&\leq \frac{1}{\lambda}\left[\lambda\left(D^{2\beta}\partial z, F(w)\right) - \int_{\mathbb{T}}(G(z) - G(v))\,dx\right] \\
&= (D^{2\beta}\partial z, F(w)) - \frac{1}{\lambda}\int_{\mathbb{T}}\left(\int_0^1 F(\tau z + (1-\tau)v)d\tau\right)(z-v) \\
&= \frac{1}{\lambda}(z - v - \lambda\partial F(w), F(w)) - \frac{1}{\lambda}(z-v, F(v)) \\
&\quad - \frac{1}{\lambda}\int_{\mathbb{T}}\left(\int_0^1 F(\tau z + (1-\tau)v)d\tau - F(v)\right)(z-v)\,dx \\
&\leq \frac{1}{|\lambda|}|z-v||F(w) - F(v)| \\
&\quad + \frac{1}{|\lambda|}\int_{\mathbb{T}}\left|\int_0^1 [F(\tau z + (1-\tau)v) - F(v)]d\tau\right||z-v|\,dx \\
&\leq \frac{1}{|\lambda|}\frac{\varepsilon}{2\theta_3}|\lambda|\theta_3 + \frac{1}{|\lambda|}\frac{\varepsilon}{2\theta_3}|\lambda|\theta_3 = \varepsilon.
\end{aligned}$$

Thus we conclude
$$\varphi_1(z) \leq \varphi_1(v) + \varepsilon |\lambda| < \alpha_1.$$

The only remaining estimate is for $\varphi_2(z)$. Consider an arbitrary number $\gamma \in [0,1]$ and let

$$\varphi(z) := \frac{1}{2}|D^{2\beta}z|^2 - \gamma(F(z), D^{2\beta}z) + \gamma(I(z), 1)$$

(we will obtain the value $\gamma = \frac{4\beta+1}{4\beta+2}$ to be the useful one, thus $\varphi = \varphi_2$).

Rewrite (65) in the form

$$z - v = \lambda D^{2\beta}\partial z - \lambda \partial F(w).$$

Then,

$$\begin{aligned}
|D^{2\beta}z|^2 &= (D^{2\beta}z, D^{2\beta}v) + (D^{2\beta}z, D^{2\beta}z - D^{2\beta}v) \\
&= (D^{2\beta}z, D^{2\beta}v) - (D^{2\beta}\partial z, \partial^{-1}D^{2\beta}(z-v)) \\
&= (D^{2\beta}z, D^{2\beta}v) - \frac{1}{\lambda}(z - v + \lambda \partial F(w), \partial^{-1}D^{2\beta}(z-v)) \\
&= (D^{2\beta}z, D^{2\beta}v) - (\partial F(w), \partial^{-1}D^{2\beta}(z-v)) \\
&= (D^{2\beta}z, D^{2\beta}v) + (D^{2\beta}F(w), (z-v)) \\
&= (D^{2\beta}z, D^{2\beta}v) + (D^{2\beta}F(w), \lambda D^{2\beta}\partial z - \lambda \partial F(w)) \\
&= (D^{2\beta}z, D^{2\beta}v) + \lambda(D^{2\beta}F(w), D^{2\beta}\partial z) \\
&\leq \frac{1}{2}|D^{2\beta}z|^2 + \frac{1}{2}|D^{2\beta}v|^2 + \lambda(D^{2\beta}F(z), D^{2\beta}\partial z) + |\lambda|\varepsilon_1.
\end{aligned}$$

where ε_1 is as in (60). Hence

$$\frac{1}{2}|D^{2\beta}z|^2 - \frac{1}{2}|D^{2\beta}v|^2 \leq \lambda(D^{2\beta}F(z), D^{2\beta}\partial z) + |\lambda|\varepsilon_1. \quad (69)$$

On the other hand,

$$(F(z), D^{2\beta}z) - (F(v), D^{2\beta}v) =$$
$$= (F(z) - F(v), D^{2\beta}z) + (F(v), D^{2\beta}z - D^{2\beta}v)$$
$$= \left(\int_0^1 F'(\tau z + (1-\tau)v)d\tau D^{2\beta}z, (z-v)\right) + (D^{2\beta}F(v), z-v)$$
$$\geq (F'(z)D^{2\beta}z, z-v) - |\lambda|\varepsilon_2 - |D^{2\beta}F(v) - D^{2\beta}F(w)||z-v|$$
$$+ (D^{2\beta}F(w), z-v)$$
$$\geq (F'(z)D^{2\beta}z, \lambda D^{2\beta}\partial z - \lambda \partial F(w)) + (D^{2\beta}F(w), \lambda D^{2\beta}\partial z) - |\lambda|(\varepsilon_1 + \varepsilon_2)$$
$$\geq \lambda (F'(z)D^{2\beta}z, D^{2\beta}\partial z) - \lambda (F'(z)D^{2\beta}z, F'(z)\partial z)$$
$$+ (\lambda D^{2\beta}F(z), D^{2\beta}\partial z) - |\lambda|(2\varepsilon_1 + 2\varepsilon_2),$$

where ε_2 is as in (61).

Also,

$$\begin{aligned}(I(z), 1) - (I(v), 1) &= \int_\mathbb{T} (I(z) - I(v))\, dx \\ &= \int_\mathbb{T} \int_0^1 I'(\tau z + (1-\tau)v)d\tau (z-v)dx \\ &= \left(\int_0^1 I'(\tau z + (1-\tau)v)d\tau, \lambda D^{2\beta}\partial z - \lambda \partial F(w)\right) \\ &= -\lambda \left(\int_0^1 I''(z_\tau)(\partial z_\tau)\, d\tau, D^{2\beta}z - F(w)\right), \\ &\quad \text{where } z_\tau = \tau z + (1-\tau)v, \\ &\leq -\lambda (I''(z)\partial z, D^{2\beta}z - F(w)) + |\lambda|\varepsilon_3 \\ &\leq -\lambda (I''(z)\partial z, D^{2\beta}z) + \lambda (I''(z)\partial z, F(z)) + |\lambda|(\frac{\varepsilon}{2} + \varepsilon_3) \\ &= -\lambda (F'(z)\partial z, F'(z)D^{2\beta}z) + |\lambda|(\frac{\varepsilon}{2} + \varepsilon_3),\end{aligned}$$

where ε_3 is as in (62).

Putting all the above estimates together, we obtain

$\frac{1}{\lambda}\left(\varphi\left(z\right)-\varphi\left(v\right)\right) \leq$

$$\begin{aligned}
&\leq \left(D^{2\beta}F(z), D^{2\beta}\partial z\right) - \gamma\left(F'(z)D^{2\beta}z, D^{2\beta}\partial z\right) + \gamma\left(F'(z)D^{2\beta}z, F'(z)\partial z\right) \\
&\quad -\gamma\left(D^{2\beta}F(z), D^{2\beta}\partial z\right) - \gamma\left(F'(z)\partial z, F'(z)D^{2\beta}z\right) \\
&\quad +(1+2\gamma)\varepsilon_1 + 2\gamma\varepsilon_2 + \gamma\varepsilon_3 + \gamma\frac{\varepsilon}{2} \\
&\leq (1-\gamma)\left(D^{2\beta}F(z), D^{2\beta}\partial z\right) - \gamma\left(F'(z)D^{2\beta}z, D^{2\beta}\partial z\right) + \varepsilon \\
&= -(1-\gamma)\left(D^{2\beta}\partial F(z), D^{2\beta}z\right) - \gamma\left(F'(z)D^{2\beta}\partial z, D^{2\beta}z\right) + \varepsilon.
\end{aligned}$$

Thus,

$\frac{1}{\lambda}\left(\varphi\left(z\right)-\varphi\left(v\right)\right) \leq$

$$\begin{aligned}
&\leq (\gamma-1)\left[\left(D^{2\beta}\partial F(z), D^{2\beta}z\right) - \left(F'(z)D^{2\beta}\partial z, D^{2\beta}z\right)\right] \\
&\quad - \left(F'(z)D^{2\beta}z, \partial D^{2\beta}z\right) + \varepsilon \\
&= (\gamma-1)\left[\left(D^{2\beta}\partial F(z), D^{2\beta}z\right) - \left(F'(z)D^{2\beta}\partial z, D^{2\beta}z\right)\right] \\
&\quad -\frac{1}{2}\left(F'(z), \partial\left(D^{2\beta}z\right)^2\right) + \varepsilon. \\
&= (\gamma-1)\left[\left(D^{2\beta}\partial F(z), D^{2\beta}z\right) - \left(F'(z)D^{2\beta}\partial z, D^{2\beta}z\right)\right] \\
&\quad +\frac{1}{2}\left(\partial F'(z)D^{2\beta}z, D^{2\beta}z\right) + \varepsilon \\
&= (\gamma-1)\left(D^{2\beta}\left(F'(z)\partial z\right) - F'(z)D^{2\beta}\partial z - (2\beta+1)\partial F'(z)D^{2\beta}z, D^{2\beta}z\right) + \varepsilon,
\end{aligned}$$

provided that

$$-(\gamma-1)(2\beta+1) = \frac{1}{2}$$

which is satisfied precisely for $\gamma = \frac{4\beta+1}{4\beta+2}$, when $\varphi = \varphi_2$.

Using Lemma 13 we get the estimate

$$\frac{1}{\lambda}\left(\varphi_2\left(z\right) - \varphi_2\left(v\right)\right) \leq g(\varphi_2(z)) + \varepsilon.$$

This implies (using (54) in conjunction with (20)

$$\varphi_2(z) \leq m_{g_\varepsilon}(\lambda, \varphi_2(v)) < \alpha_2. \tag{70}$$

This concludes the proof of (64). The Schauder-Tichonov theorem applied to $\Gamma : K \to K$ gives us the desired fixed point, $u = \Gamma u$, so that

$$u - \lambda D^{2\beta}\partial u - \lambda \partial F(u) = v. \tag{71}$$

The estimates for $\varphi_j(z)$ in terms of $\varphi_j(v)$ imply (50),(51),(52).

What remains to be proven is the estimate (53) for $\varphi_3(z)$. We will make use of the dissipativity of $A + B_R - \omega_R$, for some $\omega_R > 0$.

For v and u as above we know that $\varphi_j(u) \leq \alpha_j, j = 0, 1, 2$. Then

$$\begin{aligned}
\omega_R |u - v|^2 &\geq (Au + B(u) - Av - B(v), u - v) \\
&= (D^{2\beta}\partial u - \partial F(u) - D^{2\beta}\partial v + \partial F(v), u - v) \\
&= \lambda \left| D^{2\beta}\partial u - \partial F(u) \right|^2 + (D^{2\beta}\partial v + \partial F(v), u - v) \\
&\geq \lambda \varphi_3(u)^2 - \varphi_3(v)|u - v|.
\end{aligned}$$

Thus

$$\lambda \varphi_3(u)^2 \leq \varphi_3(v)|u - v| + \omega_R |u - v|^2.$$

But $|u - v| = |\lambda| \varphi_3(u)$, so we conclude

$$\varphi_3(u) \leq \varphi_3(v) + \lambda \omega_R \varphi_3(u),$$

or, equivalently,

$$\varphi_3(u) \leq \frac{\varphi_3(v)}{1 - \lambda \omega_R}.$$

This completes the proof of Theorem 14. $\square$

Having all the hypothesis in place, we can apply the abstract Theorem 3 and conclude the proof of the well-posedness for the Cauchy Problem (NDE) in the space $H^s(\mathbb{T})$, where $s = max\{2\beta, \frac{3}{2} + \varepsilon\}$, for some $\varepsilon > 0$ as in Theorem 1.

References

[1] L. Abdelouhab, J.L. Bona, M. Felland, J.-C. Saut: Nonlocal models for nonlinear dispersive waves, *Physica D* **40** (1989), 360-392.

[2] R.A. Adams: *Sobolev Spaces*, Academic Press, New York, 1975.

[3] V. Barbu: *Nonlinear Semigroups and Differential Equations in Banach Spaces*, Noordhoff, Leyden, 1976

[4] Ph. Bénilan: *Equations d'évolution dans un espace de Banach quelconque et applications.* PhD dissertation, Université de Paris, Orsay, 1972.

[5] J.L. Bona, V.A. Dougalis, O.A. Karakashian, W.R. McKinney: Conservative, high-order numerical schemes for the generalized Korteweg-de Vries equations, *Phil. Trans. Roy. Soc. London Ser. A* **351** (1995), 107-164.

[6] J.L. Bona, V.A. Dougalis, O.A. Karakashian, W.R. McKinney: Numerical simulation of singular solutions of the generalized Korteweg-de Vries equation. Mathematical problems in the theory of water waves, *Contemp. Math.* **200** (1996), 17-29

[7] J.L. Bona, M. Scialom: The effect of change in the nonlinearity and the dispersion relation of model equations for long waves, *Canad. Appl. Math. Quart.* **3** (1995), 1-41.

[8] J.L. Bona, R. Smith: The initial value problem for the Korteweg-de Vries equation, *Royal Society of London, Ser. A* **278** (1978), 555-601.

[9] J.L. Bona, P.E. Souganidis, W.A. Strauss: Stability and instability of solitary waves of Korteweg-de Vries type, *Proc. Royal Soc. London, Ser. A* **411** (1987), 395-412.

[10] J. Bourgain: On the Cauchy problem for periodic KdV-type equations, Proceedings of the Conference in Honor of Jean-Pierre Kahane, *J. Fourier Anal. Appl.*, Special Issue, (1995), 17-86.

[11] J. Bourgain: On the growth in time of higher Sobolev norms of smooth solutions of hamiltonian PDE, *Internat. Math. Res. Notices* **6** (1996), 277-304.

[12] R.C. Cascaval: Global Well-Posedness for a Class of Dispersive Equations, PhD Thesis, University of Memphis, May 2000.

[13] R.C. Cascaval, J.A. Goldstein: A semigroup approach to dispersive waves, *Proc. of the Bad-Herrenalb Conference*, G. Lumer, L. Weis (ed.), 225-233, Marcel Dekker, New York, 2000.

[14] P. Constantin, J.-C. Saut: Local smoothing properties of dispersive equations, *J. Amer. Math. Soc.* **1** (1988), 413-439.

[15] M.G. Crandall, T.M. Liggett: Generation of semigroups of nonlinear transformations on general Banach spaces, *Amer. J. Math.* **93** (1971), 265-298.

[16] D. Dix: The dissipation of nonlinear dispersive waves: the case of asymptotically weak nonlinearity, *Comm. Partial Differential Equations* **17** (1992), 1665-1693.

[17] L.C. Evans: Nonlinear Evolution Equations in Banach Spaces, Math. Res. Center Tech. Summary Report No. 1568, Madison, Wisconsin, 1975.

[18] L.C. Evans: Nonlinear evolution equations in an arbitrary Banach space, *Israel J. Math.* **26** (1977), 1-42.

[19] J. Ginibre, Y. Tsutsumi: Uniqueness for the generalized Korteveg-de Vries equation, *SIAM J. Math. Anal.* **20** (1989), 1388-1425.

[20] J. Ginibre, G. Velo: Existence and uniqueness of solutions for the generalized Korteweg de Vries equation, *Math. Z.* **203** (1990), 9-36.

[21] J. Ginibre, G. Velo: Smoothing properties and existence of solutions for the generalized Benjamin-Ono equation, *J. Differential Equations* **93** (1991), 150-212.

[22] J.A. Goldstein: Approximation of nonlinear semigroups and evolution equations, *J. Math. Soc. Japan* **24** (1972), 558-573.

[23] J.A. Goldstein: The KdV equation via semigroups, in *"Theory and Applications of Nonlinear Operators of Accretive and Monotone Type"*, 107-114; Lecture Notes in Pure and Appl. Math. **178**, M. Dekker, New York, 1996.

[24] J.A. Goldstein: *Semigroups of Linear Operators and Applications*, Oxford U. Press, Oxford and New York, 1985.

[25] J.A. Goldstein, S. Oharu, T. Takahashi: A class of locally Lipschitzian semigroups and its application to generalized Korteweg-de Vries equations, (unpublished manuscript), 1994.

[26] J.A. Goldstein, S. Oharu, T. Takahashi: Semilinear Hille-Yosida theory: the approximation theorem and group of operators, *Nonlinear Analysis, Theory, Methods and Applications* **13** (1989), 325-339.

[27] T. Kato: Quasi-linear equations of evolution, with applications to partial differential equations. *Spectral Theory and Differential*

Equations, 27-50; Lecture Notes in Math. **448**, Springer, New York, 1975.

[28] T. Kato: On the Korteweg-de Vries equation, *Manuscripta Math.* **28** (1979), 89-99.

[29] T. Kato: On the Cauchy problem for the (generalized) Korteweg-de Vries equation, in *Studies in Applied Mathematics*, 93-128; Adv. Math. Suppl. Stud. **8**, Academic Press, New York, 1983.

[30] T. Kato, G. Ponce: Commutator estimates and the Euler and Navier-Stokes equations, *Comm. Pure Appl. Math.* **41** (1988), 891-907.

[31] T. Kato: Abstract evolution equations, linear and quasilinear, revisited, in *Functional Analysis and Related Topics*, 103-125; Lecture Notes in Math. **1540**, Springer, Berlin, 1993.

[32] C. Kenig, G. Ponce, L. Vega: On the (generalized) Korteweg-de Vries equation, *Duke Math. J.* **59** (1989), 585-610.

[33] C. Kenig, G. Ponce, L. Vega: Well-posedness of the initial value problem for the Korteweg-de Vries equation, *J. Amer. Math. Soc.* **4** (1991), 323-347.

[34] C. Kenig, G. Ponce, L. Vega: Well-posedness and scattering results for the generalized Korteweg-de Vries equation via the contraction principle, *Comm. Pure Appl. Math.* **46** (1993), 527-620.

[35] C. Kenig, G. Ponce, L. Vega: On the generalized Benjamin-Ono equation, *Trans. Amer. Math. Soc.* **342** (1994), 155-172.

[36] Y. Kobayashi: Difference approximation of Cauchy problems for quasi-dissipative operators and generation of nonlinear semigroups, *J. Math. Soc. Japan* **27** (1975), 640-665.

[37] Y. Kobayashi, S. Oharu: Semigroups of locally Lipshitzian operators and applications, in *Functional Analysis and Related Topics*, 191-211; Lecture Notes in Math. **1540**, Springer Verlag, 1993.

[38] P.I. Naumkin, I.A. Shishmarev: *Nonlinear Nonlocal Equations in the Theory of Waves*, Translations of Mathematical Monographs **133**, Amer. Math. Soc., Providence, RI, 1994.

[39] S. Oharu, T. Takahashi: Locally Lipshitz continuous perturbations of linear dissipative operators and nonlinear semigroups, *Proc. Amer. Math. Soc.* **100** (1987), 187-194.

[40] S. Oharu, T. Takahashi: Characterization of nonlinear semigroups associated with semilinear evolution equations *Trans. Amer. Math. Soc.* **311** (1989), 593-679.

[41] J.-C. Saut, R. Temam: Remarks on the Korteweg-de Vries equation, *Israel J. Math.* **24** (1976), 78-87.

[42] J.-C. Saut: Sur quelques generalisations de l'equation de Korteweg-de Vries, *J. Math. Pure Appl.* **58** (1979), 21-61.

[43] G. Staffilani: On the generalized Korteweg-de Vries- type Equations, *Differential and Integral Equations* **10** (1997), 777-796.

[44] G. Staffilani: On solutions for periodic generalized KdV equations, *Internat. Math. Res. Notices* **18** (1997), 899-917.

Viable Domains for Differential Equations Governed by Carathéodory Perturbations of Nonlinear m-Accretive Operators

Ovidiu Cârjă "Al.I. Cuza" University of Iaşi, Iaşi, Romania

Ioan I. Vrabie "Al.I. Cuza" University of Iaşi, Iaşi, Romania

ABSTRACT. Let X be a real Banach space, $A : D(A) \subset X \to 2^X$ an m-accretive operator and let $S(t) : \overline{D(A)} \to \overline{D(A)}$, $t \geq 0$ be the semigroup of nonexpansive mappings generated by $-A$. Let D be a locally closed subset in $\overline{D(A)}$ and $f : [a,b) \times D \to X$ a Carathéodory function. We assume that either $S(t)$ is compact for each $t > 0$, or that the inclusion $D \subset X$ is compact and we prove:

THEOREM. *Under the assumptions above, a necessary and sufficient condition in order that for each $(\tau, \xi) \in [a,b) \times D$ there exists at least one local mild solution $u : [\tau, T] \to D$ of $u'(t) + Au(t) \ni f(t, u(t))$ satisfying $u(\tau) = \xi$ is the tangency condition*

$$(\mathcal{TC}) \quad \begin{cases} \text{There is a negligible subset } \mathcal{Z} \text{ of } [a,b) \text{ so that, for each } (t,\xi) \in ([a,b) \setminus \mathcal{Z}) \times D, \\ \liminf_{h \downarrow 0} \tfrac{1}{h} d(u(t+h, t, \xi, f(t, \xi)), D) = 0. \end{cases}$$

Here $u(t+h, t, \xi, f(t, \xi)) = y(t+h)$ where y is the unique mild solution of the problem

$$\begin{cases} y'(s) + Ay(s) \ni f(t, \xi) \\ y(t) = \xi. \end{cases}$$

An interesting application concerning the existence of monotone solutions and some extensions to the case f multivalued are also included.

1. INTRODUCTION

Our main goal in this paper is to prove a necessary and sufficient condition in order that a given subset of a Banach space X be a viable domain for a strongly nonlinear nonautonomous differential equation. Namely, let X be a Banach space, $A : D(A) \subset X \to 2^X$ an m-accretive operator with $-A$ generating a semigroup of nonexpansive mappings $S(t) : \overline{D(A)} \to \overline{D(A)}$, for $t \geq 0$, D a nonempty subset in $\overline{D(A)}$ and $f : [a,b) \times D \to X$ a function. We consider the nonlinear perturbed differential equation

$$u'(t) + Au(t) \ni f(t, u(t)) \tag{$\mathcal{DE}$}$$

and we are interested in finding necessary and sufficient conditions in order that D be a *viable domain* for $(\mathcal{DE})$.

DEFINITION 1.1. We say that D is a *viable domain* for $(\mathcal{DE})$ if for each $(\tau, \xi) \in [a,b) \times D$ there exists at least one mild solution $u : [\tau, T] \to D$, $T < b$, of $(\mathcal{DE})$ satisfying the initial condition

$$u(\tau) = \xi. \tag{$\mathcal{IC}$}$$

We recall that the function $u : [\tau, T] \to D$ is a *mild solution* of $(\mathcal{DE})$ and $(\mathcal{IC})$ if u satisfies $(\mathcal{IC})$, it renders the function $g(\cdot) = f(\cdot, u(\cdot))$ integrable on $[\tau, T]$ and it is a mild solution on $[\tau, T]$ of the equation

$$u'(t) + Au(t) \ni g(t)$$

in the sense of Definition 1.7.5, p. 25 in [31].

The viability problem has been studied by many authors by using various frameworks and techniques. We start by reviewing the state of the art in the "continuous" case, i.e. in the case in which f is continuous. First we discuss the semilinear case, i.e., the case in which $-A$ is the generator of a C_0-semigroup. In this respect it should be noted the pioneering work of Nagumo [22] who considered X finite dimensional and $A = 0$. In this context he showed that a necessary and sufficient condition in order that D be a viable domain for $(\mathcal{DE})$ is the tangency condition below:

$$\liminf_{h \downarrow 0} \frac{1}{h} d(\xi + hf(t,\xi), D) = 0 \tag{1.1}$$

for each $(t,\xi) \in [a,b] \times D$. Here and thereafter $d(x, C)$ denotes the distance from the point $x \in X$ to the subset C in X. As far as we know, Nagumo's result (or variants of it) has been independently rediscovered several times in the seventies by Brézis [9], Crandall [16], Hartman [19] and Martin [21].

In the infinite dimensional setting with $A = 0$ and $f(t, \cdot)$ dissipative, we recall the results of Martin [21]. The semilinear "continuous" case, i.e. the case in which $-A$ is the infinitesimal generator of a C_0-compact semigroup and f is continuous has been studied by Pavel [23]. We emphasize Pavel's main contribution who, to our knowledge, was the first who formulated the corresponding tangency condition applying to the semilinear case by means of the generated C_0-semigroup, in such a way to work also for points ξ that do not belong to the domain of A. More precisely, Pavel [23] shows that, whenever A generates a compact C_0-semigroup and f is continuous on $[a,b] \times D$ with D locally closed, a sufficient condition for viability is:

$$\lim_{h \downarrow 0} \frac{1}{h} d(S(h)\xi + hf(t,\xi), D) = 0 \tag{1.2}$$

for each $(t, \xi) \in [a, b] \times D$. We notice that, whenever $\xi \in D \cap D(A)$, (1.2) is equivalent to

$$\lim_{h \downarrow 0} \frac{1}{h} d(\xi + h(A\xi + f(t,\xi)), D) = 0$$

which is nothing else than the classical Nagumo's tangency condition (1.1) with $A + f$ instead of f. However, there exist situations in which D is not included in $D(A)$, or even $D \cap D(A)$ is empty and in these cases we can use only (1.2). For instance this happens if D is the trajectory of a nowhere differentiable mild solution of $(\mathcal{DE})$. For subsequent developements, allowing f to be multivalued, see Pavel-Vrabie [24] and [25], Shi Shuzhong [27], Cârjă-Vrabie [13] and the references therein.

The fully nonlinear case, i.e. the case in which both A and f are nonlinear, with A unbounded but f still continuous, has been considered for the first time by Vrabie [30]. We notice that Vrabie [30] introduced the suitable tangency condition to apply also for points of D which do not belong to $D(A)$. Namely, the tangency condition introduced in [30] is

$$\lim_{h \downarrow 0} \frac{1}{h} d(u(t+h,t,\xi,f(t,\xi)),D) = 0, \tag{1.3}$$

Viable Domains

where $u(t + \cdot, t, \xi, y) = v(\cdot)$ is the unique mild solution of the Cauchy problem

$$\begin{cases} v'(s) + Av(s) \ni y \\ v(t) = \xi. \end{cases}$$

More precisely, Vrabie [30] proved that if $-A$ is the generator af a compact semigroup of nonexpansive operators and (1.3) holds *uniformly* with respect to (t, ξ) in $[a, b] \times D$, then D is a viable domain for $(\mathcal{DE})$. We emphasize that, whenever A is linear, (1.3) is equivalent to

$$\lim_{h \downarrow 0} \frac{1}{h} d \left(S(h)\xi + \int_t^{t+h} S(t-s)f(t,\xi)\,ds, D \right) = 0$$

which in its turn reduces to (1.2). Subsequent contributions in this context are due to Bothe [5] who allowed D to depend on t as well. In particular, in case D independent of t, Bothe [5] showed that (1.3) is necessary and sufficient for viability. We also mention Bothe [6], Bressan-Staicu [8] and Cârjă-Vrabie [14] who considered the case in which f is multivalued and satisfies a certain continuity-like condition. While Bothe [6] and Bressan-Staicu [8] consider a tangency condition which reduces to (1.3) whenever f is single valued and an $\varepsilon - \delta$ upper semicontinuity condition and respectively a lower semicontinuity on f, Cârjă-Vrabie [14] allow f to be strongly-weakly upper semicontinuous but use a tangency condition expressed in the terms of the weak topology on X and which, in certain situations, is stronger than (1.3).

Concerning the Carathéodory case, again when X is finite dimensional and $A = 0$, we have to mention the work of Ursescu [29] who was the first able to overcome the difficulties of defining a suitable a.e.-tangency condition in order to obtain a corresponding Nagumo-type theorem applicable in this general frame. More precisely, Ursescu proved that a necessary and sufficient condition for viability under Carathéodory conditions on f is the Nagumo's tangency condition (1.1) but satisfied only for each $(t, \xi) \in ([a, b] \setminus \mathcal{Z}) \times D$, where $\mathcal{Z}$ is a negligible set.

The true semilinear Carathéodory case, i.e. A unbounded and f measurable in t and continuous in ξ has been considered for the first time in Cârjă-Monteiro Marques [11] by using a "lim inf" variant of Pavel's tangency condition (1.2), again satisfied only for each $(t, \xi) \in ([a, b] \setminus \mathcal{Z}) \times D$, where $\mathcal{Z}$ is as above. We note in passing that, whenever f is single-valued, (1.2) and respectively (1.3) is equivalent with the corresponding condition obtained from (1.2) and respectively from (1.3) by substituting "lim" with "lim inf".

In the present paper we consider the fully nonlinear Carathéodory case, i.e. the case in which A is nonlinear, possible unbounded and f is a Carathéodory function. We assume that D is locally closed in the sense that for each $\xi \in D$ there exists $r > 0$ such that $D \cap B(\xi, r)$ is closed in X, where, as usual, $B(\xi, r)$ denotes the closed ball with center ξ and radius r and we prove that under some extra-conditions either on the semigroup generated by $-A$ or on the subset D, a necessary and sufficient condition for the viability of D is (1.3) satisfied only for each $(t, \xi) \in ([a, b] \setminus \mathcal{Z}) \times D$, where $\mathcal{Z}$ is a negligible subset in $[a, b]$. See Theorems 3.1 and 3.2 below. As a matter of fact, it should be mentioned that our main result here is a nontrivial nonlinear extension of that in Cârjă-Monteiro Marques [11] and, inasmuch as the specific "linear arguments" used in [11] fail in this new context, our proofs herein are distinct from their semilinear counterparts and are essentially new.

We also characterize *admissibility of a preorder* with respect to the differential equation $(\mathcal{DE})$ through a tangency condition of type (1.3). We recall that a preorder "$\preceq$" on a subset M of D is admissible with respect to $(\mathcal{DE})$ if for each $(\tau, \xi) \in [a, b] \times M$ there exists at least one mild solution $u : [\tau, T] \to M$ satisfying $u(\tau) = \xi$ and $u(s) \preceq u(t)$ for each $\tau \leq s \leq t \leq T$. We adopt the approach proposed for the first time in Cârjă-Ursescu [12], where, in the finite dimensional setting, the admissibility of a preorder "$\preceq$" with respect to $(\mathcal{DE})$ is described in terms of viability of the sets $P(\xi) = \{\eta \in M \,;\, \xi \preceq \eta\}$ with respect to the same equation. That approach has been extended recently to the infinite dimensional case by Chiş-Şter [15] in the case of a continuous perturbation f. Usually, the authors get admissibility of preorders reconstructing, and thus increasing in amount, the proofs of viability of sets. See, e.g., [18], [20]. The interested reader in this area, as well as in viability problems, is refered to Aubin-Cellina [3].

Finally, as direct applications of our main theorems, we include some results concerning the *invariance* of a given subset with respect to a differential inclusion governed by a Carathéodory multivalued perturbation of an m-accretive operator. We recall that a subset D is *invariant* with respect to $(\mathcal{DE})$ if f, as a function of its second argument, is defined on a subset in X which is larger than D, for instance if $f : [a,b) \times \overline{D(A)} \to X$, and for each $(\tau, \xi) \in [a, b) \times D$, either there is no solution of $(\mathcal{DE})$ and $(\mathcal{IC})$, or each mild solution $u : [\tau, T) \to \overline{D(A)}$ of the Cauchy problem above satisfies $u(t) \in D$ for each $t \in [\tau, T)$.

The paper is divided into seven sections, the second one being merely concerned with some necessary background material. In section three we state our main results, i.e. Theorems 3.1 and 3.2, while in section four we present several auxiliary results. Section five contains the complete proof of both Theorem 3.1 and 3.2 and section six is devoted to the statement and proof of a necessary and sufficient condition of admissibility of a preorder with respect to $(\mathcal{DE})$. Section seven includes a necessary condition in order that a given subset be invariant with respect to a differential inclusion governed by a measurable $\times$ lower semicontinuous perturbation of an m-accretive operator. We also show that the condition in question is also sufficient for viability, and, in certain specific, but important cases, we prove that it is sufficient even for invariance.

Acknowledgements. We express our warmests thanks to Professor C. Ursescu for the very careful reading of a previous version of this paper, for his helpful critical remarks and suggestions concerning the presentation and especially for calling to our attention a possibility to simplify our initial proof.

2. Preliminaries

We assume that the reader is familiar with the basic concepts and results concerning m-accretive operators and nonautonomous differential equations in abstract Banach spaces and we refer to Barbu [4] and Vrabie [31] for details.

However, we recall for easy references some basic concepts and results we will use in the sequel. Let $A : D(A) \subset X \to 2^X$ be an m-accretive operator, $\xi \in \overline{D(A)}$ and $f \in L^1(a, b; X)$ and let us consider the differential equation

$$u'(t) + Au(t) \ni f(t). \tag{2.1}$$

In all that follows we denote by $u(\cdot, a, \xi, f) : [a, b] \to \overline{D(A)}$ the unique mild solution of (2.1) satisfying $u(a, a, \xi, f) = \xi$ and by $S(t) : \overline{D(A)} \to \overline{D(A)}$, $t \geq 0$, the semigroup of nonexpansive mappings generated by $-A$, i.e. $S(t)\xi = u(t, 0, \xi, 0)$ for each $t \geq 0$ and $\xi \in \overline{D(A)}$. We recall that, for each $f, g \in L^1(a, b; X)$ each $\xi, \eta \in \overline{D(A)}$ and each $a \leq t \leq b$,

Viable Domains

we have
$$\|u(t, a, \xi, f) - u(t, a, \eta, g)\| \leq \|\xi - \eta\| + \int_a^t \|f(\tau) - g(\tau)\| \, d\tau. \tag{2.2}$$

Moreover, we note that, for each $a \leq \nu \leq t \leq b$, we have
$$u(t, a, \xi, f) = u(t, \nu, u(\nu, a, \xi, f), f|_{[\nu, b]}). \tag{2.3}$$

See Vrabie [31]. We also recall that the semigroup $S(t) : \overline{D(A)} \to \overline{D(A)}$, $t \geq 0$ is *compact* if for each $t > 0$ $S(t)$ is a compact operator. A subset G in $L^1(a, b; X)$ is *uniformly integrable* if, for each $\varepsilon > 0$ there exists $\delta(\varepsilon) > 0$ such that, for each measurable subset E in $[a, b]$ whose Lebesgue measure $\lambda(E) < \delta(\varepsilon)$, we have
$$\int_E \|g(s)\| \, ds \leq \varepsilon,$$
uniformly for $g \in G$.

REMARK 2.1. It is easy to see that, whenever $\ell \in \mathcal{L}^1(a, b; \mathbb{R}_+)$, the set
$$G_\ell = \{g \in L^1(a, b; X); \|g(t)\| \leq \ell(t), \text{ a.e. for } t \in [a, b]\}$$
is uniformly integrable.

We include for easy reference the following two compactness results which are the main ingredients in the proof of Theorems 3.1 and 3.2.

THEOREM 2.1. (VRABIE) *Let X be a real Banach space, $A : D(A) \subset X \to 2^X$ an m-accretive operator, $\xi \in \overline{D(A)}$ and G an uniformly integrable subset in $L^1(a, b; X)$. Then the following conditions are equivalent:*

(i) *the set $\{u(\cdot, a, \xi, g); g \in G\}$ is relatively compact in $C([a, b]; X)$;*
(ii) *there exists a dense subset E in $[a, b]$ such that, for each $t \in E$, $\{u(t, a, \xi, g); g \in G\}$ is relatively compact in X.*

See Theorem 2.3.1, p. 45, in Vrabie [31]. A very useful consequence of Theorem 2.1 is:

THEOREM 2.2. (BARAS) *Let X be a real Banach space, $A : D(A) \subset X \to 2^X$ an m-accretive operator such that $-A$ generates a compact semigroup, let $\xi \in \overline{D(A)}$ and G an uniformly integrable subset in $L^1(a, b; X)$. Then the set $\{u(\cdot, a, \xi, g); g \in G\}$ is relatively compact in $C([a, b]; X)$.*

See Theorem 2.3.3, p. 47, in Vrabie [31].

We are now able to introduce the tangency condition we are going to use in the sequel. We begin with the tangency concept.

DEFINITION 2.1. *Let $A : D(A) \subset X \to 2^X$ an m-accretive operator and D be a nonempty subset in $\overline{D(A)}$. We say that $y \in X$ is A-tangent to D at $\xi \in D$ if for each $\delta > 0$ and each $r > 0$ there exist $t \in (0, \delta)$ and $p \in B(0, r)$ such that*
$$u(t, 0, \xi, y) + tp \in D.$$

The set of all A-tangent elements to D at $\xi \in D$ is denoted by $\mathcal{T}_D^A(\xi)$.

REMARK 2.2. It should be noticed that whenever $A = 0$, $\mathcal{T}_D^A(\xi)$ is nothing else than the tangent cone at $\xi \in D$ in the sense of Bouligand [7] and Severi [26]. Moreover, if A is linear and $A \neq 0$, $\mathcal{T}_D^A$ is the tangency concept used by Shi Shouzhong [27] who considered the case when the perturbation is a multifunction not depending on t. It is easy to see that

$$\mathcal{T}_D^A(\xi) = \{y \in X; \liminf_{h \downarrow 0} \frac{1}{h} d(u(h,0,\xi,y), D) = 0\}.$$

Moreover, this concept can be defined equivalently by means of sequences. Namely, $y \in X$ is A-tangent to D at $\xi \in D$ if and only if there exist two sequences, $(t_n)_{n \in \mathbb{N}^*}$ decreasing to 0 and $(p_n)_{n \in \mathbb{N}^*}$ convergent to 0, such that

$$u(t_n, 0, \xi, y) + t_n p_n \in D \tag{2.4}$$

for each $n \in \mathbb{N}$.

We conclude this section by recalling that the *inclusion $D \subset X$ is compact* if each bounded subset in D is relatively compact in X.

3. THE MAIN RESULTS

We begin by recalling:

DEFINITION 3.1. A function $f : [a,b) \times D \to X$ is called a *Carathéodory function* if it satisfies:

(C_1) for every $x \in D$, the function $f(\cdot, x)$ is measurable on $[a,b)$;
(C_2) for almost every $t \in [a,b)$, the function $f(t, \cdot)$ is continuous on D;
(C_3) for every $r > 0$ there exists a locally integrable function $\ell_r : [a,b) \to \mathbb{R}$ such that $\|f(t,x)\| \leq \ell_r(t)$ for almost every $t \in [a,b)$ and for every $x \in D \cap B(0,r)$.

We are now ready to state the main results of this paper.

THEOREM 3.1. *Let X be a real Banach space, $A : D(A) \subset X \to 2^X$ an m-accretive operator with $-A$ the infinitesimal generator of a compact semigroup $S(t) : \overline{D(A)} \to \overline{D(A)}$, $t \geq 0$, D a nonempty, locally closed subset in $\overline{D(A)}$, and $f : [a,b) \times D \to X$ a Carathéodory function. Then a necessary and sufficient condition in order that the set D be viable with respect to $(\mathcal{DE})$ is the tangency condition:*

($\mathcal{T}$) *there exists a negligible subset $\mathcal{Z}$ in $[a,b)$ such that, for each $(t,\xi) \in ([a,b) \setminus \mathcal{Z}) \times D$, we have $f(t,\xi) \in \mathcal{T}_D^A(\xi)$.*

THEOREM 3.2. *Let X be a real Banach space, $A : D(A) \subset X \to 2^X$ an m-accretive operator, D a nonempty locally closed subset in $\overline{D(A)}$ with the inclusion $D \subset X$ compact and $f : [a,b) \times D \to X$ a Carathéodory function. Then a necessary and sufficient condition in order that the set D be viable with respect to $(\mathcal{DE})$ is the tangency condition ($\mathcal{T}$).*

Concerning the existence of saturated, i.e. noncontinuable mild solutions of $(\mathcal{DE})$ and $(\mathcal{JC})$, using a standard argument based on the Brézis - Browder's countable version of Zorn's Lemma, see Theorem 4.2 below, we deduce:

THEOREM 3.3. *Under the hypotheses of either Theorem 3.1, or Theorem 3.2, a necessary and sufficient condition in order that for each $\xi \in D$ there exists at least one saturated mild solution of $(\mathcal{DE})$ satisfying $(\mathcal{JC})$ is the tangency condition $(\mathcal{T})$. If D is closed, then each saturated solution is global, i.e. defined on $[\tau, b)$.*

4. SOME AUXILIARY RESULTS

We begin with the following variant of some general "Lebesgue-type" theorem established in [28] and [11].

THEOREM 4.1. *Assume that X is a real Banach space, D is a nonempty and separable subset in $\overline{D(A)}$ and $f : [a,b) \times D \to X$ is a Carathéodory function. Then there exists a negligible subset $\mathcal{Z}$ of $[a,b)$ such that, for every $t \in [a,b) \setminus \mathcal{Z}$, we have*

$$\lim_{h \downarrow 0} \frac{1}{h} \int_t^{t+h} \|f(s,y(s)) - f(t,y(t))\| \, ds = 0 \qquad (4.1)$$

for all continuous functions $y : [a,b) \to D$.

To prove Theorem 4.1 we have only to repeat the same routine as that in the proof of Lemma 1 in [28] or that in the proof of Theorem 2.3 in [11], in the latter case by taking $S(t) = I$, the identity on X, for each $t \geq 0$.

The following (perhaps known) extension of a simple remark due to Vrabie will prove useful later. See [11] Proposition 2.1.

PROPOSITION 4.1. *If X is a Banach space for which there exists a family $\{S(t)\,;\, t \geq 0\}$ of compact operators such that*

$$\lim_{t \downarrow 0} S(t)x = x$$

for each $x \in X$, then X is separable.

Proof. Take a sequence $(t_n)_{n \in \mathbb{N}}$ which is decresing to 0. Since, for each $n \in \mathbb{N}$, $S(t_n)B(0,n)$ is precompact there exists a finite family of points D_n in $B(0,n)$ such that for every $x \in B(0,n)$ there exists $x_n \in D_n$ satisfying

$$\|S(t_n)x - S(t_n)x_n\| \leq t_n.$$

Let $x \in X$ and $\varepsilon > 0$ and choose $n \in \mathbb{N}$ such that $t_n \leq \varepsilon$, $\|x - S(t_n)x\| \leq \varepsilon$ and $\|x\| \leq n$. Taking $x_n \in D_n$ as above, we have

$$\|x - S(t_n)x_n\| \leq \|x - S(t_n)x\| + \|S(t_n)x - S(t_n)x_n\| \leq 2\varepsilon.$$

So $D = \cup_n S(t_n)D_n$ (which obviously is countable) is dense in X and this completes the proof. $\square$

One of the main tools in the proof of both Theorem 3.1 and 3.2 is the following characterization of the tangency condition (J).

PROPOSITION 4.2. *Let X be a real Banach space, D a nonempty and separable subset in $\overline{D(A)}$, $A : D(A) \subset X \to 2^X$ an m-accretive operator and $f : [a,b) \times D \to X$ a Carthéodory function. Then, the tangency condition (J) is equivalent to the condition (J̃) below:*

(J̃) *There is a negligible subset $\mathcal{Z}$ of $[a,b)$ such that for every $(t,\xi) \in ([a,b) \setminus \mathcal{Z}) \times D$ there exist two sequences, $(h_n)_{n \in \mathbb{N}^*}$ decreasing to 0 and $(p_n)_{n \in \mathbb{N}^*}$ convergent to 0, such that*

$$u(t+h_n, t, \xi, f(\cdot, \xi)) + h_n p_n \in D \qquad (4.2)$$

for each $n \in \mathbb{N}^*$.

Proof. From (2.2) we have

$$\|u(t+h,t,\xi,f(t,\xi)) - u(t+h,t,\xi,f(\cdot,\xi))\| \leq \int_t^{t+h} \|f(s,\xi) - f(t,\xi)\| \, ds$$

for each $(t,\xi) \in [a,b] \times D$ and $h > 0$ with $t + h < b$. The conclusion follows directly from (2.4) and Theorem 4.1. □

We end this section by recalling a general principle on ordered sets due to H. Brézis and F. Browder [10, p. 356]. It will be used in the next sections in order to obtain some "maximal" elements in an ordered set. The existence of maximal elements is usually derived by using the well-known Zorn's Lemma, an ordering principle which is equivalent to the Axiom of Choice. The Brézis-Browder's Ordering Principle is based on an axiom which is weaker than the Axiom of Choice, i.e. the Axiom of Dependent Choices [17]. See also [12, p. 16] for other applications.

THEOREM 4.2. *Let $\mathcal{X}$ be a nonempty set, $\leq$ a preorder on $\mathcal{X}$ and $\Psi : \mathcal{X} \to \mathbb{R} \cup \{+\infty\}$ an increasing function. Suppose that each increasing sequence in $\mathcal{X}$ is majorated in $\mathcal{X}$. Then, for each $x_0 \in \mathcal{X}$ there exists $\bar{x} \in \mathcal{X}$ with $x_0 \leq \bar{x}$ such that $\bar{x} \leq x$ implies $\Psi(\bar{x}) = \Psi(x)$.*

Note that, in the paper by Brézis and Browder, the function S is supposed to be finite and bounded from above, but, as remarked in [12], this restriction can be easily removed by replacing the function Ψ by the function $x \mapsto \arctan(\Psi(x))$.

5. Poof of Theorems 3.1 and 3.2

The necessity of both Theorems 3.1 and 3.2 is an immediate consequence of the next result which is interesting by itself.

THEOREM 5.1. *Let X be a real Banach space, $A : D(A) \subset X \to 2^X$ an m-accretive operator, D a locally closed and separable subset in X and $f : [a,b) \times D \to X$ a Carathéodory function. Then, a necessary condition in order that D be a viable domain for $(\mathcal{DE})$ is the tangency condition $(\mathcal{T})$.*

Proof. Let $\mathcal{Z}$ be given by Theorem 4.1, let $\tau \in [a,b) \setminus \mathcal{Z}$, let $\xi \in D$, choose a solution u to $(\mathcal{DE})$ and $(\mathcal{IC})$, which is defined on a subinterval $[\tau, T]$ of $[a,b)$ and take a continuous function $z : [a,b) \to D$ which coincides with u on $[\tau, T]$. We have

$$u(\tau + h, \tau, \xi, f(\cdot, z(\cdot))) \in D$$

for each $h \in [0, T - \tau]$. On the other hand, by (2.2), we get

$$\|u(\tau+h,\tau,\xi,f(\cdot,z(\cdot))) - u(\tau+h,\tau,\xi,f(\tau,\xi))\| \leq \int_\tau^{\tau+h} \|f(s,z(s)) - f(\tau,\xi)\| \, ds$$

for each $h \in [0, T - \tau]$. By (4.1) we know that

$$\lim_{h \downarrow 0} \frac{1}{h} \int_\tau^{\tau+h} \|f(s,z(s)) - f(\tau,\xi)\| \, ds = 0$$

and therefore

$$\lim_{h \downarrow 0} \frac{1}{h} d(u(\tau+h,\tau,\xi,f(\tau,\xi)), D) \leq$$

$$\leq \lim_{h \downarrow 0} \frac{1}{h} \|u(\tau+h,\tau,\xi,f(\tau,\xi)) - u(\tau+h,\tau,\xi,f(\cdot,z(\cdot)))\| = 0.$$

Viable Domains

Hence $f(\tau,\xi) \in \mathcal{T}_D^A(\xi)$ for each $(\tau,\xi) \in ([a,b] \times D) \setminus \mathcal{Z}$ and this completes the proof of Theorem 5.1. □

We are now in the position to prove the necessity of both Theorems 3.1 and 3.2. To do this we have only to observe that in both cases, i.e. $S(t)$ is compact for each $t > 0$, or the inclusion $D \subset X$ is compact, D is separable and therefore the hypotheses of Theorem 5.1 are satisfied. See Proposition 4.1.

REMARK 5.1. The proof of Theorem 5.1 shows that, even in a more general frame than that assumed either in Theorem 3.1, or in Theorem 3.2, a necessary condition for the viability of D with respect to $(\mathcal{DE})$ is a tangency condition which, in general is stronger that $(\mathcal{T})$. More precisely, we proved that such a necessary condition is:

(S) there exists a negligible subset $\mathcal{Z}$ in $[a,b)$ such that, for each $(t,\xi) \in ([a,b) \setminus \mathcal{Z}) \times D$, we have $f(t,\xi) \in \mathcal{S}_D^A(\xi)$.

Here
$$\mathcal{S}_D^A(\xi) = \{y \in X \,;\, \lim_{h \downarrow 0} \frac{1}{h} d(u(h,0,\xi,y), D) = 0\}$$

which is included in $\mathcal{T}_D^A(\xi)$. Indeed, $\mathcal{S}_D^A(\xi) \subset \mathcal{T}_D^A(\xi)$ simply because the former is defined by means of "lim", while the latter is defined in a very similar way but by means of "lim inf". See Remark 2.2.

The proof of the sufficiency consists in showing that the tangency condition $(\mathcal{T})$ along with Brézis-Browder Ordering Principle, i.e. Theorem 4.2 above, imply that for each (τ,ξ) in $[a,b) \times D$ there exists at least one sequence of "approximate solutions" of $(\mathcal{DE})$, defined on the same interval, $v_n : [\tau,T] \to X$, satisfying $(\mathcal{IC})$ for each $n \in \mathbb{N}^*$ and such that (v_n) converges in some sense to a mild solution of $(\mathcal{DE})$ satisfying $(\mathcal{IC})$.

The next lemma represents an existence result concerning "approximate solutions" of $(\mathcal{DE})$ satisfying $(\mathcal{IC})$ and it is a nonlinear version of Lemma 3.1 in [11]. Its proof relies on an interplay between some techniques developed in [5] and [11].

LEMMA 5.1. *Let X be a real Banach space, $A : D(A) \subset X \to X$ an m-dissipative operator, D a nonempty, locally closed subset in $\overline{D(A)}$ and $f : [a,b) \times D \to X$ a Carathéodory function satisfying the tangency condition $(\mathcal{T})$. Then for each $(\tau,\xi) \in [a,b) \times D$ there exist $r > 0$, $t_0 \in [a,b) \setminus \mathcal{Z}$ and $T \in (\tau,b)$ such that $D \cap B(\xi,r)$ is closed and for each $n \in \mathbb{N}^*$ and for each open set $\mathcal{L}$ of $\mathbb{R}$ with $\mathcal{Z} \subset \mathcal{L}$ and $\lambda(\mathcal{L}) < \frac{1}{n}$ (λ is the Lebesgue measure), there exist a family of nonempty and pairwise disjoint intervals: $\mathcal{P} = \{[t_m, s_m)\,;\, m \in \mathcal{J}\}$ and two measurable functions $g \in \mathcal{L}^1(\tau,T;X)$ and $v : [\tau,T] \to X$ satisfying*

(i) $\bigcup_{m \in \mathcal{J}} [t_m, s_m) = [\tau, T)$ *and* $s_m - t_m \leq \frac{1}{n}$ *for each* $m \in \mathcal{J}$;

(ii) *if* $t_m \in \mathcal{L}$ *then* $[t_m, s_m) \subset \mathcal{L}$;

(iii) $v(t_m) \in D \cap B(\xi,r)$ *for each* $m \in \mathcal{J}$, $v(T) \in D \cap B(\xi,r)$ *and* $v([\tau,T])$ *is precompact*;

(iv) $g(s) = f(s, v(t_m))$ *a.e. on* $[t_m, s_m)$ *if* $t_m \notin \mathcal{L}$, *and* $g(s) = f(t_0, v(t_m))$ *a.e. on* $[t_m, s_m)$ *if* $t_m \in \mathcal{L}$;

(v) $\|v(t) - u(t, t_m, v(t_m), g)\| \leq \frac{t - t_m}{n}$ *for each* $m \in \mathcal{J}$ *and each* $t \in [t_m, T]$.

Proof. Let $(\tau,\xi) \in [a,b) \times D$ be arbitrary and choose $r > 0$ so that $D \cap B(\xi,r)$ is closed and there exists a locally integrable function $\ell(\cdot)$ such that $\|f(t,x)\| \leq \ell(t)$ for almost every

$t \in [a, b)$ and for every $x \in D \cap B(\xi, r)$. This is always possible because D is locally closed and f satisfies (C_3) in Definition 3.1. Fix $t_0 \notin \mathcal{Z}$ and $T \in (\tau, b)$ such that

$$\sup_{\tau \leq t \leq T} \|S(t-\tau)\xi - \xi\| + K + T - \tau \leq r, \tag{5.1}$$

where $K = \max\left\{(T-\tau)\ell(t_0), \int_\tau^T \ell(s)ds\right\}$.

We prove first that the conclusion of Lemma 5.1 remains true if we replace T as above with a possible smaller number $\mu \in (\tau, T]$ which, at this stage, is allowed to depend on $n \in \mathbb{N}^*$ and then, by using the Brézis-Browder Ordering Principle Theorem 4.2, we will prove that we can take $\mu = T$ independent of $n \in \mathbb{N}^*$.

For $n \in \mathbb{N}^*$ take an open set $\mathcal{L}$ of $\mathbb{R}$ with $\mathcal{Z} \subset \mathcal{L}$ and whose Lebesgue measure $\lambda(\mathcal{L}) < \frac{1}{n}$.

Case 1. In case $\tau \in \mathcal{L}$, since $f(t_0, \xi)$ is A-tangent to D at ξ it is easy to see that there exist $\delta \in (0, \frac{1}{n})$ and $p \in X$ with $\|p\| \leq \frac{1}{n}$ such that $[\tau, \tau + \delta) \subset \mathcal{L}$ and such that

$$u(\tau + \delta, \tau, \xi, f(t_0, \xi)) + \delta p \in D.$$

Now, let us define $g : [\tau, \tau + \delta] \to X$ and $v : [\tau, \tau + \delta] \to X$ by $g(t) = f(t_0, \xi)$ and respectively by

$$v(t) = u(t, \tau, \xi, g) + (t - \tau)p \tag{5.2}$$

for each $t \in [\tau, \tau + \delta]$.

Let us observe that the family $\mathcal{P}_{\tau+\delta} = \{[\tau, \tau + \delta)\}$ and the functions g and v satisfy (i)-(v) with T substituted by $\tau + \delta$.

Case 2. In case $\tau \notin \mathcal{L}$, we have $\tau \notin \mathcal{Z}$ and in view of Proposition 4.2 there exist $\delta \in (0, \frac{1}{n})$ and $p \in X$ with $\|p\| \leq \frac{1}{n}$ such that

$$u(\tau + \delta, \tau, \xi, f(\cdot, \xi)) + \delta p \in D.$$

Setting $g(s) = f(s, \xi)$ and defining v by (5.2), we can easily see that, again, the family $\mathcal{P}_{\tau+\delta} = \{[\tau, \tau + \delta)\}$ and the functions g and v satisfy (i)-(v) with T substituted by $\tau + \delta$.

Next, we show that there exists at least one triplet $(\mathcal{P}, g, v)$ satisfying (i)-(v) with T given by (5.1). To this aim we shall use the Brézis-Browder Ordering Principle (see Theorem 4.2) as follows. Let $\mathcal{U}$ be the set of all triplets $(\mathcal{P}_\mu, g_\mu, v_\mu)$ with $\mu \leq T$ and satisfying (i)-(v) with μ instead of T. This set is clearly nonempty as we already proved. On $\mathcal{U}$ we introduce a partial order as follows. We say that

$$(\mathcal{P}_{\mu_1}, g_{\mu_1}, v_{\mu_1}) \leq (\mathcal{P}_{\mu_2}, g_{\mu_2}, v_{\mu_2}),$$

where $\mathcal{P}_{\mu_k} = \{[t_m^k, s_m^k); m \in \mathcal{J}_k\}$, $k = 1, 2$, if

(O_1) $\mu_1 \leq \mu_2$ and if $\mu_1 < \mu_2$ there exists $i \in \mathcal{J}_2$ such that $\mu_1 = t_i^2$;
(O_2) for each $m_1 \in \mathcal{J}_1$ there exists $m_2 \in \mathcal{J}_2$ such that $t_{m_1}^1 = t_{m_2}^2$ and $s_{m_1}^1 = s_{m_2}^2$;
(O_3) $g_{\mu_1}(s) = g_{\mu_2}(s)$ and $v_{\mu_1}(s) = v_{\mu_2}(s)$ for each $s \in [\tau, \mu_1]$.

Let us define the function $\Psi : \mathcal{U} \to \mathbb{R}$ by

$$\Psi((\mathcal{P}_\mu, g_\mu, v_\mu)) = \mu.$$

It is clear that Ψ is increasing on $\mathcal{U}$. Let us take now an increasing sequence

$$((\mathcal{P}_{\mu_j}, g_{\mu_j}, v_{\mu_j}))_{j \in \mathbb{N}}$$

in $\mathcal{U}$ and let us show that it is majorated in $\mathcal{U}$. We define a majorant as follows. First, set

$$\mu^* := \sup\{\mu_j; j \in \mathbb{N}\}.$$

Viable Domains

If $\mu^* = \mu_j$ for some $j \in \mathbb{N}$, $(\mathcal{P}_{\mu_j}, g_{\mu_j}, v_{\mu_j})$ is clearly a majorant. If $\mu_j < \mu^*$ for each $j \in \mathbb{N}$, let us observe first that the intervals in the family $\mathcal{P}_{\mu^*} = \{[t_m^j, s_m^j); j \in \mathbb{N}, m \in \mathcal{J}_j\}$ are pairwise disjoint and so this family is at most countable. For our latter purposes, it is important to emphasize that in fact $\mathcal{P}_{\mu^*}$ is countable. Indeed, by (O_1), we have that for each $j \in \mathbb{N}$ there exists $m \in \mathcal{J}_j$ such that $\mu_j = t_m^j$. On the other hand, the set $\{\mu_j; j \in \mathbb{N}\}$ is clearly countable because $\mu^* = \sup\{\mu_j; j \in \mathbb{N}\}$ and $\mu_j < \mu^*$ for each $j \in \mathbb{N}$. Hence $\mathcal{P}_{\mu^*}$ can be written in the form $\mathcal{P}_{\mu^*} = \{[t_m, s_m); m \in \mathbb{N}\}$. We define

$$g_{\mu^*}(t) = g_{\mu_j}(t), \quad v_{\mu^*}(t) = v_{\mu_j}(t)$$

for $j \in \mathbb{N}$ and every $t \in [\tau, \mu_j]$. Now let us observe that $(\mathcal{P}_{\mu^*}, g_{\mu^*}, v_{\mu^*})$, where $\mathcal{P}_{\mu^*}$, g_{μ^*} and v_{μ^*} are defined as above, satisfies (i), (ii), (iv) with T replaced with $\tau + \mu^*$. Notice that (v) is also satisfied but only on $[\tau, \mu^*)$. Obviously we have $v_{\mu^*}(t_m) \in D \cap B(\xi, r)$ for each $m \in \mathbb{N}$. To see that $(\mathcal{P}_{\mu^*}, g_{\mu^*}, v_{\mu^*})$ satisfies also (iii) we have to check first that $v_{\mu^*}([\tau, \mu^*))$ is precompact in X and next to show how to define $v_{\mu^*}(\mu^*)$. By (ii) and (C_3) we know that $g \in \mathcal{L}^1(\tau, \mu^*; X)$ and so, for each $j \in \mathbb{N}$, the function $u(\cdot, \mu_j, v_{\mu^*}(\mu_j), g) : [\mu_j, \mu^*] \to \overline{D(A)}$ is continuous. Therefore $C_j = u([\mu_j, \mu^*], \mu_j, v_{\mu^*}(\mu_j), g)$ is precompact. On the other hand, by (iii), for each $j \in \mathbb{N}$, we know that $K_j = v_{\mu^*}([\tau, \mu_j])$ is precompact too. By (v) and (O_1) we deduce that, for each $j \in \mathbb{N}$,

$$v_{\mu^*}([\tau, \mu^*)) \subset C_j \cup K_j + \frac{\mu^* - \mu_j}{n} B(0, 1).$$

Let $\varepsilon > 0$ be arbitrary and fix $j \in \mathbb{N}$ such that

$$(\mu^* - \mu_j)\frac{1}{n} \leq \frac{\varepsilon}{2}.$$

Since $C_j \cup K_j$ is precompact, there exists a finite family $\{x_1, x_2, \ldots, x_{n(\varepsilon)}\}$ such that, for each $x \in C_j \cup K_j$, there exists $k \in \{1, 2, \ldots, n(\varepsilon)\}$ such that

$$\|x - x_k\| \leq \frac{\varepsilon}{2}.$$

¿From the last two inequalities and the inclusion above, we get $v_{\mu^*}([\tau, \mu^*)) \subset \cup_{k=1}^{n(\varepsilon)} B(x_k, \varepsilon)$ and accordingly $v_{\mu^*}([\tau, \mu^*))$ is precompact. Now, take any limit point v^* of $v_{\mu^*}(\mu_j)$ as j tends to $+\infty$ and set $v_{\mu^*}(\mu^*) = v^*$. Clearly $v_{\mu^*}(\mu^*) \in D \cap B(\xi, r)$. So, with $v_{\mu^*} : [\tau, \mu^*] \to X$, defined as above, we obviously have that $(\mathcal{P}_{\mu^*}, g_{\mu^*}, v_{\mu^*})$ satisfies (i), (ii), (iii) and (iv). It is also easy to see that (v) holds for each $m \in \mathbb{N}$ and each $t \in [t_m, \mu^*)$. To check (v) for $t = \mu^*$, we have to fix any $m \in \mathbb{N}$, to take $t = \mu_j$ with $\mu_j > t_m$ in (v) and to pass to the limit for j tending to $+\infty$ both sides in (v) on that subsequence on which $(v_j(\mu_j))_{j \in \mathbb{N}}$ tends to $v^* = v_{\mu^*}(\mu^*)$. So $(\mathcal{F}_{\mu^*}, \mathcal{P}_{\mu^*}, g_{\mu^*}, v_{\mu^*})$ is a majorant for $((\mathcal{F}_{\mu_j}, \mathcal{P}_{\mu_j}, g_j, v_j))_{j \in \mathbb{N}}$ and consequently the set $\mathcal{U}$ endowed with the partial order $\leq$ and the function Ψ satisfy the hypotheses of Theorem 4.2. Accordingly there exists at least one element $(\mathcal{P}_\nu, g_\nu, v_\nu)$ in $\mathcal{U}$ such that, if $(\mathcal{P}_\nu, g_\nu, v_\nu) \leq (\mathcal{P}_\sigma, g_\sigma, v_\sigma)$ then $\nu = \sigma$.

We show next that $\nu = T$, where T satisfies (5.1). To this aim let us assume by contradiction that $\nu < T$ and let $\xi_\nu := v_\nu(\nu)$ which belongs to $D \cap B(\xi, r)$. In view of (2.2) and (i)-(v) we have

$$\|\xi_\nu - \xi\| \leq \|v_\nu(\nu) - u(\nu, \tau, \xi, g_\nu)\| + \|u(\nu, \tau, \xi, g_\nu) - S(\nu - \tau)\xi\| + \|S(\nu - \tau)\xi - \xi\| \leq$$

$$\leq \frac{\nu - \tau}{n} + \|S(\nu - \tau)\xi - \xi\| + \int_\tau^\nu \|g_\nu(s)\| ds \leq$$

$$\leq \frac{\nu - \tau}{n} + \sup_{\tau \leq t \leq \nu} \|S(t-\tau)\xi - \xi\| + \max\left\{(\nu-\tau)\ell(t_0), \int_\tau^\nu \ell(s)\,ds\right\}.$$

Recalling that $\nu < T$, from (5.1) we get

$$\|\xi_\nu - \xi\| < r. \tag{5.3}$$

There are two possibilities: either $\nu \in \mathcal{L}$, or $\nu \notin \mathcal{L}$.

If $\nu \in \mathcal{L}$ we act as in *Case* 1 above with ν instead of τ and with ξ_ν instead of ξ. So from the tangency condition (T) combined with (5.3) we infer that there exist $\delta \in (0, \frac{1}{n}]$ with $\nu + \delta \leq T$, $[\nu, \nu + \delta) \subset \mathcal{L}$ and $p \in X$ satisfying $\|p\| \leq \frac{1}{n}$, such that

$$u(\nu + \delta, \nu, \xi_\nu, f(t_0, \xi_\nu)) + \delta p \in D \cap B(\xi, r).$$

If $\nu \notin \mathcal{L}$ we act as in *Case* 2 above with ν instead of τ and with ξ_ν instead of ξ. So from Proposition 4.2 combined with (5.3) we infer that there exist $\delta \in (0, \frac{1}{n}]$ with $\nu + \delta \leq T$ and $p \in X$ satisfying $\|p\| \leq \frac{1}{n}$, such that

$$u(\nu + \delta, \nu, \xi_\nu, f(\cdot, \xi_\nu)) + \delta p \in D \cap B(\xi, r).$$

We define $\mathcal{P}_{\nu+\delta} = \mathcal{P}_\nu \cup \{[\nu, \nu+\delta)\}$, $g_{\nu+\delta} : [\tau, \nu+\delta] \to X$ and $v_{\nu+\delta} : [\tau, \nu+\delta] \to X$ by

$$g_{\nu+\delta}(t) = \begin{cases} g_\nu(t) & \text{if } t \in [\tau, \nu] \\ f(t_0, \xi_\nu) & \text{if } t \in (\nu, \nu+\delta] \end{cases}$$

in case $\nu \in \mathcal{L}$ and

$$g_{\nu+\delta}(t) = \begin{cases} g_\nu(t) & \text{if } t \in [\tau, \nu] \\ f(t, \xi_\nu) & \text{if } t \in (\nu, \nu+\delta] \end{cases}$$

in case $\nu \notin \mathcal{L}$ and respectively by

$$v_{\nu+\delta}(t) = \begin{cases} v_\nu(t) & \text{if } t \in [\tau, \nu] \\ u(t, \nu, \xi_\nu, g_{\nu+\delta}) + (t-\nu)p & \text{if } t \in (\nu, \nu+\delta]. \end{cases}$$

Since $v_{\nu+\delta}(\nu+\delta) \in D$ and by (2.2) and (5.1)

$$\|v_{\nu+\delta}(t) - \xi\| \leq \|v_{\nu+\delta}(t) - u(t, \tau, \xi, g_{\nu+\delta})\| + \|u(t, \tau, \xi, g_{\nu+\delta}) - S(t-\tau)\xi\| + \|S(t-\tau)\xi - \xi\| \leq$$

$$\leq (t-\tau)\frac{1}{n} + \int_\tau^t \|g_{\nu+\delta}(s)\|\,ds + \sup_{\tau \leq t \leq T} \|S(t-\tau)\xi\| \leq$$

$$\leq \sup_{\tau \leq t \leq T} \|S(t-\tau)\xi - \xi\| + K + T - \tau \leq r,$$

for each $t \in [\nu, \nu+\delta]$, $(\mathcal{P}_{\nu+\delta}, g_{\nu+\delta}, v_{\nu+\delta})$ satisfies (i), (ii), (iii) and (iv). with T replaced by $\nu + \delta$. Clearly (v) holds for each t_m and t satisfying $t_m \leq t \leq \nu$, or $t_m = \nu \leq t$. The only case we have to check is that in which $t_m \leq \nu < t \leq \nu + \delta$ To this aim, let us observe that, by virtue of (2.3) and (2.2) and (v), we have

$$\|v_{\nu+\delta}(t) - u(t, t_m, v_{\nu+\delta}(t_m), g_{\nu+\delta})\| =$$
$$= \|u(t, \nu, v_{\nu+\delta}(\nu), g_{\nu+\delta}) + (t-\nu)p - u(t, \nu, u(\nu, t_m, v_{\nu+\delta}(t_m), g_{\nu+\delta}))\| \leq$$
$$\leq \|v_{\nu+\delta}(\nu) - u(\nu, t_m, v_{\nu+\delta}(t_m), g_{\nu+\delta})\| + (t-\nu)\|p\| \leq$$
$$\leq \frac{\nu - t_m}{n} + \frac{t-\nu}{n} = \frac{t-t_m}{n}.$$

So (v) holds for each $m \in \mathbb{N}$ and each $t \in [t_m, \nu+\delta]$.

Viable Domains

Thus, $(\mathcal{P}_{\nu+\delta}, g_{\nu+\delta}, v_{\nu+\delta}) \in \mathcal{U}$, $(\mathcal{P}_\nu, g_\nu, v_\nu) \leq (\mathcal{P}_{\nu+\delta}, g_{\nu+\delta}, v_{\nu+\delta})$ and $\nu < \nu + \delta$. This contradiction can be eliminated only if $\nu = T$ and this completes the proof of Lemma 5.1. □

DEFINITION 5.1. Let $(\tau, \xi) \in [a, b] \times D$, $n \in \mathbb{N}^*$ and the set $\mathcal{L}$ as in Lemma 5.1. A triplet $(\mathcal{P}, g, v)$ satisfying (i)-(v) is called an n-$\mathcal{L}$-*approximate solution* of $(\mathcal{DE})$ and $(\mathcal{IC})$ on $[\tau, T]$.

We are now prepared to complete the proof of the sufficiency of Theorems 3.1 and 3.2.

Proof. Let $(\mathcal{L}_n)$ be a decreasing sequence of open subsets in $\mathbb{R}$ such that $\mathcal{Z} \subset \mathcal{L}_n$ and $\lambda(\mathcal{L}_n) < \frac{1}{n}$ for every $n \in \mathbb{N}^*$. Take $\mathcal{L} := \cap_{n \geq 1} \mathcal{L}_n$ and a sequence of n-$\mathcal{L}_n$-approximate solutions $((\mathcal{P}_n, g_n, v_n))_{n \in \mathbb{N}^*}$ of $(\mathcal{DE})$ and $(\mathcal{IC})$ on $[\tau, T]$. From (iii) and condition (C_3) we know that $\{g_n; n \in \mathbb{N}^*\}$ is a uniformly integrable subset in $L^1(\tau, T; X)$. See Remark 2.1. Under the hypotheses of Theorem 3.1, since the semigroup $S(t) : X \to X$, $t \geq 0$, is compact, by Theorem 2.2 it follows that there exists $v \in C([\tau, T]; X)$ such that, on a subsequence at least, we have

$$\lim_n u(t, \tau, \xi, g_n) = v(t) \tag{5.4}$$

uniformly for $t \in [\tau, T]$. We shall prove now that, under the hypotheses of Theorem 3.2, (5.4) still holds true. First, let us remark that $D \cap B(\xi, r)$ is compact. Next, by (v) we have

$$\|u(t_m^k, \tau, \xi, g_k) - v_k(t_m^k)\| \leq \frac{t_m^k - \tau}{k} \leq \frac{T - \tau}{k}$$

for each $k \in \mathbb{N}^*$ and $m \in \mathcal{J}_k$. For $n \in \mathbb{N}^*$ let us denote

$$C_n = \{u(t, \tau, \xi, g_k);\ k = \overline{1, n},\ t \in [\tau, T]\} \quad \text{and} \quad K = \{v_k(t_m^k);\ k \in \mathbb{N}^*,\ m \in \mathcal{J}_k\}$$

and let us observe that, in view of the inequality above, we have

$$\{u(t_m^k, \tau, \xi, g_k);\ k \in \mathbb{N}^*,\ m \in \mathcal{J}_k\} \subset C_n \cup K + \frac{T - \tau}{n} B(0, 1) \tag{5.5}$$

for each $n \in \mathbb{N}^*$. But, for each $n \in \mathbb{N}^*$, C_n and K are precompact, the former because each function $u(\cdot, \tau, \xi, g_k)$ is continuous on $[\tau, T]$ for $k = \overline{1, n}$, and the latter as a subset of $D \cap B(\xi, r)$ which in its turn is compact as already mentioned. This remark along with (5.5) shows that $\{u(t_m^k, \tau, \xi, g_k);\ k \in \mathbb{N}^*,\ m \in \mathcal{J}_k\}$ is precompact too. Now let us observe that, by (i), for each $t \in [\tau, T)$ and each $k \in \mathbb{N}^*$ there exists $m \in \mathcal{J}_k$ such that $t \in [t_m^k, s_m^k)$ and $s_m^k - t_m^k \leq \frac{1}{k}$. As a consequence $\{t_m^k;\ k \in \mathbb{N}^*,\ m \in \mathcal{J}_k\}$ is dense in $[\tau, T]$. Thus we are in the hypotheses of Theorem 2.1 and accordingly there exists $v \in C([\tau, T]; X)$ such that, on a subsequence at least, we have (5.4). So, in both cases, i.e. under the hypotheses of either Theorem 3.1, or Theorem 3.2, (5.4) holds. By (v) and (5.4), on the same subsequence, we also have

$$\lim_n v_n(t) = v(t) \tag{5.6}$$

uniformly for $t \in [\tau, T]$. Next, recalling once again that the set $\{t_m^k;\ k \in \mathbb{N}^*,\ m \in \mathcal{J}_k\}$ is dense in $[\tau, T]$ and $v_k(t_m^k)$ belongs to $D \cap B(\xi, r)$ for each $k \in \mathbb{N}^*$ and $m \in \mathcal{J}_k$, from (5.4) and (5.6) we conclude that $v(t) \in D \cap B(\xi, r)$ for each $t \in [\tau, T]$. Indeed, this is clearly the case if $t = \tau$. So take $t \in (\tau, T]$, $k \in \mathbb{N}^*$, $m \in \mathcal{J}_k$ and let us denote (for the sake of simplicity) $s = t_m^k$. Assume that $s < t$ and let us observe that, by virtue of (v) and (2.2), we have

$$\|v(t) - v_k(s)\| \leq \|v(t) - u(t, s, v_k(s), g_k)\| + \|u(t, s, v_k(s), g_k) - S(t-s)v_k(s)\| +$$

$$+ \|S(t-s)v_k(s) - v_k(s)\| \leq \frac{t-s}{k} + \int_s^t \|g_k(\theta)\|\, d\theta + \sup_{\eta \in C} \|S(t-s)\eta - \eta\| \leq$$

$$\leq \frac{t-s}{k} + \int_s^t \tilde{\ell}(\theta)\, d\theta + \sup_{\eta \in C} \|S(t-s)\eta - \eta\|,$$

where $\tilde{\ell}(\theta) = \max\{\ell(t_0), \ell(\theta)\}$ a.e. for $\theta \in [\tau, T]$ and

$$C = \{v_k(t);\ k \in \mathbb{N}^*,\ t \in [\tau, T]\}.$$

Since, due to (5.6), C is precompact in X, we have

$$\lim_{\delta \downarrow 0} \sup_{\eta \in C} \|S(\delta)\eta - \eta\| = 0.$$

Recalling that $\tilde{\ell} \in L^1(\tau, T; \mathbb{R}_+)$, that s denote a generic element t_m^k with $t_m^k < t$, and using the relation above, we easily deduce that

$$v(t) \in \overline{\{v_k(t_m^k);\ k \in \mathbb{N},\ m \in \mathcal{J}_k\}},$$

where the latter is included in $D \cap B(\xi, r)$. So, we necessarily have $v(t) \in D \cap B(\xi, r)$ for each $t \in [\tau, T]$.

Now, let us observe that if $s \notin \mathcal{L}$ there exists $n(s) \in \mathbb{N}^*$ such that, for each $n \geq n(s)$, $s \notin \mathcal{L}_n$. Hence, by (i) and (iii) we have $g_n(s) = f(s, v_n(t_m^n))$ for each $n \geq n(s)$ and for some $m \in \mathcal{J}_n$ with $|s - t_m^n| \leq \frac{1}{n}$. Therefore by condition (C_2) we get

$$\lim_n g_n(s) = f(s, v(s))$$

for almost every $s \in [\tau, T]$. ¿From (C_3) and Lebesgue Dominated Convergence Theorem we deduce that

$$\lim_n g_n = f(\cdot, v(\cdot))$$

in $L^1(\tau, T; X)$. In view of (2.2) we then have $v(t) = u(t, \tau, \xi, g)$ for each $t \in [\tau, T]$, where g satisfies $g(s) = f(s, u(s, \tau, \xi, g))$ for almost every $s \in [\tau, T]$. But this means that v is a mild solution of $(\mathcal{DE})$ and $(\mathcal{IC})$ and this completes the proof. □

REMARK 5.2. We notice that the above proof may be easily adapted to handle a slightly more general result obtained from Theorem 3.2, by replacing the hypothesis "*the inclusion $D \subset X$ is compact*" with "*D is locally compact and separable*".

6. ADMISSIBLE PREORDERS

Let M be a nonempty subset of D.

DEFINITION 6.1. The set M is *admissible* with respect to $(\mathcal{DE})$ if for each $(\tau, \xi) \in [a, b) \times M$ there exists at least one mild solution $u : [\tau, T] \to M$, $T < b$, of $(\mathcal{DE})$ and $(\mathcal{IC})$.

Applying either Theorem 3.1, or Theorem 3.2 to the differential equation

$$u'(t) + Au(t) \ni f_M(t, u(t)),$$

where f_M stands for the restriction of f to $[a, b) \times M$, we get

THEOREM 6.1. *Under the hypotheses of either Theorem 3.1, or Theorem 3.2, assume that M is a closed subset of D. Then M is admissible with respect to $(\mathcal{DE})$ if and only if there exists a negligible subset $\mathcal{Z}$ in $[a, b]$ such that, for each $(t, \xi) \in ([a, b) \setminus \mathcal{Z}) \times M$, we have $f(t, \xi) \in \mathcal{T}_M^A(\xi)$*

We also have

Viable Domains

COROLLARY 6.1. *Under the hypotheses of Theorem 6.1, a necessary and sufficient condition in order that for each $\xi \in M$ there exists at least one saturated mild solution $u : [\tau, T) \to M$, $T \leq b$, of $(\mathcal{DE})$ satisfying $(\mathcal{JC})$ is the tangency condition $(\mathcal{T})$ with D replaced with M.*

Now, let $\preceq$ be a preorder on M, i.e. a reflexive and transitive binary relation on M. It is convenient to identity the preorder $\preceq$ on M with the multifunction $P : M \to 2^M$ defined by
$$P(\xi) = \{\eta \in M \, ; \, \xi \preceq \eta\}$$
for all $\xi \in M$.

DEFINITION 6.2. *The preorder $P : M \to 2^M$ is admissible with respect to $(\mathcal{DE})$ if for every $(\tau, \xi) \in [a, b) \times M$ there exists a mild solution $u : [\tau, T] \to M$ to $(\mathcal{DE})$ and $(\mathcal{JC})$ such that for every $s \in [\tau, T]$ and for every $t \in [s, T]$, $u(t) \in P(u(s))$.*

Our main goal in the sequel is to characterize the admissibility of a given preorder P with respect to the differential inclusion $(\mathcal{DE})$.

THEOREM 6.2. *Under the hypotheses either of Theorem 3.1, or of Theorem 3.2, assume that M is closed in D and the graph of P is closed in $D \times D$. Then a necessary and sufficient condition in order that P be admissible with respect to $(\mathcal{DE})$ is the tangency condition below.*

$(\mathcal{G})$ *There is a negligible subset $\mathcal{Z}$ of $[a, b)$ such that for each $(t, \xi) \in ([a, b) \setminus \mathcal{Z}) \times M$ we have $f(t, \xi) \in \mathcal{T}^A_{P(\xi)}(\xi)$.*

Proof. The proof of the necessity follows the same lines as that of the necessity of either Theorem 3.1, or of Theorem 3.2 and so we are not going to give details. We only note that we have to use the simple fact that the admissibility of P implies the admissibility of $P(\xi)$ for each $\xi \in M$. The main point is that, the converse is also true, as Proposition 6.1 below shows, and this is the essential clue to conclude the proof of the sufficiency. Indeed, since $P(\eta) \subset P(\xi)$ for all $\xi \in M$ and for all $\eta \in P(\xi)$, it follows that $\mathcal{T}^A_{P(\eta)}(\eta) \subset \mathcal{T}^A_{P(\xi)}(\eta)$ for all $\xi \in M$ and for all $\eta \in P(\xi)$. Thus, assuming condition $(\mathcal{G})$ we get that, for every $\xi \in M$, $P(\xi)$ satisfies the tangency condition in Theorem 6.1. Therefore, for each $\xi \in M$, $P(\xi)$ is admissible with respect to $(\mathcal{DE})$ and Proposition 6.1 below comes into play to deduce that P is admissible with respect to $(\mathcal{DE})$. $\square$

PROPOSITION 6.1. *Under the hypotheses of Theorem 6.2 the preorder P is admissible with respect to the differential equation $(\mathcal{DE})$ if and only if, for every $\xi \in M$, the set $P(\xi)$ is admissible with respect to the differential equation $(\mathcal{DE})$.*

Proof. Clearly, if P is admissible with respect to $(\mathcal{DE})$, then, for all $x \in M$, $P(\xi)$ is admissible with respect to $(\mathcal{DE})$. To show the converse, assume that, for each $\xi \in M$, $P(\xi)$ is admissible with respect to $(\mathcal{DE})$. Let $\xi \in M$ and $\tau \in [a, b)$. We shall show that there exists at least one solution $u : [\tau, T] \to M$ to $(\mathcal{DE})$, with $u(\tau) = \xi$ and such that $u([s, T]) \subset P(u(s))$ for each $s \in [\tau, T]$. To this aim we proceed in several steps.

In the first step we note that, reasoning as in Lemma 3.1 [13], one can show that there exists $\sigma \in (\tau, b)$ such that for every noncontinuable solution $u : [\tau, T) \to M$ to $(\mathcal{DE})$ with $u(\tau) = \xi$ we have $\sigma < T$. According to Corollary 6.1, there exists a solution $u : [\tau, T) \to M$ to $(\mathcal{DE})$ with $u(\tau) = \xi$ and $u([\tau, \sigma]) \subset P(\xi)$.

In the second step we observe that, for every solution $v : [\tau, \sigma] \to M$ to $(\mathcal{DE})$, with $v(\tau) = \xi$ and $v([\tau, \sigma]) \subset P(\xi)$, and for every $\nu \in [\tau, \sigma)$, there exists a solution $w : [\tau, \sigma] \to M$ to $(\mathcal{DE})$ such that w equals v on $[\tau, \nu]$ and $w([\nu, \sigma]) \subset P(w(\nu))$.

In the third step we observe that, according to the first two steps, for every nonempty and finite subset S of $[\tau, \sigma]$, with $\tau \in S$, there exists a solution $u : [\tau, \sigma] \to M$ to $(\mathcal{DE})$ and $(\mathcal{IC})$ with $u([s, \sigma]) \subset P(u(s))$ for all $s \in S$.

In the fourth step we consider a sequence $(S_n)_{n \in \mathbb{N}}$ of nonempty finite subsets of $[\tau, \sigma]$ such that: $\tau \in S_n$ and $S_n \subset S_{n+1}$ for each $n \in \mathbb{N}$; the set $S = \cup_{n \in \mathbb{N}} S_n$ is dense in $[\tau, \sigma]$. For example we can take

$$S_n = \{\tau + (i/2^n)(\sigma - \tau); i \in \{0, 1, ..., 2^n\}\}.$$

Further we apply the third step to get a sequence of solutions $(u_n : [\tau, \sigma] \to M)_{n \in \mathbb{N}}$ to $(\mathcal{DE})$ and $(\mathcal{IC})$ such that $u_n([s, \sigma]) \subset P(u_n(s))$ and for each $n \in \mathbb{N}$ and each $s \in S_n$. Now, applying either Theorem 2.2, or Theorem 2.1, we can assume, taking a subsequence if necessary, that the sequence $(u_n)_{n \in \mathbb{N}}$ converges uniformly on $[\tau, \sigma]$ to a solution $u \in C([\tau, \sigma]; X)$ of $(\mathcal{DE})$. Clearly $u(\tau) = \xi$.

In the fifth step we show that $u([s, \sigma]) \subset P(u(s))$ for all $s \in S \cap [\tau, \sigma]$. Indeed, given s as above, there exists $n \in \mathbb{N}$ such that $s \in S_n$. Then $s \in S_m$ and $u_m([s, \sigma]) \subset P(u_m(s))$ for all $m \in \mathbb{N}$ with $n \leq m$. At this point, the closedness of the graph of P in $D \times D$ implies that $u([s, \sigma]) \subset P(u(s))$.

In the sixth and final step, taking into account that $S \cap [\tau, \sigma]$ is dense in $[\tau, \sigma]$, u is continuous on $[\tau, \sigma]$ and the graph of P is closed in $D \times D$, we conclude that the preceding relation holds for every $s \in [\tau, \sigma]$ and this completes the proof. □

7. Some Problems of Invariance

Let us consider the differential inclusion

$$u' \in Au + F(t, u), \qquad (\mathcal{DI})$$

where $A : D(A) \subset X \to 2^X$ is an m-accretive operator, $D \subset \overline{D(A)}$ is locally closed, while $F : [a, b) \times \overline{D(A)} \to 2^X$ is a given multifunction. By a *mild solution* of $(\mathcal{DI})$ on $[\tau, T]$ we mean a continuous function $u : [\tau, T] \to \overline{D(A)}$ for which there exists $g \in L^1(\tau, T; X)$ such that $g(t) \in F(t, u(t))$ a.e. for $t \in [\tau, T]$ and u is a mild solution on $[\tau, T]$ of $u' \in Au + g$ in the sense of Definition 1.7.5, p. 25 in Vrabie [31] .

We say that D is *invariant* with respect to $(\mathcal{DI})$ if for each $(\tau, \xi) \in [a, b) \times D$ each mild solution $u : [\tau, T] \to \overline{D(A)}$ of $(\mathcal{DI})$ satisfying $(\mathcal{IC})$, i.e. $u(\tau) = \xi$, satisfies also $u(t) \in D$ for each $t \in [\tau, T]$.

Let Y be a nonempty and closed subset in X. We recall that a multivalued mapping Q from Y into 2^X is called *lower semicontinuous* on Y if for each open subset U in X the inverse $Q^-(U) = \{t \in [a, b); F(t) \cap U \neq \emptyset\}$ is open in Y.

Let $(\Omega, \mathcal{M})$ be a measure space. A multivalued mapping F from Ω into 2^X is called *measurable* is for each open subset U in X the inverse $F^-(U) = \{\omega \in \Omega; F(\omega) \cap U \neq \emptyset\}$ is measurable, i.e. $F^-(U) \in \mathcal{M}$.

DEFINITION 7.1. A multivalued mapping F from $[a, b) \times Y$ into 2^X is called a *Michael-mapping* (*M-mapping* for short) if for each closed subset K in $[a, b) \times Y$ for which $F_{|K}$ is lower semicontinuous, each closed subset C in K and each continuous selection $f_C : C \to X$ of $F_{|C}$, there exists at least one continuous selection $f_K : K \to Y$ of $F_{|K}$ such that $f_C(t, y) = f_K(t, y)$ for every $(t, y) \in K$.

Viable Domains

We note that whenever X is separable and Q is nonempty, closed, convex valued then Q is an M-mapping. This is consequence of the well-known Michael' Selection Theorem. See Theorem 9.1.2, p. 355 in Aubin-Frankowska [2].

DEFINITION 7.2. A multivalued mapping $F : [a,b) \times Y \to 2^X$ is called a *Carathéodory mapping* if it has nonempty values and satisfies:

(CM_1) the mapping F is jointly measurable on $[a,b) \times Y$ endowed with the Lebesgue×Borel measure space structure;

(CM_2) for almost every $t \in [a,b)$, the mapping $F(t, \cdot)$ is lower semicontinuous on Y;

(CM_3) for every $r > 0$ there exists a locally integrable function $\ell_r : [a,b) \to \mathbb{R}$ such that $\sup\{\|y\|;\ y \in F(t,x)\} \leq \ell_r(t)$ for almost every $t \in [a,b)$ and for every $x \in Y \cap B(0,r)$.

In our setting, the next result is a slight extension of Theorem 3.2 in Artstein-Prikry [1]. Its proof, which is not evident, is inspired from that of Theorem 2.3 in Cârjă-Monteiro Marques [11].

THEOREM 7.1. *Let X be a separable real Banach space, Y a nonempty and closed subset in X and $F : [a,b) \times Y \to 2^X$ a Carathéodory M-mapping. Then, there exists a negligible subset $\mathcal{Z}$ in $[a,b)$ such that, for each $(\tau,\xi) \in ([a,b) \setminus \mathcal{Z}) \times Y$ and each $\eta \in F(\tau,\xi)$, there exists at least one Carathéodory function $f : [a,b) \times Y \to X$ with $f(t,x) \in F(t,x)$ for each $(t,x) \in ([a,b) \setminus \mathcal{Z}) \times Y$, $f(\tau,\xi) = \eta$ and*

$$\lim_{h \downarrow 0} \frac{1}{h} \int_\tau^{\tau+h} \|f(s,u(s)) - \eta\| \, ds = 0 \qquad (7.1)$$

for each continuous function $u : [\tau, T) \to Y$ satisfying $u(\tau) = \xi$.

Proof. First let us observe that we may assume without loss of generality that $b < +\infty$. By Theorem 2.1 in Artstein-Prikry [1] we have that, there exists a sequence $(K_n)_{n \in \mathbb{N}^*}$ of compact subsets in $[a,b)$ such that, for each $n \in \mathbb{N}^*$ we have:

(i) $\lambda([a,b) \setminus K_n) \leq \frac{1}{n}$;
(ii) $F_{|K_n \times Y}$ is lower semicontinuous;
(iii) $K_n \subset K_{n+1}$.

Next, let L_n be the set of all density points of K_n which, at the same time, are Lebesgue points of all the functions belonging to the family $\{\psi_n;\ n \in \mathbb{N}^*\}$, where, for each $n \in \mathbb{N}^*$, $\psi_n(t) = \ell_n(t)\chi_{|[a,b) \setminus K_n}(t)$ for $t \in [a,b)$ and ℓ_n is given by (CM_3). Since almost all points of a measurable set are density points, it is easy to see that $\lambda(L_n) = \lambda(K_n)$ and, by the definition of L_n, we have

$$\lim_{h \downarrow 0} \frac{\lambda(K_n \cap [t,t+h])}{h} = 1 \qquad (7.2)$$

for each $t \in L_n$. Since, for each $n \in \mathbb{N}^*$, L_n contains only Lebesgue points of ψ_n and $\psi_{n|K_n} = 0$, from (7.2) we deduce

$$\lim_{h \downarrow 0} \frac{1}{h} \int_{[t,t+h] \setminus K_n} \psi_n(s) \, ds = 0 \qquad (7.3)$$

for every $t \in L_n$. Set

$$\mathcal{Z} = [a,b) \setminus \bigcup_{n \in \mathbb{N}^*} L_n$$

and let us observe that $\mathcal{Z}$ is negligible. Let $(\tau, \xi) \in ([a,b) \setminus \mathcal{Z}) \times Y$. By (iii) and the definition of L_n it follows that $L_n \subset L_{n+1}$ for each $n \in \mathbb{N}^*$. Therefore, there exists $n_{\tau,\xi} \in \mathbb{N}^*$ such that $\tau \in L_n \subset K_n$ and $\|\xi\| \leq n-1$ for each $n \geq n_{\tau,\xi}$. Since F is an M-mapping, for each $\xi \in X$ and each $\eta \in F(\tau, \xi)$, there exists a sequence of continuous functions $(f_n)_{n \geq n_{\tau,\xi}}$, $f_n : K_n \times Y \to X$ such that $f_n(t, x) \in F(t, x)$ for each $(t, x) \in K_n \times Y$, $f_n(\tau, \xi) = \eta$, $f_n(t, x) = f_{n+1}(t, x)$ for each $n \geq n_{\tau,\xi}$ and $(t, x) \in K_n \times Y$. Indeed, for $n = n_{\tau,\xi}$, in Definition 7.1 we consider the set $K = K_n \times Y$ and $C = \{(\tau, \xi)\}$ and $f_C(\tau, \xi) = \eta$. We then obtain a continuous selection f_{K_n} of $F_{|K_n \times Y}$, denoted for simplicity by $f_n : K_n \times Y \to X$, such that $f_n(\tau, \xi) = \eta$. From this point we proceed inductively. Namely, given the continuous selection f_m of $F_{|K_m \times Y}$, $m \geq n_{\tau,\xi}$, we extend it to a continuous selection f_{m+1} of $F_{|K_{m+1} \times Y}$. By induction this clearly proves the existence of the sequence $(f_n)_{n \geq n_{\tau,\xi}}$ with all the properties mentioned above.

Let us define $f : [a,b) \times Y \to X$, by
$$f(t,x) = \begin{cases} f_n(t,x) & \text{if } t \in L_n \\ 0 & \text{if } t \in \mathcal{Z}. \end{cases}$$

Obviously f is a Carathéodory function, $f(t,x) \in F(t,x)$ for each $(t,x) \in ([a,b) \setminus \mathcal{Z}) \times Y$ and $f(\tau, \xi) = \eta$.

To prove (7.1), let $u : [\tau, T] \to Y$ be a continuous function. We may assume with no loss of generality that T is small enough so that $\|u(t)\| \leq n$ for each $t \in [\tau, T]$. Recall that $\|\xi\| = \|u(\tau)\| \leq n - 1$. Let $\varepsilon > 0$ be arbitrary and let us observe that inasmuch as $\tau \in L_n$, by (7.2) and (7.3), there exists $\mu(\varepsilon, n) > 0$ such that

$$\frac{\lambda([\tau, \tau + h] \setminus K_n)}{h} \psi_n(\tau) \leq \frac{\varepsilon}{3} \quad \text{and} \quad \frac{1}{h} \int_{[\tau, \tau+h] \setminus K_n} \psi_n(s)\, ds \leq \frac{\varepsilon}{3} \qquad (7.4)$$

for each $h \in [0, \mu(\varepsilon, n))$. Since the restriction of f to $[a,b) \times K_n$ is continuous there exists $\delta(\varepsilon, n) > 0$ such that

$$\|f(s, u(s)) - \eta\| \leq \frac{\varepsilon}{3} \qquad (7.5)$$

for each $s \in [\tau, \tau + \delta(\varepsilon, n)] \cap K_n$. Then, by (7.3), (7.4) and (7.5) we have

$$\frac{1}{h} \int_\tau^{\tau+h} \|f(s, u(s)) - \eta\|\, ds = \frac{1}{h} \int_{[\tau,\tau+h] \setminus K_n} \|f(s, u(s)) - \eta\|\, ds +$$
$$+ \frac{1}{h} \int_{[\tau,\tau+h] \cap K_n} \|f_n(s, u(s)) - \eta\|\, ds \leq$$
$$\leq \frac{1}{h} \int_{[\tau,\tau+h] \setminus K_n} \|f(s, u(s))\|\, ds + \frac{1}{h} \int_{[\tau,\tau+h] \setminus K_n} \|\eta\|\, ds + \frac{1}{h} \int_{[\tau,\tau+h] \cap K_n} \|f_n(s, u(s)) - \eta\|\, ds \leq$$
$$\leq \frac{1}{h} \int_{[\tau,\tau+h] \setminus K_n} \psi_n(s))\, ds + \frac{\lambda([\tau, \tau + h] \setminus K_n)}{h} \psi_n(\tau) + \frac{1}{h} \int_{[\tau,\tau+h] \cap K_n} \|f_n(s, u(s)) - \eta\|\, ds \leq$$
$$\leq \frac{\varepsilon}{3} + \frac{\varepsilon}{3} + \frac{\varepsilon}{3} = \varepsilon$$

for each $n \in \mathbb{N}^*$ and each $h \in (0, \delta(\varepsilon, n)) \cap (0, \mu(\varepsilon, n))$ and this completes the proof. □

DEFINITION 7.3. An element $(\tau, \xi) \in [a, b) \times \overline{D(A)}$ is called an *existence point* for $(\mathcal{DI})$ if, for each $\eta \in F(\tau, \xi)$ and each Carathéodory selection $f : [\tau, T] \times \overline{D(A)} \to X$ of F satisfying $f(\tau, \xi) = \eta$, there exists at least one mild solution $u : [\tau, T] \to \overline{D(A)}$ of $u' \in Au + f(t, u)$ satisfying $u(\tau) = \xi$.

Viable Domains

REMARK 7.1. If $-A$ generates a compact semigroup each element $(\tau, \xi) \in [a, b) \times \overline{D(A)}$ is an existence point. See Theorem 3.8.1 of Vrabie in [31], p. 131. Therefore, whenever X is finite dimensional, $[a, b) \times \overline{D(A)}$ contains only existence points.

The first result concerning the invariance of D with respect to $(\mathcal{DJ})$ is:

THEOREM 7.2. *Let X be a separable real Banach space, $A : D(A) \subset X \to 2^X$ an m-accretive operator, D a locally closed subset in $\overline{D(A)}$ and $F : [a, b) \times \overline{D(A)} \to 2^X$ a Carathéodory M-mapping. Then, a necessary condition in order that D be invariant for $(\mathcal{DE})$ is the tangency condition:*

$(\mathcal{TJ})$ *there exists a negligible subset $\mathcal{Z}$ in $[a\,b)$ such that, for each $(\tau, \xi) \in ([a, b) \setminus \mathcal{Z}) \times D$ which is an existence point of $(\mathcal{DJ})$, we have*

$$F(t, \xi) \subset \mathcal{T}_D^A(\xi).$$

Proof. Let D be invariant with respect to $(\mathcal{DJ})$. Since F is a Carathéodory M-mapping, by Theorem 7.1, there exists a negligible subset $\mathcal{Z}$ in $[a, b)$ such that, for each $(\tau, \xi) \in ([a, b) \setminus \mathcal{Z}) \times \overline{D(A)}$ and for each $\eta \in F(\tau, \xi)$ there exists at least one Carathéodory function $f : [a, b) \times \overline{D(A)} \to X$ such that $f(t, x) \in F(t, x)$ for each $(t, x) \in ([a, b) \setminus \mathcal{Z}) \times \overline{D(A)}$, $f(\tau, \xi) = \eta$ and

$$\lim_{h \downarrow 0} \frac{1}{h} \int_\tau^{\tau + h} \|f(s, u(s)) - \eta\| \, ds = 0$$

for each continuous function $u : [\tau, T) \to \overline{D(A)}$ satisfying $u(\tau) = \xi$. Let $(\tau, \xi) \in [a, b) \setminus \mathcal{Z}$ an existence point of $(\mathcal{DJ})$ and let $u : [\tau, T) \to \overline{D(A)}$ be a mild solution of the problem

$$\begin{cases} u' \in Au + f(t, u) \\ u(\tau) = \xi, \end{cases}$$

where $f : [a, b) \times \overline{D(A)} \to X$ is a Carathéodory function satisfying all the conditions mentioned above. Obviously u is a mild solution of $(\mathcal{DJ})$, and since D is invariant with respect to $(\mathcal{DJ})$, we necessarily have $u(t) \in D$ for each $t \in [\tau, T)$. From this point the proof follows exactly the same lines as those in the proof of Theorem 5.1. □

REMARK 7.2. We do not know whether or not the condition $(\mathcal{TJ})$ is sufficient for invariance in this general setting. More than this, we do not know whether or not the stronger tangency condition below

$(\mathcal{SJ})$ there exists a negligible subset $\mathcal{Z}$ in $[a, b)$ such that, for each $(\tau, \xi) \in ([a, b) \setminus \mathcal{Z}) \times D$, we have

$$F(t, \xi) \subset \mathcal{T}_D^A(\xi)$$

is sufficient for invariance even under the extra-hypothesis that, either $-A$ generates a compact semigroup, or D is compactly embedded in X. The only thing we are able to prove in this specific setting is that, whenever F belongs to a quite narrow but important class of multivalued mappings, the tangency condition $(\mathcal{SJ})$ is necessary and sufficient for invariance. We will present later on such a particular case. See Theorem 7.4. We emphasize however, that the condition $(\mathcal{SJ})$ is sufficient for the viability of the set D even if F is a Carathéodory M-mapping as Theorem 7.3 below shows.

THEOREM 7.3. *Let X be a real Banach space, $A : D(A) \subset X \to 2^X$ an m-accretive operator, D a closed subset in $\overline{D(A)}$ and $F : [a,b) \times D \to 2^X$ a Carathéodory M-mapping. Assume that either the semigroup generated by $-A$ is compact, or that X is separable and D is locally compact. Then, a sufficient condition in order that D be viable with respect to $(\mathcal{DE})$ is the tangency condition $(\mathcal{SJ})$.*

Proof. The necessity follows from Theorem 7.2 combined with Proposition 4.1. To prove the sufficiency, take $(\tau,\xi) \in ([a,b) \setminus \mathcal{Z}) \times D$, $\eta \in F(t,\xi)$ and let $f : [a,b) \times D \to X$ be the Carathéodory selection given by Theorem 7.1. Clearly f satisfies the tangency condition in both Theorems 3.1 and 3.2. The conclusion follows from Theorem 3.1 and Remark 5.2. $\square$

We finally consider a special class of multivalued mappings F, for which, under some extra general hypotheses, $(\mathcal{SJ})$ is even a sufficient condition for invariance of D with respect to $(\mathcal{DJ})$. More precisely, we introduce:

DEFINITION 7.4. *Let Y a nonempty subset in X. We say that $F : [a,b) \times Y \to 2^X$ is a superposition mapping if there exist a complete, separable metric space $\mathcal{V}$, a lower semicontinuous mapping $G : [a,b) \to 2^{\mathcal{V}}$ with nonempty, convex and closed values and a function $f : [a,b) \times \mathcal{V} \times Y \to X$ such that:*

(SP_1) *for each $(v,u) \in \mathcal{V} \times Y$ the function $f(\cdot,v,u)$ is measurable;*
(SP_2) *for almost each $t \in [a,b)$, the function $f(t,\cdot,\cdot)$ is continuous;*
(SP_3) *for each $(t,u) \in [a,b) \times Y$, we have*

$$F(t,u) = \bigcup_{v \in G(t)} f(t,v,u).$$

DEFINITION 7.5. *We say that the function $f : [a,b) \times \mathcal{V} \times Y \to X$ has the uniqueness property if there exists a continuous function $\omega : \mathbb{R}_+ \to \mathbb{R}_+$ such that*

$$(u_1 - u_2, f(t,v,u_1) - f(t,v,u_2))_+ \leq \omega(\|u_1 - u_2\|)\|u_1 - u_2\|$$

a.e. for $t \in [a,b)$, for each $v \in \mathcal{V}$ and each $u_1, u_2 \in Y$ and such that the differential inequality $x' \leq \omega(x)$, $x(0) = 0$ has only the solution $x \equiv 0$.

THEOREM 7.4. *Let X be a real Banach space, $A : D(A) \subset X \to 2^X$ an m-accretive operator with $-A$ generating a compact semigroup, D a closed subset in $\overline{D(A)}$ and $F : [a,b) \times \overline{D(A)} \to 2^X$ a superposition mapping satisfying condition (CM_3) in Definition 7.2. Assume that the function f in Definition 7.4 has the uniqueness property. Then a necessary and sufficient condition in order that D be viable and invariant for $(\mathcal{DE})$ is the tangency condition $(\mathcal{SJ})$.*

Proof. Necessity. Let $(\tau,\xi) \in [a,b) \times D$, $\eta \in F(\tau,\xi)$ and let us observe that, by (SP_3) in Definition 7.4 there exists $v_\eta \in G(\tau)$ such that $f(\tau,v_\eta,\xi) = \eta$. Since G is lower semicontinuous with nonempty, convex and closed values, by Michael' Selection Theorem 9.1.2, p. 355 in Aubin-Frankowska [2] it follows that there exists a continuous function $v : [\tau,b) \to \mathcal{V}$ such that $v(\tau) = v_\eta$ and $v(t) \in G(t)$ for each $t \in [\tau,b)$. Clearly, $g : [\tau,b) \times \overline{D(A)} \to X$, defined by $g(t,u) = f(t,v(t),u)$ is a Carathéodory function. Since the semigroup generated by $-A$ is compact, we are in the hypotheses of Theorem 3.8.1 of Vrabie, [31], p. 131. Accordingly, the Cauchy problem

$$\begin{cases} u' \in Au + g(t,u) \\ u(\tau) = \xi \end{cases} \qquad (\mathcal{CP})$$

has at least one mild local solution $u : [\tau, T) \to \overline{D(A)}$. Obviously u is a mild solution of $(\mathcal{DI})$ and, inasmuch as D is invariant with respect to $(\mathcal{DI})$, we necessarily have $u(t) \in D$ for each $t \in [\tau, T)$. From this point the proof follows exactly the same lines as that of Theorem 5.1 and therefore we do not enter into details.

Sufficiency. We begin by showing that D is viable. To this aim let $v : [a, b) \to V$ be a continuous selection of G. We define $g : [\tau, b) \times \overline{D(A)} \to X$ by

$$g(t, u) = f(t, v(t), u)$$

for each $(t, u) \in [a, b) \times \overline{D(A)}$. Clearly g is a Carathéodory function. Let now $\mathcal{Z}$ be the negligible subset in $[a, b)$ which corresponds to $g_D = g_{|[a,b) \times D}$ by means of Theorem 4.1. Since F satisfies $(\mathcal{SI})$, we easily conclude that g_D satisfies $(\mathcal{I})$. So, by Theorem 3.1, for each $(\tau, \xi) \in [a, b) \times D$, the Cauchy problem $(\mathcal{CP})$ with g replaced by g_D has at least one saturated solution $u : [\tau, T) \to D$ which obviously is a mild solution of $(\mathcal{DI})$.

To complete the proof, i.e. to prove the invariance of D, we have merely to show that each saturated mild solution $u : [\tau, T) \to \overline{D(A)}$ of $(\mathcal{DI})$ satisfies $u(t) \in D$ for each $t \in [\tau, T)$. Thus, let u be such a solution and let $g_u \in L^1(\tau, T; X)$ be such that u is a mild solution of

$$\begin{cases} u' \in Au + g_u \\ u(\tau) = \xi \end{cases}$$

and $g_u(t) \in F(t, u(t))$ a.e. for $t \in [\tau, T)$. At this point, let us remark that, since G is obviously measurable, by virtue of Fillipov's Theorem 8.2.10, p. 316 in Aubin-Frankowska [2], there exists a measurable function $v : [a, b) \to \mathcal{V}$ such that $g_u(t) = f(t, v(t), u(t))$ for almost every $t \in [a, b)$. Next, with v as above, let us consider the function $g : [a, b) \times D \to X$, defined by $g(t, x) = f(t, v(t), x)$. By the sufficiency part of Theorem 3.1 we know that the Cauchy problem $(\mathcal{CP})$ has at least one saturated solution $w : [\tau, T_1) \to D$. Inasmuch as f has the uniqueness property and D is closed, it follows that $T_1 = T$ and w coincides with u on $[\tau, T)$. This completes the proof. □

References

[1] Z. ARSTEIN AND K. PRIKRY, Carathéodory selections and the Scorza Dragoni property, *J. Math. Anal. Appl.*, 127(1987), 540-547.

[2] J.-P. AUBIN AND H. FRANKOWSKA, *Set-Valued Analysis*, Birkhäuser, Basel, 1990.

[3] J. P. AUBIN AND A. CELLINA, *Differential Inclusions*, Springer Verlag, Berlin-Heidelberg-New York-Tokyo, 1984.

[4] V. BARBU, *Nonlinear Semigroups and Differential Equations in Banach Spaces*, Noordhoff, Leyden, 1976.

[5] D. BOTHE, Flow invariance for perturbed nonlinear evolution equations, *Abstr. Appl. Anal.*, 1(1996), 417-433.

[6] D. BOTHE, Reaction diffusion systems with discontinuities. A viability approach, *Nonlinear Anal.*, 30(1997), 677-686.

[7] H. BOULIGAND, Sur les surfaces dépourvues de points hyperlimités, *Ann. Soc. Polon. Math.*, 9(1930), 32-41.

[8] A. BRESSAN AND V. STAICU, On nonconvex perturbations of maximal monotone differential inclusions, *Set-Valued Anal.*, 2(1994), 415-437.

[9] H. BRÉZIS, On a characterization of flow-invariant sets, *Comm. Pure Appl. Math.*, 23(1970), 261-263.

[10] H. BRÉZIS AND F. BROWDER, A general principle on ordered sets in nonlinear functional analysis, *Advances in Math.*, 21(1976), 355-364.

[11] O. CÂRJĂ AND M. D. P. MONTEIRO MARQUES, Viability for nonautonomous semilinear differential equations, *J. Differential Equations*, 165(2000), 000-000.

[12] O. Cârjă and C. Ursescu, The characteristics method for a first order partial differential equation, *An. Ştiinţ. Univ. Al. I. Cuza Iaşi Secţ. I a Mat.*, **39**(1993), 367-396.

[13] O. Cârjă and I. I. Vrabie, Some new viability results for semilinear differential inclusions, *NoDEA Nonlinear Differential Equations Appl.*, **4**(1997), 401-424.

[14] O. Cârjă and I. I. Vrabie, Viability results for nonlinear perturbed differential inclusions, *Panamer. Math. J.*, **9**(1999), p. 63-74.

[15] I. Chiş-Şter, Monotone solutions for single-valued perturbed nonlinear evolution equations, *Commun. Appl. Anal.*, to appear.

[16] M. G. Crandall, A generalization of Peano's existence theorem and flow-invariance, *Proc. Amer. Math. Soc.*, **36**(1972), 151-155.

[17] S. Feferman, Independence of the axiom of choice from the axiom of dependent choices, *J. Symbolic Logic*, **29**(1964), 226.

[18] G. Haddad, Monotone trajectories of differential inclusions and functional differential inclusions with memory, *Israel J. Math.*, **39**(1981), 83-100.

[19] P. Hartman, On invariant sets and on a theorem of Ważewski, *Proc. Amer. Math. Soc.*, **32**(1972), 511-520.

[20] L. Malaguti, Monotone trajectories of differential inclusions in Banach spaces, *J. Convex Anal.*, **3**(1996), 269-281.

[21] R. H. Martin Jr., Differential equations on closed subsets of a Banach space, *Trans. Amer. Math. Soc.*, **179**(1973), 399-414.

[22] M. Nagumo, Über die Lage der Integralkurven gewönlicher Differentialgleichungen, *Proc. Phys. Math. Soc. Japan*, **24**(1942), 551-559.

[23] N. H. Pavel, Invariant sets for a class of semi-linear equations of evolution, *Nonlinear Anal.*, **1**(1977), 187-196.

[24] N. H. Pavel and I. I. Vrabie, Equations d'évolution multivoques dans des espaces de Banach, *C. R. Acad. Sci. Paris Sér. I Math.*, **287**(1978), 315-317.

[25] N. H. Pavel and I. I. Vrabie, Semilinear evolution equations with multivalued right-hand side in Banach spaces, *An. Ştiinţ. Univ. Al. I. Cuza Iaşi Secţ. I a Mat.*, **25**(1979), 137-157.

[26] F. Severi, Su alcune questioni di topologia infinitesimale, *Ann. Polon. Soc. Math.*, **9**(1930), 97-108.

[27] Shi Shuzhong, Viability theorems for a class of differential operator inclusions, *J. Differential Equations*, **79**(1989), 232-257.

[28] C. Ursescu, Carathéodory solutions of ordinary differential equations on locally compact sets in Fréchet spaces, *"Al. I. Cuza" University of Iaşi, Preprint Series in Mathematics of "A. Myller" Mathematical Seminar*, **18**(1982), p. 1-27.

[29] C. Ursescu, Carathéodory solutions of ordinary differential equations on locally closed sets in finite dimensional spaces, *Math. Japon.*, **31** (1986), 483-491.

[30] I. I. Vrabie, Compactness methods and flow-invariance for perturbed nonlinear semigroups, *An. Ştiinţ. Univ. Al. I. Cuza Iaşi Secţ. I a Mat.*, **27** (1981), 117-125.

[31] I. I. Vrabie, *Compactness Methods for Nonlinear Evolutions*, Second Edition, Pitman Monographs and Surveys in Pure and Applied Mathematics 75, Addison-Wesley and Longman, 1995.

Faculty of Mathematics, "Al. I. Cuza" University of Iaşi, Iaşi 6600, Romania
E-mail address: ocarja@uaic.ro

Faculty of Mathematics, "Al. I. Cuza" University of Iaşi, Iaşi 6600, Romania
Current address: P. O. Box 180, Ro, Iş 1, Iaşi 6600, Romania
E-mail address: ivrabie@uaic.ro

Almost Periodic Solutions to Neutral Functional Equations

C. Corduneanu   University of Texas at Arlington, Arlington, Texas

One problem oftenly encountered in the applications of functional equations is the existence of periodic or almost periodic (in time) solutions.

For basic definitions and properties concerning almost periodic functions we shall refer the reader to our book [1].

We shall consider the functional equation

$$(Vx)(t) = (Wx)(t), \quad t \in R, \tag{1}$$

where V and W stand for operators (not causal, in general) on the space $AP(R, R^n)$, consisting of Bohr almost periodic functions on R, with values in R^n.

Let us first assume that V in (1) is a linear operator on $AP(R, R^n)$, and rewrite the equation (1) in the form

$$(Lx)(t) = (Nx)(t), \quad t \in R, \tag{2}$$

with N standing, in general, for a nonliner operator on the space $AP(R, R^n)$.

The special case of (2)

$$(Vx)(t) = f(t), \quad t \in R, \tag{3}$$

with $f \in AP(R, R^n)$, is solvable in case, and only in case,

$$\sup |(Lx)(t)| \geq m \sup |x(t)|, \quad t \in R, \tag{4}$$

for some positive m, and any $x \in AP(R, R^n)$. Condition (1) is the well known condition for the invertibility (with bounded inverse) of the linear continuous operator L, taking into account that supremum is the norm in $AP(R, R^n)$.

Based on the solvability of the equation (3), under condition (4), we can proceed to the discussion of the equation (2). It turns out that (2) is also uniquely solvable in $AP(R, R^n)$, if N is Lipschitz continuous, with a sufficiently small Lipschitz constant K:

$$|Nx - Ny|_{AP} \leq K|x - y|_{AP}. \tag{5}$$

Indeed, from (2) we obtain

$$\begin{aligned}
m \sup |x(t) - y(t)| &\leq \sup |(Lx)(t) - (Ly)(t)| \\
&\leq \sup |(Nx)(t) - (Ny)(t)| \\
&\leq K \sup |x(t) - y(t)|.
\end{aligned}$$

The above estimates show that the iteration process defined by

$$(Lx^m)(t) = (Nx^{m-1})(t), \quad m \geq 1, \tag{6}$$

with $x^0(t)$ arbitrary in $AP(R, R^n)$ is convergent in this space when $K < m$. We can obviously write

$$\sup \left|x^{m+1}(t) - x^m(t)\right| \leq Km^{-1} \sup \left|x^m(t) - x^{m-1}(t)\right|, \tag{7}$$

and since $Km^{-1} < 1$, the assertion is proved. The uniqueness of the solution of (2) is obtained from

$$m \sup |x(t) - y(t)| \leq K \sup |x(t) - y(t)|,$$

which has been established above.

The discussion conducted so far, leads to the following result:

Proposition 1. *Consider the equation (2) in the space $AP(R, R^n)$, and assume that the operators L and N satisfy (4), resp. (5). If $K < m$, then the iteration process defined by (6) is convergent in $AP(R, R^n)$ to the unique solution of (2).*

Remark. The existence of the solution to equations (1) and (2), can be interpreted as existence of coincidence points to the couple of operators (V, W), resp. (L, N).

Example. As an illustration to Proposition 1, we shall consider the equation

$$x(t) + \int_R k(t-s)x(s)ds = f(t, x(t), x(t+h)), \tag{8}$$

in which $k = (k_{ij})_{n\times m}$ is integrable on R, while f is almost periodic in the first argument and Lipschitz continuous with aspect to last two arguments. One chooses $h \in R$ arbitrarily.

The condition equivalent to (4) can be written as

$$\left|\det[I + \tilde{k}(i\omega)]\right| \geq \mu > 0, \ \omega \in R, \tag{9}$$

where

$$\tilde{k}(i\omega) = \int_R k(t)e^{-i\omega t}dt, \ \omega \in R, \tag{10}$$

is the Fourier transform of k.

According to Proposition 1, equation (8) has a unique solution in $AP(R, R^n)$ if (9) is verified and the Lipschitz constant for f is sufficiently small.

We shall consider again equation (2), and notice that under condition (4) it can be rewritten in the form

$$x(t) = L^{-1}((Nx)(t)), \ t \in R. \tag{11}$$

In this form, the application of fixed point method appears to be appropriate. We shall make such assumptions that will allow us to obtain existence of a solution by means of Schauder fixed point theorem for compact operators.

In order to secure the compactness of the operator $L^{-1}N$ it suffices to assume that N is a compact operator on $AP(R, R^n)$. Since (4) implies that $|L^{-1}| \leq m^{-1}$, one can write

$$\left|L^{-1}(Nx)(t)\right| \leq m^{-1}|(Nx)(t)|, \ t \in R. \tag{12}$$

Denote
$$a(r) = \sup |(Nx)(t)|, \ |x(t)| \leq r, \tag{13}$$

assuming, of course, that the supremum in (13) is finite for each $r > 0$. From (12) and (13) we derive

$$\left|L^{-1}(Nx)(t)\right| \leq m^{-1}a(r) \ \text{ for } |x(t)| \leq r. \tag{14}$$

Therefore, the operator $L^{-1}N$ will take the ball of radius r, centered at the zero element of $AP(R, R^n)$, into itself, if for this r one has

$$m^{-1}a(r) \leq r. \tag{15}$$

Let us point out the fact that we need *only one value* of $r > 0$, such that (15) be valid. Such values for r do exist, for instance, if we assume

$$\limsup_{r \to \infty} \frac{a(r)}{r} < m. \qquad (16)$$

In particular, when $a(r)$ grows slower than r at infinity, condition (16) is verified.

Summarizing the discussion carried out above, the following existence (only!) result can be stated:

Proposition 2. *Consider equation (2), with L and N continuous operators on $AP(R, R^n)$. Moreover, assume that L is linear and invertible, while N is compact from $AP(R, R^n)$ into itself, and such that (15) or (16) is valid. Then equation (2) has a solution in $AP(R, R^n)$.*

Remark. The compactness of a set $S \subset AP(R, R^n)$ is equivalent to the following conditions: a) S is bounded, i.e., there exists $M > 0$ such that $|x(t)| \leq M$, $t \in R$, for each $x \in S$; b) S is equi–continuous, i.e., for each $\varepsilon > 0$, there exists $\delta(\varepsilon) > 0$, such that $t, s \in R$, $|t - s| < \delta$ implies $|x(t) - x(s)| < \varepsilon$ for any $x \in S$; c) S is equi–almost periodic, i.e., for each $\varepsilon > 0$, there exists $\ell(\varepsilon) > 0$, such that $|x(t + \tau) - x(t)| < \varepsilon$, $t \in R$, for al least one τ in any interval $(a, a + \ell) \subset R$, and any $x \in S$.

Let us return now to the equation (1), and consider the case similar to (3), namely

$$(Vx)(t) = f(t), \ t \in R, \qquad (17)$$

in which V is acting on $AP(R, R^n)$ and $f \in AP(R, R^n)$. Since V is, in general, a nonlinear operator, we do not have a condition of the form (4), to guarantee the existence of the inverse operator V^{-1}.

Following E. Zeidler [2], we shall impose on the operator V in (17) a condition of monotonicity:

$$m|x(t) - y(t)|^2 \leq\ <(Vx)(t) - (Vy)(t), x(t) - y(t)>. \qquad (18)$$

In (18), $m > 0$ is fixed, while $x, y \in AP$ are arbitrary.

As we shall see, condition (18) assures the existence of the inverse operator, which means that equation (17) is solvable in $AP(R, R^n)$.

Actually, an iteration process can be applied in order to obtain the existence of the solution for (17).

Let us consider the auxiliary operator

$$(T_\lambda x)(t) = x(t) - \lambda[(Vx)(t) - f(t)], \ t \in R, \tag{19}$$

where λ is a positive number. It is obvious that any fixed point of T_λ in $AP(R, R^n)$ is a solution to the equation (17). It will be shown now that we can find $\lambda > 0$, such that T_λ is a contraction on $AP(R, R^n)$. One more condition will be required for V, namely

$$|Vx - Vy|_{AP} \leq M|x - y|_{AP}, \tag{20}$$

where $M > 0$ is fixed and $x, y \in AP(R, R^n)$ are arbitrary.

The following equality follows easily from the properties of the scalar product (in R^n!):

$$|T_\lambda x - T_\lambda y|_{AP}^2 = |x - y|_{AP}^2 - 2\lambda <Vx - Vy, x - y> + \lambda^2 |Vx - Vy|_{AP}^2,$$

which leads to the inequality

$$|T_\lambda x - T_\lambda y|_{AP}^2 \leq (1 - 2m\lambda + M^2 \lambda^2)|x - y|_{AP}^2, \tag{21}$$

if we take into account (18) and (20).

¿From (21) we derive that T_λ is a contraction, if we can achieve $1 - 2m\lambda + M^2 \lambda^2 < 1$ for some positive λ. This is obvious if we choose $0 < \lambda < 2mM^{-2}$. Therefore, with such λ, the operator T_λ is a contraction.

Summarizing the above discussion on equation (17), we can state the following result.

Proposition 3. *Consider equation (17), with V acting on $AP(R, R^n)$ and $f \in AP(R, R^n)$ arbitrary. If V satisfies the monotonicity condition (18), then (17) has a unique solution in $AP(R, R^n)$. This solution can be obtained by the iteration process $x^{m+1}(t) = (T_\lambda x^m)(t)$, $m \geq 0$, $0 < \lambda < 2MM^{-2}$, starting with an arbitrary $x^0(t) \in AP(R, R^n)$.*

We shall consider now the general equation (1), under the basic assumption that a solution of this equation does exist. Since this solution is automatically almost periodic, it is interesting to establish some connection between the almost periods of the solution and those of the data.

Assume that the operator V satisfies the monotonicity condition (18). If $x(t)$ is a solution of (1), then the following inequality holds:

$$m|x(t+\tau) - x(t)|^2 \leq \, <(Wx)(t+\tau) - (Wx)(t), x(t+\tau) - x(t)>. \quad (22)$$

Equation (22) leads to

$$m|x(t+\tau) - x(t)|^2 \leq |(Wx)(t+\tau) - (Wx)(t)| \cdot |x(t+\tau) - x(t)|,$$

from which we derive, based on the inequality $2ab \leq \varepsilon a^2 + \varepsilon^{-1} b^2$, with $\varepsilon = m$,

$$|x(t+\tau) - x(t)| \leq m^{-1}|(Wx)(t+\tau) - (Wx)(t)|, \quad (23)$$

for $t \in R$ and τ a fixed real number.

The inequality (23) can be easily dealt with to find out the connection between the almost periods of $x(t)$ and those of the equi–almost periodic set $\{Wy\}$, where $y \in AP(R, R^n)$ is such that $|y(t)| \leq \sup |x(t)|$, while W is assumed compact on $AP(R, R^n)$. Namely, one reads from (23) that any $m\varepsilon$–almost periodic for the functions in $\{Wy\}$, $|y(t)| \leq \sup |x(t)|$, $t \in R$, is an ε–almost period for $x(t)$.

Let us point out the fact that the almost periods of the functions in $\{Wy\}$, $|y(t)| \leq \sup |x(t)|$, $t \in R$, depend only of the properties of the operator W.

These remarks are useful if we look for solutions of equation (1), in the form

$$\sum_k a_k \exp\{i\lambda_k t\}.$$

Finally, in concluding this Appendix, let us consider an alternate approach in regard to the almost periodicity of solutions of functional equations, such that the case of functional differential equations can be covered.

Let us assume that in equation (2), the operator L is a differential operator of the form

$$(\mathcal{L}x)(t) = \dot{x}(t) - (Lx)(t), \ t \in R. \quad (24)$$

Of course, it is necessary to choose another underlying space than $AP(R, R^n)$. It appears natural to consider the space $AP^{(1)}(R, R^n)$, consisting of all functions such that $x(t), \dot{x}(t) \in AP(R, R^n)$, the natural norm being $\sup(|x(t)| + |\dot{x}(t)|)$, $t \in R$. Endowed with this norm, $AP^{(1)}(R, R^n)$ becomes a Banach space.

The invertibility of $\mathcal{L}$, given by (24), in the space $AP^{(1)}(R, R^n)$ can be discussed by means of the equation $(\mathcal{L}x)(t) = f(t)$. The general case, to the best of our knowledge, has not been investigated in the literature. The case $(Lx)(t) = A(t)x(t)$ is throroughly investigated in [3].

The case when L is a causal operator may be dealt with on the same lines as the case mentioned above. This idea can be motivated by the fact that the equation $(\mathcal{L}x)(t) = f(t)$, with L causal, possesses an integral representation of the solutions (see Ch.3).

REFERENCES

[1] C. Corduneanu (Almost Periodic Functions) (Second English edition, Chelsea Publ. Co., New York, 1989; currently distributed by American Math. Society).

[2] E. Zeidler (Nonlinear Functional Analysis and Its Applications), II. Springer, Berlin, 1983.

[3] M.A. Krasnoselskii et al. (*Nonlinear Almost Periodic Oscillations*, John Wiley, New York, 1973).

The One Dimensional Wave Equation with Wentzell Boundary Conditions

ANGELO FAVINI, GISÈLE RUIZ GOLDSTEIN,
JEROME A. GOLDSTEIN, AND SILVIA ROMANELLI

DIPARTIMENTO DI MATEMATICA, UNIVERSITA' DI BOLOGNA, PIAZZA DI PORTA S.DONATO 5, 40127 BOLOGNA, ITALY
E-mail address: favini@dm.unibo.it

CERI, UNIVERSITY OF MEMPHIS, MEMPHIS, TENNESSEE 38152
E-mail address: gisele@ceri.memphis.edu

DEPARTMENT OF MATHEMATICAL SCIENCES, UNIVERSITY OF MEMPHIS, MEMPHIS, TENNESSEE 38152
E-mail address: goldstej@msci.memphis.edu

DIPARTIMENTO INTERUNIVERSITARIO DI MATEMATICA, UNIVERSITA' DI BARI, VIA E.ORABONA 4, 70125 BARI, ITALY
E-mail address: romans@pascal.dm.uniba.it

ABSTRACT. We prove that in $C[0,1]$ the mixed problem for the wave equation $\partial^2 u/\partial t^2 = c^2 \partial^2 u/\partial x^2$, $u(x,0) = f(x)$, $\partial u/\partial t(x,0) = g(x)$, $\partial^2 u/\partial x^2(j,t) = 0$, for $x \in [0,1]$, $j = 0,1$ and $t \in \mathbf{R}$ is governed by a cosine function whose generator is the operator $Au = d^2/dx^2$ with domain including Wentzell boundary conditions $(Au(j) = 0$, for $j = 0,1)$. Relations with squares of first order operators are also considered.

1. INTRODUCTION AND MOTIVATION

Consider the parabolic problem

$$\frac{\partial u}{\partial t} = \Delta u + h(x,t) \quad (x \in \Omega \subset\subset \mathbf{R}^n, t \geq 0)$$
$$u(x,0) = f(x) \quad (x \in \Omega)$$
$$u(x,t) = 0 \quad (x \in \partial\Omega, t \geq 0).$$

A standard approach to this problem is to write it as

$$\frac{du}{dt} = A_0 u + h(t) \quad (t \geq 0), \quad u(0) = f$$

where $u : [0,\infty) \to X$, X being a Banach space of functions on Ω and A_0 is the realization on X of the Dirichlet Laplacian. Focussing on the maximum principle leads to the supremum norm, and demanding that the problem is governed by a strongly continuous (or (C_o)) semigroup [13] requires that A_0 is densely defined. Thus we are led to take $X := C_0(\Omega)$, the space of all continuous func-

---

*Work supported by G.N.A.F.A. (I.N.D.A.M.) and by University of Bari Research Funds.

tions on Ω that vanish on the boundary $\partial \Omega$. But then $h(t) \in X$ requires that $h(x,t) = 0$ for all $x \in \partial \Omega$, which is too restrictive. We prefer to assume $h \in C(\overline{\Omega} \times [0,\infty))$, or that $h \in C([0,\infty); C(\overline{\Omega}))$. Thus we want to extend the semigroup $\mathcal{T}_0 := \{T_0(t) : t \geq 0\}$ generated by the Dirichlet Laplacian A_0 to $\mathcal{T}$, a (C_o) contraction semigroup on $C(\overline{\Omega})$. Its generator A will be an extension of A_0; which one is it? Now take $X := C(\overline{\Omega})$. Let A be the Laplacian defined on $D(A) := \{u \in C(\overline{\Omega}) : $ the distributional Laplacian Δu is in $C_0(\Omega)\}$. The Wentzell boundary condition [18] implicit in this definition is $\Delta u = 0$ on $\partial \Omega$. For the problem

$$\frac{\partial u}{\partial t} = \Delta u \qquad x \in \Omega, t \geq 0$$
$$u(x,0) = f(x) \qquad x \in \Omega$$
$$\Delta u = 0 \qquad x \in \partial\Omega, t \geq 0,$$

the boundary condition $\Delta u = 0$ on $\partial \Omega$, coupled with $u_t = \Delta u$ implies $u_t = 0$ on $\partial \Omega$; hence $u(x,t) = f(x)$ for $x \in \partial \Omega$ and $t \geq 0$ (assuming $f \in C(\overline{\Omega})$). This is the homogeneous Dirichlet boundary condition when $f \in C_0(\Omega)$, but it is more general and ties an inhomogeneous Dirichlet boundary condition to the initial condition in a linear way. This leads to the definition of a linear operator A which is m-dissipative and densely defined on $C(\overline{\Omega})$; the semigroup $\mathcal{T}$ that A generates is the desired extension of $\mathcal{T}_0$.

We have recently made a systematic study of parabolic problems involving Wentzell boundary conditions and generalized Wentzell boundary conditions [1]-[12]. The case of Wentzell boundary conditions for the wave equation is much trickier. There are a few simple results and enticing calculations but lots of open questions. These results and problems will be the focus of this paper.

We shall work in one dimensional space for the following reason. W. Littman [15] showed that the wave equation $u_{tt} = \Delta u$ is wellposed in an $L^p(\mathbf{R}^n)$ context for an $n \geq 2$ *if and only if* $p = 2$, whereas when $n = 1$, d'Alembert's formula gives the solution in terms of translation operators and translation is continuous in many norms, including the norms of $L^p(\mathbf{R})$, $1 \leq p < \infty$ and $BUC(\mathbf{R})$ (= the space of all real-valued bounded uniformly continuous functions defined on $\mathbf{R}$). So we shall work in the context of $\Omega := (0,1)$ and $X := C(\overline{\Omega}) = C[0,1]$.

2. THE WAVE EQUATION

Of concern is the mixed problem for the wave equation

$$(2.1) \quad \begin{cases} (WE) & \frac{\partial^2 u}{\partial t^2} = c^2 \frac{\partial^2 u}{\partial x^2} & 0 \leq x \leq 1, t \in \mathbf{R} \\ (IC) & u(x,0) = f(x), \quad \frac{\partial u}{\partial t}(x,0) = g(x) \quad x \in [0,1] \\ (BC) & \frac{\partial^2 u}{\partial x^2}(j,t) = 0 & j = 0,1, \quad t \in \mathbf{R}. \end{cases}$$

We study the problem in the space $C[0,1]$. The function $v := \frac{\partial^2 u}{\partial x^2}$ satisfies

$$(2.2) \quad \begin{cases} \frac{\partial^2 v}{\partial t^2} = c^2 \frac{\partial^2 v}{\partial x^2}, \\ v(x,0) = f''(x), \quad \frac{\partial v}{\partial t}(x,0) = g''(x), \\ v(j,t) = 0 \quad \text{for} \quad j = 0,1. \end{cases}$$

Wave Equation with Wentzell Boundary Conditions

The mixed (Dirichlet) problem is well known to have a a unique solution given by d'Alembert's formula, namely

$$(2.3) \qquad v(x,t) = \frac{1}{2}\left[\widetilde{f''}(x+ct) + \widetilde{f''}(x-ct)\right] + \frac{1}{2c}\int_{x-ct}^{x+ct} \widetilde{g''}(s)\,ds,$$

where $\widetilde{h}$ is the odd, 1-periodic extension of $h \in C[0,1]$. For our original problem, we assume $f, g \in C^2[0,1]$, so that $f'', g'' \in C[0,1]$. In order that $\widetilde{f''}, \widetilde{g''}$ be continuous on $\mathbf{R}$, we must further assume that $f''(j) = g''(j) = 0$ for $j = 0, 1$, so that

$$f'', g'' \in C_0(0,1) := \{w \in C[0,1] : w(0) = w(1) = 0\}.$$

When the initial value problem for $u'' = Au, u(0) = f, u'(0) = 0$ is well posed for A a densely defined linear operator on a Banach space X, then the unique solution is given by

$$u(t) = C(t)f,$$

where $C := \{C(t) : t \in \mathbf{R}\}$ is the cosine function generated by A. (Cf. e.g. [13].) If $w'' = Aw, w(0) = 0, w'(0) = g$, then $z = w'$ satisfies $z'' = Az, z(0) = g, z'(0) = w''(0) = Aw(0) = 0$, whence

$$z(t) = C(t)g$$
$$(2.4) \qquad w(t) = S(t)g := \int_0^t C(s)g\,ds.$$

Thus the unique solution of

$$u'' = Au, \quad u(0) = f, \quad u'(0) = g$$

is given by

$$u(t) = C(t)f + S(t)g,$$

where S is given by (2.4). Hence the solution (2.3) of (2.2) can be rewritten as

$$v(\cdot, t) = C_0(t)\widetilde{f''} + S_0(t)\widetilde{g''}$$

where C_0 is the cosine function generated by $\frac{d^2}{dx^2}$ on $C_0(0,1)$ and S_0 is the corresponding sine function, defined as in (2.4) with the subscript zero added.

Now, let P be the canonical projection from $C[0,1]$ onto $C_0(0,1)$. Thus for $f \in C[0,1]$,

$$(Pf)(x) = f(x) - f(0) - (f(1) - f(0))x$$

is the unique function in $C_0(0,1)$ satisfying

$$(Pf)'' = f'', \quad \text{in case} \quad f \in C^2[0,1];$$

equivalently Pf differs from f by a linear function and $Pf(0) = Pf(1) = 0$.

Theorem 2.1. *The mixed problem (2.1) is well posed in $C[0,1]$. It is governed by a cosine function C and a corresponding sine function S given by (2.4); the unique* solution of (2.1) is

$$u(x,t) = [C(t)f + S(t)g](x)$$
$$= C_0(t)\widetilde{P}f + S_0(t)\widetilde{P}g + (f(1) - f(0))x + f(0)$$
$$+ t[(g(1) - g(0))x + g(0)].$$

The proof is now merely a straightforward computation; we omit the details.

Corollary 2.1. *Let λ be the identity function $\lambda(x) := x$. The cosine and sine functions associated with (2.1) are given by (see (2.4))*

$$C(t)f = C_0(t)\widetilde{P}f + (f(1) - f(0))\lambda + f(0),$$
$$S(t)g = S_0(t)\widetilde{P}g + t[(g(1) - g(0))\lambda + g(0)].$$

Corollary 2.2. *The operator $A = \frac{d^2}{dx^2}$ with Wentzell boundary conditions on $C[0,1]$ (with*

$$D(A) := \{u \in C^2[0,1] : u''(0) = u''(1) = 0\})$$

generates a strongly continuous uniformly bounded cosine function on $C[0,1]$. Also, A generates a semigroup analytic in the right half plane.

Proof. The first sentence is an immediate consequence of Theorem 2.1. Since $\|C_0(t)\| \leq 1$, we get the estimate

$$\sup_{t \in \mathbf{R}} \|C(t)\| \leq 3,$$

which gives the uniform boundedness. The last assertion follows from Romanov's formula [17], [13, Theorem 8.7, p. 120], and the semigroup generated by A is given by

$$T(t)f = \frac{1}{\sqrt{\pi t}} \int_0^\infty e^{-\frac{s^2}{4t}} C(s)f \, ds$$

for $t > 0$, $f \in X$. Clearly this is a well defined analytic function for $\operatorname{Re} t > 0$.

3. The First Order System

Let us write (2.1) as a system

(3.1) $\qquad U' = \mathcal{A}U, \quad U(0) = F,$

where

$$U(t) := \begin{pmatrix} u(t) \\ u'(t) \end{pmatrix}, \quad \mathcal{A} := \begin{pmatrix} 0 & I \\ c^2 D^2 & 0 \end{pmatrix}, \quad F := \begin{pmatrix} f \\ g \end{pmatrix}, \quad \text{and} \quad D := \frac{d}{dx}, ' := \frac{d}{dt}.$$

The Wentzell boundary condition for this system is

(3.2) $\qquad \mathcal{A}U(t) = 0 \quad \text{on} \quad \partial\Omega = \{0,1\}.$

This becomes

$$u_t(j,t) = u_{xx}(j,t) = 0 \quad \text{for all} \quad t \in \mathbf{R} \quad \text{and} \quad x = 0,1.$$

This introduces an extra boundary condition, namely

$$u_t(0,t) = u_t(1,t) = 0.$$

Wave Equation with Wentzell Boundary Conditions

This requires $g \in C_0(0,1)$, which is an inessential requirement, according to the results of the previous Section. Thus problem (2.1) is *not* equivalent to problem (3.1), (3.2). This observation complements a classical result of B. Nagy [16] and J. Kisynski [14] concerning cosine functions on $C[0,1]$. In many cases, the generator A of a cosine function is of the form $A = G^2 + H$ where G generates a (C_0) group and $D(H) \supset D(G)$ (or perhaps H is bounded or even a multiple of the identity). We show this to be the case for an example related to problem (2.1), working at a heuristic level.

4. Squares of First Order Operators

Define B by

(4.1) $$Bf(x) := c\frac{d}{dx}f(1-x) = cf'(1-x)$$

(where $c \in \mathbf{R}\setminus\{0\}$) with boundary condition

(4.2) $$\alpha f(0) + \beta f'(0) = 0.$$

Thus

$$D(B) := \{f \in C^1[0,1] : \alpha f(0) + \beta f'(0) = 0\}.$$

Clearly, for f smooth enough,

$$B^2 f(x) = c^2 f''(x),$$
$$B^{2n} f(x) = c^{2n} f^{(2n)}(x),$$
$$B^{2n+1} f(x) = c^{2n+1} f^{(2n+1)}(1-x),$$

thus for all polynomials (and functions analytic on $|z| \leq 1$), g, we can calculate $g(B)$. In particular,

$$e^{tB}f(x) = \sum_{n=0}^{\infty} \frac{t^n B^n f}{n!} = \sum_{k=0}^{\infty} \frac{c^{2k}(D^{2k}f)(x)}{(2k)!} + \sum_{k=0}^{\infty} \frac{c^{2k+1}(D^{2k+1}f)(1-x)}{(2k+1)!}$$
$$= [cosh(cD)f](x) + [sinh(cD)f](1-x);$$

thus

$$e^{tB} = cosh(cD) + R\,sinh(cD)$$

where $D := \frac{d}{dx}$ and $Rg(x) := g(1-x)$. Recall that $e^{\pm tcD}f(x) = f(x \pm ct)$, and these calculations make sense when we determine how to extend functions f in $C[0,1]$ to all of $\mathbf{R}$ (say to $F \in BUC(\mathbf{R})$) in a consistent manner with the boundary conditions. In particular we should have $\|F\|_\infty = \|f\|_\infty$.

CASE I. $\alpha = 0, \beta = 1$. Then $A = B^2 = c^2 D^2$ has boundary conditions $f'(0) = 0$, $f''(1) = 0$ for $f(\in C^2[0,1])$ in $D(A)$. Thus we have a Neumann condition at 0 and a Wentzell condition at 1. According to the Neumann condition, f in $D(A)$

should be an *even* function. Thus we are led to the space $X := C_e[-1, 1]$ of even continuous functions on $[-1, 1]$. On this space, $A = c^2 D^2$ with Wentzell boundary conditions. Except for inessential changes, this is the case treated in Section 1.

CASE II. $\alpha = 1$, $\beta = 0$. The boundary conditions for $A = B^2 = c^2 D^2$ are

$$f(0) = 0, \quad f'(1) = 0.$$

Thus functions in $C[0, 1]$ should be extended to be odd about 0, even about 1, and periodic of period 2. This is a well understood case which has nothing to do with the Wentzell boundary condition.

5. OPEN PROBLEMS

The main case of $\alpha \neq 0$, $\beta \neq 0$ leads us to (for $A = B^2$)

$$\alpha f(0) + \beta f'(0) = 0, \quad \alpha f'(1) + \beta f''(1) = 0.$$

This is very difficult to analyze and represents an open problem. If we again define B by (4.1) but replace (4.2) by

$$\alpha f(0) + \beta f(1) = 0,$$

then the second boundary condition for B^2 becomes

$$\alpha f'(1) + \beta f'(0) = 0.$$

Taking $\alpha = -\beta$ leads to 1− periodic boundary conditions. The most general game to play in this context is to replace (4.2) by

$$\alpha f(0) + \beta f'(0) + \gamma f(1) + \delta f'(1) = 0,$$

where $(\alpha, \beta, \gamma, \delta)$ is a nonzero vector in $\mathbf{R}^4$.

Another open problem is as follows. Consider the wave equation with generalized Wentzell boundary conditions on $[0, 1]$, namely

$$\frac{\partial^2 u}{\partial t^2} = c^2 \frac{\partial^2 u}{\partial x^2}, \quad 0 \leq x \leq 1, \quad t \in \mathbf{R},$$

$$u(x, 0) = f(x), \quad \frac{\partial u}{\partial t}(x, 0) = g(x), \quad 0 \leq x \leq 1,$$

$$\frac{\partial^2 u}{\partial x^2}(j) + \beta_j \frac{\partial u}{\partial x}(j) + \gamma_j u(j) = 0, \quad \text{at} \quad j = 0, 1$$

where $\gamma_0, \gamma_1 \geq 0$, $\beta_1 > 0 > \beta_0$. Is this problem well posed in $C[0, 1]$? The corresponding heat problem is known to be well posed [7].

References

1. V. Barbu and A. Favini, *The analytic semigroup generated by a second order degenerate differential operator in $C[0,1]$*, Suppl. Rend. Circolo Matem. Palermo **52** (1998), 23-42.
2. A. Favini, G.R. Goldstein, J.A. Goldstein and S. Romanelli, *C_0-semigroups generated by second order differential operators with general Wentzell boundary conditions*, Proc. Am. Math. Soc. **128** (2000), 1981-1989.
3. _____, *Nonlinear boundary conditions for nonlinear second order operators on $C[0,1]$*, Arch. der Mathem. (to appear).
4. _____, *On some classes of differential operators generating analytic semigroups*, Evolution Equations and their Applications in Physical and Life Sciences (G. Lumer and L. Weis eds.), M. Dekker, 2000, pp. 99-114.
5. _____, *Generalized Wentzell boundary conditions and analytic semigroups in $C[0,1]$*, Proceedings of the 1st International Conference on Semigroups of Operators: Theory and Applications, Newport Beach, California, December 14-18, 1998, Birkhäuser (to appear).
6. _____, *Degenerate second order differential operators generating analytic semigroups in L^p and $W^{1,p}$*, (submitted).
7. _____, *The heat equation with generalized Wentzell boundary conditions*, (in preparation).
8. A. Favini, J.A. Goldstein and S. Romanelli, *An analytic semigroup associated to a degenerate evolution equation*, Stochastic Processes and Functional Analysis (J.A. Goldstein, N.A. Gretsky and J. Uhl, eds.) M. Dekker, New York, 1997, pp. 85-100.
9. _____, *Analytic semigroups on $L^p_w(0,1)$ and on $L^p(0,1)$ generated by some classes of second order differential operators*, Taiwanese J. Math. **3**, No.2 (1999), 181-210.
10. A. Favini and S. Romanelli, *Analytic semigroups on $C[0,1]$ generated by some classes of second order differential operators*, Semigroup Forum **56** (1998), 367-372.
11. _____, *Degenerate second order operators as generators of analytic semigroups on $C[0,+\infty]$ or on $L^p_{\alpha-\frac{1}{2}}(0,+\infty)$*, Approximation and Optimization, Proceedings of the International Conference on Approximation and Optimization, Cluj-Napoca, July 29-August 1, 1996 (D. Stancu, G. Coman, W.W. Breckner and P. Blaga eds.), Volume II, Transilvania Press, 1997, pp. 93-100.
12. A. Favini and A. Yagi, *Degenerate Differential Equations in Banach Spaces*, Pure and Applied Mathematics: A Series of Monographs and Textbooks 215, M. Dekker, New York, 1998.
13. J.A. Goldstein, *Semigroups of Linear Operators and Applications*, Oxford University Press, Oxford, New York, 1985.
14. J. Kisynski, *On cosine operator functions and one-parameter groups of operators*, Studia Math. **44** (1972), 93-105.
15. W. Littman, *The wave operator and L^p-norms*, J. Math. Mech. **12** (1963), 55-68.
16. B. Nagy, *Cosine operator functions and the abstract Cauchy problem*, Periodica Math. Hung. **7** (1976).
17. N.P. Romanoff, *On one parameter operator groups of linear transformations I*, Ann. Math. **48** (1947), 216-233.
18. A.D. Wentzell, *On boundary conditions for multi-dimensional diffusion processes*, Theory Prob. Appl. **4** (1959), 164-177.

On the Longterm Behaviour of a Parabolic Phase-Field Model with Memory

MAURIZIO GRASSELLI AND VITTORINO PATA

DIPARTIMENTO DI MATEMATICA "F. BRIOSCHI"

POLITECNICO DI MILANO

VIA E. BONARDI, 9 - MILANO - ITALY

maugra@mate.polimi.it, pata@mate.polimi.it

Abstract. We consider a non-conserved phase-field model with memory effects in the internal energy and in the heat flux according to the Coleman-Gurtin law. Thus, the temperature and the order parameter evolve according to a nonlinear parabolic integrodifferential coupled system. In particular, the equation derived form the energy balance contains two time-dependent convolution terms which are characterized by two relaxation kernels a and k, respectively. In a previous paper, the phase-field system was analyzed by the present authors jointly with C. Giorgi. We first proved that the system was indeed a dynamical system in a suitable phase space depending on the temperature history. Then we showed the existence of a uniform absorbing set and the existence of a uniform attractor of finite fractal dimension. All these results were obtained by assuming that the energy relaxation kernel a was smooth, bounded, and concave. This assumption is thermodynamically compatible, but in the literature a more common assumption is that a must be smooth, summable, and convex. In this case, the dissipativity of the system, i.e., the existence of an absorbing set, is more delicate to prove, since the memory term in the internal energy is somehow antidissipative. Here we provide conditions which ensure the existence of a uniform absorbing set when a is smooth, summable, and convex. Existence of an attractor of finite fractal dimension is also discussed.

1. INTRODUCTION

In [12] (cf. also [3, 4]), we proposed and analyzed a nonlinear integrodifferential coupled system describing, in absence of mechanical stresses, phase transitions in a material with memory effects, like, e.g., certain high-viscosity liquids (see, e.g., [17] and references therein). More precisely, we considered a material occupying a bounded domain $\Omega \subset \mathbb{R}^3$ with smooth boundary $\partial\Omega$, whose state at a point $x \in \Omega$, at time $t \in \mathbb{R}$, was characterized by three variables; namely, the temperature variation field $\vartheta(x,t)$, its past history $\vartheta^t(x,s) = \vartheta(x,t-s)$, $s \geq 0$, and the order parameter $\chi(x,t)$. Then, referring to

[12] for the details, we derived the following evolution system

$$\partial_t \vartheta(t) + a(0)\vartheta(t) + \int_0^\infty a'(\sigma)\vartheta(t-\sigma)\,d\sigma + \lambda'(\chi(t))\partial_t \chi_t(t) - \epsilon \Delta \vartheta(t)$$

$$- \int_0^\infty k(\sigma)\Delta\vartheta(t-\sigma)\,d\sigma = f(t) \tag{1.1}$$

$$\partial_t \chi(t) - \Delta\chi(t) + \chi^3(t) = \gamma(\chi(t)) + \lambda'(\chi(t))\vartheta(t) \tag{1.2}$$

in Ω, for any $t \in \mathbb{R}$. Here a and k are (smooth) relaxation kernels which fulfill suitable thermodynamic restrictions (see below), f basically represents the heat supply, and λ and γ are smooth given functions. The instantaneous diffusion coefficient ϵ is supposed to be positive (see [8]); for the case $\epsilon = 0$ the reader is referred to [11, 13, 14] and references therein.

In [12] we investigated the longterm dynamics described by system (1.1)-(1.2) by assuming that k and k' were summable on $(0, +\infty)$ and k was convex; while a was required to be bounded and concave with $a(0) > 0$ and a', a'' summable on $(0, +\infty)$. These assumptions are thermodynamically compatible (see [11]); however, a more common assumption on a (see, e.g., [10] and its references) is that a must be convex and summable along with a' and a'' on $(0, +\infty)$. In this case, as we noted in [11], the dissipativity of the system, i.e., according to [19], the existence of a uniform absorbing set, is harder to prove. This is basically due to the fact that the integral term characterized by a' behaves in an antidissipative way. Here we prove that the existence of an absorbing set can be obtained even in this case, provided that a is suitably dominated by k and the Poincaré inequality for ϑ can be used; in other words, ϑ must, for instance, vanish on a portion of $\partial\Omega$ with positive Lebesgue measure (compare with [11]). This result, combined with the techniques developed in [12], allows us to prove the existence of a uniform compact attractor of finite fractal dimension in the present case as well. We shall assume $a(0) > 0$ even though the results still holds when $a \equiv 0$, as it can be easily realized. It is worth pointing out that a similar analysis can also be performed in the conserved case, where equation (1.2) is replaced by a fourth-order equation (see [15]).

To introduce the dynamical system, we first specify the initial conditions at a given time $\tau \in \mathbb{R}$ for all the state variables. Thus, due to the presence of memory dependent terms, besides the values of ϑ and χ at τ, the whole past history of ϑ up to τ must be given, namely,

$$\vartheta(\tau) = \vartheta_0 \quad \text{in } \Omega$$
$$\chi(\tau) = \chi_0 \quad \text{in } \Omega$$
$$\vartheta^\tau(s) = \vartheta_0(s) \quad \text{in } \Omega, \ \forall s > 0$$

where $\vartheta_0(s)$ is the *initial past history* of ϑ.

Concerning boundary conditions, we assume

$$\partial_\mathbf{n} \chi = 0 \quad \text{on } \partial\Omega \times (\tau, +\infty)$$
$$\vartheta = 0 \quad \text{on } \partial\Omega \times \mathbb{R}$$

$\partial_\mathbf{n}$ being the usual outward normal derivative. Here we assume homogeneous Dirichlet boundary condition for ϑ just for the sake of simplicity. Indeed, it is not difficult to

Parabolic Phase-Field Model

check that mixed Neumann-Dirichlet boundary conditions for ϑ can be treated provided that the Dirichlet's hold on a portion of $\partial\Omega$ of positive Lebesgue measure so that the Poincaré inequality can be applied.

In order to operate in a history space setting, along the lines of [11, 12], we introduce the additional variable η^t (cf. [9]), which is defined by

$$\eta^t(x,s) = \int_0^s \vartheta^t(x,\tau)\,d\tau \qquad s > 0.$$

This variable η is easily seen to satisfy the equation

$$\partial_t \eta^t(s) + \partial_s \eta^t(s) = \vartheta(t) \quad \text{in } \Omega,\ (t,s) \in (\tau,+\infty) \times (0,+\infty)$$

along with the initial and boundary conditions

$$\eta^\tau = \eta_0 \quad \text{in } \Omega \times (0,+\infty)$$
$$\eta(0) = 0 \quad \text{in } \Omega \times (\tau,+\infty)$$

where

$$\eta_0(x,s) = \int_0^s \vartheta_0(x,y)\,dy$$

is the *initial summed past history* of ϑ.

Assuming a suitable and physically reasonable asymptotic behavior of kernels k and a' (see below), making a formal integration by parts, and setting

$$\mu(s) = -k'(s) \quad \text{and} \quad \nu(s) = a''(s)$$

for any $s > 0$, the above choice of variables leads to the following initial and boundary value problem, where we have set $a(0) = \epsilon = 1$.

Problem P. *Find (ϑ, χ, η) solution to the system*

$$\partial_t\big(\vartheta(t) + \lambda(\chi(t))\big) - \Delta\vartheta(t) + \vartheta(t) - \int_0^\infty \nu(\sigma)\eta^t(\sigma)\,d\sigma - \int_0^\infty \mu(\sigma)\Delta\eta^t(\sigma)\,d\sigma = f(t)$$
$$\partial_t\chi(t) - \Delta\chi(t) + \chi^3(t) = \gamma(\chi(t)) + \lambda'(\chi(t))\vartheta(t)$$
$$\partial_t\eta^t(s) + \partial_s\eta^t(s) = \vartheta(t)$$

in Ω, for any $t > \tau$ and any $s > 0$, which satisfies the initial and boundary conditions

$$\vartheta = 0 \quad \text{on } \partial\Omega \times (\tau,+\infty)$$
$$\partial_\mathbf{n}\chi = 0 \quad \text{on } \partial\Omega \times (\tau,+\infty)$$
$$\eta(0) = 0 \quad \text{on } \Omega \times (\tau,+\infty)$$
$$\vartheta(\tau) = \vartheta_0 \quad \text{in } \Omega$$
$$\chi(\tau) = \chi_0 \quad \text{in } \Omega$$
$$\eta^\tau = \eta_0 \quad \text{in } \Omega \times (0,+\infty).$$

In the next section we shall give a rigorous formulation of problem **P** and we shall state the result which ensures **P** to be a dynamical system on a suitable phase space. The main result, i.e., the existence of a uniform absorbing set, is proved in Section 3. The existence of a compact uniform attractor of finite fractal dimension is discussed in the final section.

2. THE SOLUTION PROCESS

Before introducing the variational formulation of **P** and the related well-posedness result, some notation and assumptions are needed.

We recall that $\Omega \subset \mathbb{R}^3$ is a bounded domain with smooth boundary $\partial\Omega$. We set $H = L^2(\Omega)$, $V = H^1(\Omega)$, $V_0 = H_0^1(\Omega)$, $W = H^2(\Omega)$, and $\mathcal{V}_0 = H^2 \cap H_0^1(\Omega)$, with the identification $H \equiv H^*$ (dual space). We denote the norm and the product on a space X by $\langle \cdot, \cdot \rangle_X$ and $\|\cdot\|_X$, respectively. In particular, due to the Poincaré inequality

$$\|v\|_H \leq C_P \|\nabla v\|_{H^3} \qquad \forall\, v \in V_0 \tag{2.1}$$

we can take $\|v\|_{V_0} = \|\nabla v\|_{H^3}$. The symbol $\langle \cdot, \cdot \rangle$ will stand for the duality pairing between V_0^* and V_0.

Given a positive function α defined on $\mathbb{R}^+ = (0, +\infty)$, and a real Hilbert space X, let $L_\alpha^2(\mathbb{R}^+, X)$ be the Hilbert space of X-valued functions whose norms belong to $L^2(\mathbb{R}^+)$ with respect to the measure $\alpha(s)ds$.

The assumptions on the memory kernels are the following (see [11–13])

$$\nu, \mu \in C^1(\mathbb{R}^+) \cap L^1(\mathbb{R}^+) \tag{K1}$$
$$\nu(s) \geq 0, \ \mu(s) \geq 0 \quad \forall\, s \in \mathbb{R}^+ \tag{K2}$$
$$\nu'(s) \leq 0, \ \mu'(s) \leq 0 \quad \forall\, s \in \mathbb{R}^+. \tag{K3}$$

In view of (K1)-(K2), we introduce the space $\mathcal{M} = L_\nu^2(\mathbb{R}^+, H) \cap L_\mu^2(\mathbb{R}^+, V_0)$. Furthermore, we assume

$$\gamma \in C^1(\mathbb{R}) \quad \text{and} \quad \gamma' \in L^\infty(\mathbb{R}) \tag{H1}$$
$$\lambda \in C^2(\mathbb{R}) \quad \text{and} \quad \lambda'' \in L^\infty(\mathbb{R}) \tag{H2}$$
$$f \in L^1_{\text{loc}}(\mathbb{R}, H) \tag{H3}$$
$$\vartheta_0 \in H \tag{H4}$$
$$\chi_0 \in V \tag{H5}$$
$$\eta_0 \in \mathcal{M}. \tag{H6}$$

Definition 2.1. Let (K1)-(K2) and (H1)-(H6) hold. Pick $\tau, T \in \mathbb{R}$ such that $T > \tau$ and set $I = [\tau, T]$. A triplet (ϑ, χ, η) which fulfills

$$\vartheta \in C^0(I, H) \cap L^2(I, V_0)$$
$$\partial_t \vartheta \in L^2(I, V_0^*) + L^1(I, H)$$
$$\chi \in H^1(I, H) \cap C^0(I, V) \cap L^2(I, W)$$
$$\eta \in C^0(I, \mathcal{M})$$

Parabolic Phase-Field Model

is a solution to problem **P** in the time interval I provided that

$$\langle \partial_t \vartheta, v\rangle + \langle \lambda'(\chi)\partial_t \chi, v\rangle_H + \langle \nabla\vartheta, \nabla v\rangle_{H^3} + \langle \vartheta, v\rangle_H - \int_0^\infty \nu(\sigma)\langle \eta(\sigma), v\rangle_H\, d\sigma$$

$$+ \int_0^\infty \mu(\sigma)\langle \nabla\eta(\sigma), \nabla v\rangle_{H^3}\, d\sigma = \langle f, v\rangle_H \qquad \forall\, v \in V_0, \text{ a.e. in } I \qquad (2.2)$$

$$\partial_t \chi - \Delta\chi + \chi^3 = \gamma(\chi) + \lambda'(\chi)\vartheta \qquad \text{a.e. in } \Omega \times I \qquad (2.3)$$

$$\langle \partial_t \eta + \partial_s \eta, \psi\rangle_{\mathcal{M}} = \langle \vartheta, \psi\rangle_{\mathcal{M}} \qquad \forall\, \psi \in \mathcal{M}, \text{ a.e. in } I \qquad (2.4)$$

$$\partial_\mathbf{n}\chi = 0 \qquad \text{a.e. on } \partial\Omega \times I$$

$$\vartheta(\tau) = \vartheta_0 \qquad \text{a.e. in } \Omega$$

$$\chi(\tau) = \chi_0 \qquad \text{a.e. in } \Omega$$

$$\eta^\tau = \eta_0 \qquad \text{a.e. in } \Omega \times \mathbb{R}^+.$$

Here we point out that $-\partial_s$ is the infinitesimal generator of the right-translation semigroup on $\mathcal{M}$.

Assuming (K1)-(K3) and (H1)-(H6), along the lines of [11, 12], it is possible to show that problem **P** admits a unique solution in every time interval I. In particular, **P** generates a *strongly continuous process* of continuous operators on the product Hilbert space $\mathcal{H} = H \times V_0 \times \mathcal{M}$. Indeed, denoting by $U_f(t,\tau)z_0$ the solution (ϑ, χ, η) to problem **P** at time t with source term f and initial data $z_0 = (\vartheta_0, \chi_0, \eta_0) \in \mathcal{H}$ given at time τ, the two-parameter family of operators $U_f(t,\tau)$, with $t \geq \tau$, $\tau \in \mathbb{R}$, satisfies the usual properties of a process (see, e.g., [16], Chapter 6). In particular, the crucial continuity property follows from

Theorem 2.2. *Assuming* (K1)-(K3) *and* (H1)-(H6), *there holds*

$$\|U_{f_1}(t,\tau)z_{01} - U_{f_2}(t,\tau)z_{02}\|_{\mathcal{H}} + \int_\tau^T \|\chi_1(y) - \chi_2(y)\|_V^2\, dy$$

$$\leq \Lambda\Big(\|z_{01} - z_{02}\|_{\mathcal{H}} + \|f_1 - f_2\|_{L^1(I,H)}^2\Big) \qquad \forall\, t \in I$$

for some $\Lambda > 0$ depending only on I and on the size of z_{0i} and f_i, where $\chi_i(t)$ denotes the second component of the vector $U_{f_i}(t,\tau)z_{0i}$.

Remark 2.3. Here, for sake of simplicity, we assumed $f \in L^1_{\text{loc}}(\mathbb{R}, H)$. Nonetheless, the same results hold for the more general situation $f \in L^1_{\text{loc}}(\mathbb{R}, H) + L^2_{\text{loc}}(\mathbb{R}, V^*)$.

3. Uniform Absorbing Sets

In this section we consider a family of processes $\{U_f(t,\tau), f \in \mathcal{F}\}$, where $\mathcal{F}$ is a certain functional space, and we prove the existence of a bounded, connected, invariant *uniform absorbing set* (as $f \in \mathcal{F}$) for $U_f(t,\tau)$ in $\mathcal{H}$. That is, we show that there exists a bounded connected set $\mathcal{B}_0 \subset \mathcal{H}$ such that $U_f(t,\tau)\mathcal{B}_0 \subset \mathcal{B}_0$, for every $f \in \mathcal{F}$ and every $t \geq \tau$; moreover, given any bounded set $\mathcal{B} \subset \mathcal{H}$ there is $t^* = t^*(\mathcal{B}) \geq 0$ such that

$$\bigcup_{f \in \mathcal{F}} U_f(t,\tau)\mathcal{B} \subset \mathcal{B}_0, \qquad \forall\, t \geq t^* + \tau,\ \forall\, \tau \in \mathbb{R}.$$

To this aim, we introduce the Banach space $\mathcal{T}$ of L^1_{loc}-*translation bounded* functions with values in H, namely,

$$\mathcal{T} = \left\{ f \in L^1_{\text{loc}}(\mathbb{R}, H) : \|f\|_{\mathcal{T}} = \sup_{r \in \mathbb{R}} \int_r^{r+1} \|f(y)\|_H \, dy < \infty \right\}.$$

We also need to make some additional hypotheses on the memory kernels. We set

$$a_0 = \int_0^\infty \nu(\sigma) \, d\sigma \geq 0 \quad \text{and} \quad k_0 = \int_0^\infty \mu(\sigma) \, d\sigma \geq 0$$

and we assume

$$\mu'(s) + \delta \mu(s) \leq 0 \quad \text{for some } \delta > 0, \ \forall \, s \in \mathbb{R}^+ \tag{K4}$$

$$a_0 \nu(s) \leq \frac{2}{1 + C_P^2} \delta_0 \mu(s) \quad \text{for some } \delta_0 \in (0, \delta), \ \forall \, s \in \mathbb{R}^+. \tag{K5}$$

Notice that (K4) entails the exponential decay of the kernel μ. This should be compared with the results in the literature concerning the decay of linear homogeneous systems with memory (cf. [18] and references therein). We also remark that condition (K5) can be rewritten in terms of the original kernels a and k as follows

$$a'(0) a''(s) \geq \frac{2}{1 + C_P^2} \delta_0 k'(s) \quad \text{for some } \delta_0 \in (0, \delta), \ \forall \, s \in \mathbb{R}^+.$$

In light of (K5), it is immediate to check that the spaces $L^2_\nu(\mathbb{R}^+, H) \cap L^2_\mu(\mathbb{R}^+, V_0)$ and $L^2_\mu(\mathbb{R}^+, V_0)$ coincide, and have equivalent norms. Thus in the sequel we agree to denote $\mathcal{M} = L^2_\mu(\mathbb{R}^+, V_0)$.

We have the following uniform energy estimates.

Theorem 3.1. *Let (K1)-(K5) and (H1)-(H6) hold, and let $f \in \mathcal{T}$. Then there exist $\varepsilon > 0$ and two continuous increasing functions $C_j : \mathbb{R}^+ \to \mathbb{R}^+$, $j = 1, 2$, such that*

$$\|U_f(t, \tau) z_0\|_{\mathcal{H}}^2 \leq C_1(\|z_0\|_{\mathcal{H}}) e^{-\varepsilon(t - \tau)} + C_2(\|f\|_{\mathcal{T}})$$

for every $t \geq \tau$, $\tau \in \mathbb{R}$.

Proof. We perform some a priori estimates, which clearly hold in a proper approximation scheme (see [12]). Take $v = \vartheta$ in equation (2.1); multiply equation (2.3) by χ_t and then integrate over Ω. Adding the resulting equations, we have

$$\frac{1}{2} \frac{d}{dt} \left(\|\vartheta\|_H^2 + \|\nabla \chi\|_{H^3}^2 + \frac{1}{2} \|\chi\|_{L^4(\Omega)}^4 \right) + \|\nabla \vartheta\|_{H^3}^2 + \|\vartheta\|_H^2 + \|\partial_t \chi\|_H^2$$

$$= \langle \gamma(\chi), \partial_t \chi \rangle_H + \langle f, \vartheta \rangle_H + \int_0^\infty \nu(\sigma) \langle \eta(\sigma), \vartheta \rangle_H \, d\sigma - \langle \eta, \vartheta \rangle_{\mathcal{M}}. \tag{3.1}$$

Then, multiply equation (2.3) by $\kappa \chi$, for $\kappa \in (0, 1)$ to be fixed later, and integrate over Ω, so obtaining

$$\frac{1}{2} \frac{d}{dt} \kappa \|\chi\|_H^2 + \kappa \|\chi\|_H^2 + \kappa \|\nabla \chi\|_{H^3}^2 + \kappa \|\chi\|_{L^4(\Omega)}^4 = \kappa \langle \tilde{\gamma}(\chi), \chi \rangle_H + \kappa \langle \lambda'(\chi) \vartheta, \chi \rangle_H \tag{3.2}$$

Parabolic Phase-Field Model

where $\tilde{\gamma}(r) = r + \gamma(r)$. Finally, take $\psi = \eta$ in equation (2.4) to get

$$\frac{1}{2}\frac{d}{dt}\|\eta\|_{\mathcal{M}}^2 = -\langle \partial_s \eta, \eta \rangle_{\mathcal{M}} + \langle \eta, \vartheta \rangle_{\mathcal{M}}. \tag{3.3}$$

Using (K4) and performing an integration by parts, we have that

$$\begin{aligned}
-\langle \partial_s \eta, \eta \rangle_{\mathcal{M}} &= -\frac{1}{2} \int_0^\infty \mu(\sigma) \frac{d}{d\sigma} \|\nabla \eta(\sigma)\|_{H^3}^2 \, d\sigma \\
&= \frac{1}{2} \int_0^\infty \mu'(\sigma) \|\nabla \eta(\sigma)\|_{H^3}^2 \, d\sigma \\
&\leq -\frac{\delta}{2} \|\eta\|_{\mathcal{M}}^2.
\end{aligned} \tag{3.4}$$

Set now

$$\Phi^2(t) = \|\vartheta(t)\|_H^2 + \kappa\|\chi(t)\|_H^2 + \|\nabla \chi(t)\|_{H^3}^2 + \frac{1}{2}\|\chi(t)\|_{L^4(\Omega)}^4 + \|\eta^t\|_{\mathcal{M}}^2.$$

Addition of (3.1)-(3.3), on account of (3.4), leads to

$$\begin{aligned}
\frac{d}{dt}\Phi^2 &+ 2\|\nabla\vartheta\|_{H^3}^2 + 2\|\vartheta\|_H^2 + 2\|\partial_t\chi\|_H^2 + 2\kappa\|\chi\|_H^2 + 2\kappa\|\nabla\chi\|_{H^3}^2 + 2\kappa\|\chi\|_{L^4(\Omega)}^4 + \delta\|\eta\|_{\mathcal{M}}^2 \\
&\leq 2\langle \gamma(\chi), \partial_t\chi \rangle_H + 2\kappa\langle \tilde{\gamma}(\chi), \chi \rangle_H + 2\kappa\langle \lambda'(\chi)\vartheta, \chi \rangle_H \\
&\quad + 2\langle f, \vartheta \rangle_H + 2\int_0^\infty \nu(\sigma)\langle \eta(\sigma), \vartheta \rangle_H \, d\sigma.
\end{aligned} \tag{3.5}$$

Denoting

$$\ell = 1 - \frac{2\delta_0}{\delta + \delta_0} \in (0, 1)$$

with δ_0 given by (K5), using the Young inequality and (2.1), we have the estimate

$$\begin{aligned}
2\int_0^\infty \nu(\sigma)\langle \eta(\sigma), \vartheta \rangle_H \, d\sigma \\
&\leq \frac{(1+C_P^2)(2-2\ell)}{C_P^2}\|\vartheta\|_H^2 + \frac{a_0 C_P^2}{(1+C_P^2)(2-2\ell)}\int_0^\infty \nu(\sigma)\|\nabla\eta(\sigma)\|_{H^3}^2 \, d\sigma \\
&\leq \frac{(1+C_P^2)(2-2\ell)}{C_P^2}\|\vartheta\|_H^2 + \frac{\delta_0}{1-\ell}\|\eta\|_{\mathcal{M}}^2 \\
&= \frac{(1+C_P^2)(2-2\ell)}{C_P^2}\|\vartheta\|_H^2 + \frac{\delta+\delta_0}{2}\|\eta\|_{\mathcal{M}}^2.
\end{aligned}$$

Therefore, exploiting once again (2.1),

$$\begin{aligned}
2\|\vartheta\|_H^2 &+ 2\|\nabla\vartheta\|_{H^3}^2 + \delta\|\eta\|_{\mathcal{M}}^2 - 2\int_0^\infty \nu(\sigma)\langle\eta(\sigma),\vartheta\rangle_H \, d\sigma \\
&\geq 2\ell\|\vartheta\|_H^2 + 2\ell\|\nabla\vartheta\|_{H^3}^2 + \left(\frac{\delta-\delta_0}{2}\right)\|\eta\|_{\mathcal{M}}^2
\end{aligned} \tag{3.6}$$

Hence, setting
$$\varepsilon(\kappa) = \min\left\{2\ell, 2\kappa, \frac{\delta - \delta_0}{2}\right\}$$

(recall that $\kappa, \ell < 1$) substituting (3.6) into (3.5) we are led to

$$\frac{d}{dt}\Phi^2 + \varepsilon(\kappa)\Phi^2 + 2\ell\|\nabla\vartheta\|_{H^3}^2 + 2\|\partial_t\chi\|_H^2$$
$$\leq 2\langle\gamma(\chi), \partial_t\chi\rangle_H + 2\kappa\langle\tilde{\gamma}(\chi), \chi\rangle_H + 2\kappa\langle\lambda'(\chi)\vartheta, \chi\rangle_H + 2\langle f, \vartheta\rangle_H. \quad (3.7)$$

Moreover, from (H1)-(H2), (2.1), and Young inequality, it is fairly easy to conclude that

$$2\langle\gamma(\chi), \partial_t\chi\rangle_H + 2\kappa\langle\tilde{\gamma}(\chi), \chi\rangle_H + 2\kappa\langle\lambda'(\chi)\vartheta, \chi\rangle_H + 2\langle f, \vartheta\rangle_H$$
$$\leq c + c(1 + \|f\|_H)\Phi + \kappa c\|\nabla\vartheta\|_{H^3}^2 + \|\partial_t\chi\|_H^2 + \frac{\varepsilon(\kappa)}{2}\Phi^2 \quad (3.8)$$

for some $c > 0$ depending only on Ω, μ, γ, and λ. At this point, we choose $\kappa = \ell/c$, and we put $\varepsilon = \varepsilon(\kappa)/2$. Then (3.7)-(3.8) yield the differential inequality

$$\frac{d}{dt}\Phi^2 + \varepsilon\Phi^2 + \ell\|\nabla\vartheta\|_{H^3}^2 + \|\partial_t\chi\|_H^2 \leq c + c(1 + \|f\|_H)\Phi. \quad (3.9)$$

Making use of a Gronwall-type lemma (see, for instance, Lemma 2.5 in [12]), we end up with

$$\Phi^2(t) \leq 2\Phi^2(\tau)e^{-\varepsilon(t-\tau)} + \frac{2c}{\varepsilon} + \frac{c^2 e^{2\varepsilon}}{(1-e^{-\varepsilon})^2}(1 + \|f\|_\mathcal{T})^2, \quad \forall\, t \in [\tau, +\infty)$$

which implies the thesis at once. □

Exploiting Theorem 3.1, it is now fairly easy to find the required set $\mathcal{B}_0$ for the family of processes $\{U_f(t,\tau), f \in \mathcal{F}\}$, when $\mathcal{F}$ is a bounded subset of $\mathcal{T}$. Indeed, denoting $M_\mathcal{F} = \sup_{f \in \mathcal{F}}\|f\|_\mathcal{T}$, let $\mathcal{B}$ be the ball of $\mathcal{H}$ of radius $2C_2(M_\mathcal{F})$. It is then immediate to check that the set $\mathcal{B}_0 = \cup_{f \in \mathcal{F}} \cup_{t \geq \tau} \cup_{\tau \in \mathbb{R}} \mathcal{B}$ will do.

Integrating (3.9) over $[t, t+1]$, for $t \geq \tau$, we get another estimate, which is of some importance for further asymptotic analysis (cf. [12]).

Lemma 3.2. *Let (K1)-(K6) and (H1)-(H6) hold, and let $f \in \mathcal{T}$. Then there exists a continuous function $C_3 : \mathbb{R}^+ \times \mathbb{R}^+ \to \mathbb{R}^+$, increasing in both variables, such that*

$$\sup_{\tau \in \mathbb{R}} \sup_{t \geq \tau} \int_t^{t+1} \left(\|\nabla\vartheta(y)\|_{H^3}^2 + \|\chi(y)\|_W^2\right) dy \leq C_3(\|z_0\|_\mathcal{H}, \|f\|_\mathcal{T})$$

where $\vartheta(y)$ and $\chi(y)$ are the first and the second component, respectively, of $U_f(y, \tau)z_0$.

Remark 3.3. The above results hold as well if we consider the same model without memory in the internal energy (that is, $a \equiv 0$). In this case, observe that the first equation of **P** reads (compare with [3])

$$\partial_t\big(\vartheta(t) + \lambda(\chi(t))\big) - \Delta\vartheta(t) - \int_0^\infty \mu(\sigma)\Delta\eta^t(\sigma)\,d\sigma = f(t).$$

Parabolic Phase-Field Model

4. Existence of a Uniform Attractor

In this section, we assume (K1)-(K5) and (H1)-(H6), and we study the longterm behavior of the family of processes $U_f(t,\tau)$ as $f \in \mathrm{H}(g)$, for a given translation compact function g in $L^1_{\mathrm{loc}}(\mathbb{R}, H)$.

Recall that $g \in L^1_{\mathrm{loc}}(\mathbb{R}, H)$ is said to be *translation compact* (cf. [7] and references therein) in $L^1_{\mathrm{loc}}(\mathbb{R}, H)$ if the *hull* of g, defined as

$$\mathrm{H}(g) = \overline{\{g^r\}_{r \in \mathbb{R}}}^{L^1_{\mathrm{loc}}(\mathbb{R}, H)}$$

is compact in $L^1_{\mathrm{loc}}(\mathbb{R}, H)$, where $g^r(\cdot) = g(\cdot + r)$ is the translate of g by r.

Repeating with no substantial changes the argument developed in [12], we have

Theorem 4.1. *There exists a compact set $\mathcal{K} \subset \mathcal{H}$ such that, for every $z_0 \in \mathcal{B}_0$ and every $f \in \mathrm{H}(g)$, the solution $U_f(t,\tau)z_0$ to* **P**, *for every $\tau \in \mathbb{R}$ and every $t \geq \tau$, admits a decomposition*

$$U_f(t,\tau)z_0 = z_D(t;\tau) + z_C(t;\tau)$$

such that

$$\|z_D(t;\tau)\|_{\mathcal{H}} \leq M e^{-\varepsilon_0(t-\tau)} \|z_0\|_{\mathcal{H}} \qquad \forall\, z_0 \in \mathcal{H}$$

for some $M \geq 1$ and $\varepsilon_0 > 0$, independent of z_0 and f, and

$$z_C(t;\tau) \in \mathcal{K}.$$

In particular, Theorem 4.1 says that $\mathcal{K}$ is a *uniformly attracting* compact set for the family $\{U_f(t,\tau),\, f \in \mathrm{H}(g)\}$; that is, for any $\tau \in \mathbb{R}$ and any bounded set $\mathcal{B} \subset \mathcal{H}$,

$$\lim_{t \to \infty} \left[\sup_{f \in \mathrm{H}(g)} \delta_{\mathcal{H}}(U_f(t,\tau)\mathcal{B}, \mathcal{K}) \right] = 0$$

where $\delta_{\mathcal{H}}(\mathcal{B}_1, \mathcal{B}_2) = \sup_{z_1 \in \mathcal{B}_1} \inf_{z_2 \in \mathcal{B}_2} \|z_1 - z_2\|_{\mathcal{H}}$ denotes the Hausdorff semidistance of two sets $\mathcal{B}_1$ and $\mathcal{B}_2$ in $\mathcal{H}$.

Referring to [2, 16, 19] for a detailed presentation of the theory of attractors of dynamical systems, we recall the following

Definition 4.2. A closed set $\mathcal{A} \subset \mathcal{H}$ is said to be a *uniform attractor* for the family $\{U_f(t,\tau),\, f \in \mathrm{H}(g)\}$ if it is at the same time uniformly attracting and contained in every closed uniformly attracting set.

Due to well-known results from [5, 6] (see also the monograph [20]), the existence of a uniformly attracting compact set $\mathcal{K}$, together with the continuity of $U_\bullet(t,\tau)$ as a map from $\mathcal{H} \times \mathrm{H}(g)$ to $\mathcal{H}$, for every $\tau \in \mathbb{R}$ and $t \geq \tau$ (which is a consequence of Theorem 2.2), entail

Theorem 4.3. *The family $\{U_f(t,\tau),\, f \in \mathrm{H}(g)\}$ is uniformly asymptotically compact possesses a compact and connected uniform attractor given by*

$$\mathcal{A} = \left\{ \begin{array}{l} z(0) \text{ such that } z(t) \text{ is any bounded complete} \\ \text{trajectory of } U_f(t,\tau) \text{ for some } f \in \mathrm{H}(g) \end{array} \right\}.$$

If we replace conditions (H1)-(H3) with

$$\gamma \in C^2(\mathbb{R}) \quad \text{and} \quad \gamma' \in L^\infty(\mathbb{R}) \qquad \text{(H1$'$)}$$
$$\lambda(r) = \lambda_0 r \qquad \lambda_0 \in \mathbb{R} \qquad \text{(H2$'$)}$$
$$g \text{ is quasiperiodic in } H \qquad \text{(H3$'$)}$$

then there holds (cf. [12])

Theorem 4.4. *The attractor $\mathcal{A}$ of the family $\{U_f(t,\tau),\, f \in \mathrm{H}(g)\}$ has finite fractal and Hausdorff dimensions.*

Recall that a function $g : \Omega \times \mathbb{R} \to \mathbb{R}$ is *quasiperiodic* (see, e.g., [1]) if

$$g(x,t) = G(x, \kappa t) = G(x, \kappa_1 t, \ldots, \kappa_n t)$$

where $G(\cdot, \varpi) \in C^1(\mathbf{T}^n, H)$ is a 2π-periodic function of ϖ on the n-dimensional torus $\mathbf{T}^n$ and $\kappa = (\kappa_1, \ldots, \kappa_n)$ are rationally independent numbers. Such g is translation compact in $L^1_{\mathrm{loc}}(\mathbb{R}, H)$, and $f \in \mathrm{H}(g)$ if and only if $f(x,t) = G(x, \kappa t + \varpi_0)$, for some $\varpi_0 \in \mathbf{T}^n$.

Acknowledgments

This work has been partially supported by the Italian MURST Research Project "Simmetrie, Strutture Geometriche, Evoluzione e Memoria in Equazioni a Derivate Parziali".

References

[1] L. Amerio, G. Prouse, *Abstract almost periodic functions and functional equations*, Van Nostrand, New York (1971)

[2] A.V. Babin, M.I. Vishik, *Attractors of evolution equations*, North-Holland, Amsterdam (1992)

[3] G. Bonfanti, F. Luterotti, *Global solution to a phase-field model with memory and quadratic nonlinearity*, Adv. Math. Sci. Appl. **9**, 523–538 (1999)

[4] G. Bonfanti, F. Luterotti, *Regularity and convergence results for a phase-field model with memory*, Math. Meth. Appl. Sci. **21**, 1085–1105 (1998)

[5] V.V. Chepyzhov, M.I. Vishik, *Nonautonomous evolution equations and their attractors*, Russian J. Math. Phys. **1**, 165–190 (1993)

[6] V.V. Chepyzhov, M.I. Vishik, *Attractors of non-autonomous dynamical systems and their dimension*, J. Math. Pures Appl. **73**, 279–333 (1994)

[7] V.V. Chepyzhov, M.I. Vishik, *Non-autonomous evolutionary equations with translation compact symbols and their attractor*, C.R. Acad. Sci. Paris Sér. I Math. **321**, 153–158 (1995)

[8] B.D. Coleman, M.E. Gurtin, *Equipresence and constitutive equations for rigid heat conductors*, Z. Angew. Math. Phys. **18**, 199–208 (1967)

[9] C.M. Dafermos, *Asymptotic stability in viscoelasticity*, Arch. Rational Mech. Anal. **37**, 297–308 (1970)

[10] G. Gentili, C. Giorgi, *Thermodynamic properties and stability for the heat flux equation with linear memory*, Quart. Appl. Math. **51**, 343–362 (1993)

[11] C. Giorgi, M. Grasselli, V. Pata, *Well-posedness and longtime behavior of the phase-field model with memory in a history space setting*, Quart. Appl. Math. (to appear)

[12] C. Giorgi, M. Grasselli, V. Pata, *Uniform attractors for a phase-field model with memory and quadratic nonlinearity*, Indiana Univ. Math. J. **48**, 1395–1445 (1999)

[13] M. Grasselli, V. Pata, *Upper semicontinuous attractors for a hyperbolic phase-field model with memory*, Indiana Univ. Math. J. (to appear)

[14] M. Grasselli, V. Pata, *On the dissipativity of a hyperbolic phase-field system with memory*, Nonlinear Anal. (to appear)

[15] M. Grasselli, V. Pata, F. Vegni, *Longterm dynamics of a conserved phase-field system with memory*, submitted

[16] A. Haraux, *Systèmes dynamiques dissipatifs et applications*, Coll. RMA **17**, Masson, Paris (1990)

[17] J. Jäckle, *Heat conduction and relaxation in liquids of high viscosity*, Physica A **48**, 337-404 (1990)

[18] Z. Liu, S. Zheng, *Semigroups associated with dissipative systems*, Chapman & Hall/CRC Res. Notes Math. n. **398**, Boca Raton (1999)

[19] R. Temam, *Infinite-dimensional dynamical systems in mechanics and physics*, Springer, New York (1988)

[20] M.I. Vishik, *Asymptotic behaviour of solutions of evolutionary equations*, Cambridge University Press, Cambridge (1992)

On the Kato Classes of Distributions and the *BMO*-Classes

A. GULISASHVILI Department of Mathematics, Ohio University, Athens, Ohio 45701, USA

1 Introduction

The Kato class K_n on n-dimensional Euclidean space R^n was introduced and studied by Aizenman and Simon (see [1, 20]). The definition of K_n is based on a condition considered by Kato in [11]. Similar function classes were defined by Schechter [18] and Stummel [23]. We refer the reader to [4, 10, 20] for more information concerning the Kato class and its applications. The Kato classes of order s were studied by Davies and Hinz [3]. For the generalizations of the Kato class to the case of time-dependent functions see [7, 13, 14, 15, 19].

It is known that the following conditions are equivalent for a function $V \in L^1_{loc}$: <u>Condition A</u>:

$$V \in K_n.$$

<u>Condition B</u>: $J^{-2}|V| \in L^\infty$ and

$$\lim_{\alpha \to 0+} \alpha^2 \| J^{-2}(|V|)_\alpha \|_\infty = 0. \tag{1}$$

<u>Condition C</u>:

$$J^{-2}|V| \in BUC \tag{2}$$

(see [8, 9], see also [6]). In (1) and (2), the symbol J^{-2} stands for the Bessel potential of order -2, BUC denotes the space of bounded uniformly continuous functions on R^n, and $(|V|)_\alpha(x) = |V(\alpha x)|$.

Since

$$\alpha^2 \| J^{-2}(|V|)_\alpha \|_\infty = \|(\alpha^{-2} - \Delta)^{-1}|V|\|_\infty,$$

condition B is equivalent to the following condition from [19]:
<u>Condition D</u>: $(I - \Delta)^{-1}|V| \in L^\infty$ and
$$\lim_{E \to \infty} ||(E - \Delta)^{-1}|V|||_\infty = 0.$$

It was shown in [19] that Condition D characterizes the Kato class.

In the present paper we introduce and study two scales of classes of tempered distributions on R^n (see definitions 2 and 3). We will call the scales in these definitions the scale of Kato classes of distributions and the scale of BMO-classes, respectively. Our main goal in this paper is to study whether conditions similar to conditions B and C above characterize the Kato classes of distributions and the BMO-classes. Our main results are contained in sections 2 and 3. In the proofs of these results we use the theory of Fourier multipliers and some ideas of Stein (see [21]), concerning the connections between the Bessel and the Riesz potentials.

2 The Kato classes of order s and the classes of distributions

We define the scale of the Kato classes of order $s \geq 0$ on R^n as follows:

Definition 1 *Let $V \in L^1_{loc}$ and $s > 0$. Then we say that $V \in K_{s,n}$ iff*
$$\lim_{\alpha \to 0+} \sup_x \int_{|x-y| \leq \alpha} |V(y)| g_s(x-y) dy = 0$$

where
$$g_s(x) = \begin{cases} |x|^{s-n} & \text{if } s - n \neq 0, 2, 4, \cdots \\ |x|^{s-n} \ln \frac{1}{|x|} & \text{if } s - n = 0, 2, 4, \cdots \end{cases}$$

We also put $K_{0,n} = BUC$.

Remark 1 There exists a positive constant r_s such that the function $h_s = r_s g_s$ coincides with the fundamental function for the fractional power $(-\Delta)^{\frac{s}{2}}$ of the Laplace operator. In other words,
$$(-\Delta)^{\frac{s}{2}}(h_s) = \delta$$

in the sense of distributions (see [16]). The class $K_{2,n}$ coincides with the classical Kato class on R^n. The classes $K_{s,n}$ with $0 < s < n$ were studied by Davies and Hinz in [3] where the notation K_s was used. For the explanation why we put $K_{0,n} = BUC$ see Remark 2 below.

Let S' denote the space of tempered distributions on R^n. For every $V \in S'$ and $\alpha > 0$ we define the dilation $(V)_\alpha$ of the distribution V as follows:
$$(V)_\alpha(\phi) = \alpha^{-n} V(\phi(\alpha \cdot)), \quad \phi \in S.$$
If V is a function, then $(V)_\alpha(x) = V(\alpha x)$ for almost all $x \in R^n$.

For any $r > 0$ and $x \in R^n$ we will denote by $B(x, r)$ the closed ball of radius r in R^n centered at x. Throughout the paper we will denote by Λ the class of functions $\lambda \in C_0^\infty$ such that $0 \leq \lambda \leq 1$, $supp(\lambda) \subset B(0,1)$, and λ is equal to 1 in a neighborhood of 0. For $s - n = 0, 2, 4, \cdots$ and $0 < \alpha \leq 1$, we define a function $g_{s,\alpha}$ by
$$g_{s,\alpha}(x) = |x|^{s-n} \ln \frac{\alpha}{|x|}.$$

The next definition introduces the Kato classes of tempered distributions on R^n.

Definition 2 *Let $V \in S'$ and let $s > 0$ be a number such that $s - n \neq 0, 2, 4, \cdots$ Then we say that $V \in \tilde{K}_{s,n}$ iff for every α with $0 < \alpha \leq 1$ and every $\lambda \in \Lambda$ the distribution $V \star ((\lambda)_{\frac{1}{\alpha}} g_s)$ coincides with a function from L^∞, and moreover,*
$$\lim_{\alpha \to 0+} \|V \star ((\lambda)_{\frac{1}{\alpha}} g_s)\|_\infty = 0.$$

If $s - n = 0, 2, 4, \cdots$, then we say that $V \in \tilde{K}_{s,n}$ iff for every α with $0 < \alpha \leq 1$ and every $\lambda \in \Lambda$ the distribution $V \star ((\lambda)_{\frac{1}{\alpha}} g_{s,\alpha})$ coincides with a function from L^∞, and moreover,
$$\lim_{\alpha \to 0+} \|V \star ((\lambda)_{\frac{1}{\alpha}} g_{s,\alpha})\|_\infty = 0.$$

It is not difficult to show that if $V \in L^1_{loc}$, then $|V| \in \tilde{K}_{s,n} \Leftrightarrow V \in K_{s,n}$.

Let $-\infty < s < \infty$. The Bessel potential J^s of order s is defined by
$$F(J^s \gamma) = (1 + |\xi|^2)^{s/2} F(\gamma), \quad \xi \in R^n, \quad \gamma \in S'.$$

Here $F : S' \to S'$ denotes the Fourier transform on S'. For $f \in L^1$ we have
$$F(f)(\xi) = \int_{R^n} f(x) e^{-i\xi \cdot x} dx.$$

If $s > 0$, then the Bessel potential J^{-s} is a convolution type operator,
$$J^{-s} f(x) = \int_{R^n} f(y) G_s(x - y) dy,$$

where the Bessel potential kernel G_s is defined by

$$G_s(x) = \frac{1}{(2\pi)^{\frac{n}{2}} 2^{\frac{s}{2}} \Gamma(\frac{s}{2})} \int_0^\infty \delta^{\frac{s-n}{2}} e^{-\frac{|x|^2}{2\delta}} e^{-\frac{\delta}{2}} \frac{d\delta}{\delta}. \quad (3)$$

In (3), Γ denotes the gamma function. Formula (3) can be found in [16, 21]. For more information concerning the Bessel potentials see [12, 16, 21]. Note that the definition of the Fourier transform in the present paper differs from that in [21].

Theorem 1 *Let $V \in L^1_{loc}$ and $s > 0$. Then the following are equivalent:*
(i) $V \in K_{s,n}$.
(ii) $J^{-s}|V| \in L^\infty$ and

$$\lim_{\alpha \to 0+} \alpha^s \|J^{-s}(|V|)_\alpha\|_\infty = 0.$$

(iii) $J^{-s}|V| \in BUC$.

Remark 2 In the case where $s = 2$, Theorem 1 gives the equivalence of conditions $A - D$ for the classical Kato class. Theorem 1 also clarifies why we put $K_{0,n} = BUC$. Indeed, for $s = 0$ condition (iii) in Theorem 1 is $|V| \in BUC$ which is equivalent to $V \in BUC$.

Our next result generalizes the equivalence of parts (ii) and (iii) in Theorem 1 to the case of tempered distributions.

Theorem 2 *Let $V \in S'$ and $s > 0$. Then the following are equivalent:*
(i) $V \in L^{\infty, -s}$ and

$$\lim_{\alpha \to 0+} \alpha^s \|J^{-s}(V)_\alpha\|_\infty = 0.$$

(ii) $J^{-s}V \in BUC$.

We will prove Theorem 2 first.
Proof of Theorem 2. It is not immediately clear why the condition $J^{-s}V \in L^\infty$ in part (i) of Theorem 2 implies that $J^{-s}(V)_\alpha \in L^\infty$. The next lemma shows that this implication is valid.

Lemma 1 *Let $V \in S'$, $s > 0$, and $J^{-s}V \in L^\infty$. Then for every α with $0 < \alpha < 1$ we have $J^{-s}(V)_\alpha \in L^\infty$ and moreover,*

$$\sup_{\alpha: 0 < \alpha < 1} \alpha^s \|J^{-s}(V)_\alpha\|_\infty \leq C_s \|J^{-s}V\|_\infty.$$

Proof of Lemma 1. It is easy to see that

$$\alpha^s \|J^{-s}(V)_\alpha\|_\infty = \alpha^{s-n}\|V \star (G_s)_{\frac{1}{\alpha}}\|_\infty. \tag{4}$$

We have

$$(G_s)_{\frac{1}{\alpha}} = G_s \star Y_\alpha \tag{5}$$

where

$$F(Y_\alpha)(\xi) = \alpha^n \left(\frac{1+|\xi|^2}{1+\alpha^2|\xi|^2}\right)^{\frac{s}{2}}. \tag{6}$$

Consider the expansion

$$(1-t)^{s/2} = 1 + \sum_{m=1}^\infty A_{m,s} t^m$$

where $|t| < 1$. The explicit formula for the coefficients $A_{m,s}$ is as follows:

$$A_{m,s} = \frac{(-1)^m \frac{s}{2}(\frac{s}{2}-1)\cdots(\frac{s}{2}-m+1)}{m!}$$

for all $m \geq 1$. It follows that for all $m \geq 1$

$$|A_{s,m}| \leq c_s m^{-(1+\frac{s}{2})}. \tag{7}$$

Inequality (7) can be obtained, using the Gauss formula for the gamma function (this formula can be found in, e.g., [2]). Since

$$\frac{1+|\xi|^2}{1+\alpha^2|\xi|^2} = \alpha^{-2}\left(1 - \frac{1-\alpha^2}{1+\alpha^2|\xi|^2}\right),$$

we get from (6) that

$$F(Y_\alpha)(\xi) = \alpha^{n-s} + \alpha^{n-s} \sum_{m=1}^\infty A_{m,s}(-1)^m(1-\alpha^2)(1+\alpha^2|\xi|^2)^{-m}.$$

Therefore,

$$Y_\alpha = \alpha^{n-s}\delta + \alpha^{-s}\sum_{m=1}^\infty A_{m,s}(-1)^m(1-\alpha^2)^m (G_{2m})_{\frac{1}{\alpha}} \tag{8}$$

where δ denotes the delta-measure at 0. Next, using (7), (8), and the equality $\|G_{2m}\|_1 = 1$, we get that

$$\alpha^{s-n} Y_\alpha = \delta + X_\alpha \tag{9}$$

where $X_\alpha \in L^1$ for all $0 < \alpha \leq 1$ and

$$\sup_{0<\alpha\leq 1} ||X_\alpha||_1 < \infty. \tag{10}$$

Now (4), (5), (9), and (10) show that Lemma 1 holds.

Let us continue the proof of Theorem 2.

$(i) \Longrightarrow (ii)$.

Let $V \in S'$ and suppose that part (i) in Theorem 1 holds. The following equality can be easily checked:

$$\frac{1}{(1+|\xi|^2)^{s/2}} - \frac{1}{(1+|\xi|^2)^{s/2}} \frac{(1+\alpha^2|\xi|^2)^{s/2} - \alpha^s|\xi|^s}{(1+\alpha^2|\xi|^2)^{s/2}}$$
$$= \frac{\alpha^s}{(1+\alpha^2|\xi|^2)^{s/2}} + \frac{\alpha^s|\xi|^s - \alpha^s(1+|\xi|^2)^{s/2}}{(1+|\xi|^2)^{s/2}(1+\alpha^2|\xi|^2)^{s/2}}. \tag{11}$$

It is known that there exists a function $\Phi_s \in L^1$ such that

$$\frac{(1+|\xi|^2)^{s/2} - |\xi|^s}{(1+|\xi|^2)^{s/2}} = F(\Phi_s)(\xi) \tag{12}$$

for all $\xi \in R^n$ (see [21], p. 134). It follows from (11) and (12) that

$$J^{-s}V - J^{-s}V \star (\alpha^{-n}\Phi_s(\alpha^{-1}\cdot)) = \alpha^s V \star (\alpha^{-n}G_s(\alpha^{-1}\cdot))$$
$$- \alpha^s[V \star (\alpha^{-n}G_s(\alpha^{-1}\cdot))] \star \Phi_s \tag{13}$$

Since $\Phi_s \in L^1$, the L^∞-norm of the expression on the right-hand side of (13) can be majorized by a constant multiple of the expression

$$\alpha^s ||V \star (\alpha^{-n}G_s(\alpha^{-1}\cdot))||_\infty = \alpha^s ||J^{-s}(V)_\alpha||_\infty.$$

Therefore, the left-hand side of (13) tends to 0 in L^∞ as $\alpha \to 0$. Hence, the function $J^{-s}V$ belongs to the space BUC, since it can be approximated in the L^∞-norm by an approximation of the identity.

$(ii) \Longrightarrow (i)$.

Let $V \in S'$ and $J^{-s}V \in BUC$. It is easy to check that

$$\frac{\alpha^s}{(1+\alpha^2|\xi|^2)^{s/2}} = \frac{(1+|\xi|^2)^{s/2}}{1+|\xi|^s}[\frac{1}{(1+|\xi|^2)^{s/2}}$$
$$-\frac{1}{(1+|\xi|^2)^{s/2}}\frac{(1+\alpha^2|\xi|^2)^{s/2} - \alpha^s|\xi|^s}{(1+\alpha^2|\xi|^2)^{s/2}} + \frac{\alpha^s}{(1+\alpha^2|\xi|^2)^{s/2}(1+|\xi|^2)^{s/2}}].\tag{14}$$

There exists a function $\Psi_s \in L^1$ such that

$$F(\Psi)(\xi) + 1 = \frac{(1+|\xi|^2)^{s/2}}{1+|\xi|^s} \tag{15}$$

for all $\xi \in R^n$ (see [21], p.134). Now we get from (14) and (15) that

$$\begin{aligned}
\alpha^s V \star (\alpha^{-n} G_s(\alpha^{-1} \cdot)) &= \Psi \star [J^{-s}V - J^{-s}V \star (\alpha^{-n}\Phi(\alpha^{-1}\cdot))] \\
&+ [J^{-s}V - J^{-s}V \star (\alpha^{-n}\Phi(\alpha^{-1}\cdot))] \\
&+ \Psi \star \{\alpha^s[J^{-s}V \star (\alpha^{-n}G_s(\alpha^{-1}\cdot))]\} \\
&+ \{\alpha^s[J^{-s}V \star (\alpha^{-n}G_s(\alpha^{-1}\cdot))]\}. \quad (16)
\end{aligned}$$

Since $J^{-s}V \in L^\infty$, $G_s \in L^1$, and $\Psi_s \in L^1$, the last two terms on the right-hand side of (16) tend to 0 in L^∞ as $\alpha \to 0+$. The first two terms also tend to 0, since $J^{-s}V \in BUC$, and any function from the space BUC can be approximated in L^∞ by the approximations of the identity. It follows from (16) that condition (i) in Theorem 2 holds.

This completes the proof of Theorem 2.

Proof of Theorem 1. The equivalence of parts (ii) and (iii) in Theorem 1 follows from Theorem 2. Our next goal is to prove the equivalence of parts (i) and (ii).

It is known that the local behavior of the Bessel potential kernel and the corresponding Riesz potential kernel is the same for $0 < \alpha \leq n$. It is also known that the Bessel potential kernels decay exponentially at infinity. More exactly, the following estimates hold: If $0 < s < n$, then there exist $c_s > 0$ and $\tilde{c}_s > 0$ such that

$$\tilde{c}_s |x|^{s-n} \leq G_s(x) \leq c_s |x|^{s-n}, \quad (17)$$

for all x with $0 < |x| < 1$. Moreover, there exist $c_n > 0$ and $\tilde{c}_n > 0$ such that

$$\tilde{c}_n \ln \frac{1}{|x|} \leq G_s(x) \leq c_n \ln \frac{1}{|x|} \quad (18)$$

for all $0 < |x| < 1/2$. If $s > n$, then there exist $d_s > 0$ and $\tilde{d}_s > 0$ such that

$$\tilde{d}_s \leq G_s(x) \leq d_s \quad (19)$$

for all $0 < |x| < 1$. On the other hand, for every $s > 0$ we have

$$G_s(x) \leq b_s e^{-c|x|} \quad (20)$$

for all $x \in R^n$ with $|x| > 1$. In (20), $b_s > 0$ is a constant, depending only on s, and $c > 0$ is an absolute constant (see [21], p. 132-133).

Remark 3 It is not difficult to prove, using (19), that for $s > n$ we get $K_{s,n} = L^1_{loc,u}$. Here $L^1_{loc,u}$ stands for the space of functions $f \in L^1_{loc}$ such that

$$\sup_x \int_{B(x,1)} |f(y)| dy < \infty.$$

Using (17)-(19), we get the following estimate:

$$\chi_{B(0,\alpha)} g_s \leq a_s \alpha^{s-n} (G_s)_{\frac{1}{\alpha}}, \ 0 < \alpha \leq 1, \quad (21)$$

for $s - n \neq 0, 2, 4, \cdots$. Moreover, for $s - n = 0, 2, 4, \cdots$ we have

$$\frac{1}{2}\chi_{B(0,\alpha^2)} g_s \leq \tilde{a}_s (G_s)_{\frac{1}{\alpha}}, \ 0 < \alpha \leq 1/\sqrt{2}. \quad (22)$$

Now the implication $(ii) \Rightarrow (i)$ in Theorem 1 follows from (21) and (22).

Let $0 < s \leq n$ and $V \in K_{s,n}$. In the remaining part of the proof of Theorem 1 we will denote by c_s positive constants, depending only on s, which may vary from line to line. We have from (17), (18), and (20) that

$$G_s \leq c_s(g_s \chi_{B(0,1)} + \tau) \quad (23)$$

where $\tau(x) = e^{-|x|}$. If $0 < s < n$, then for every $0 < \alpha < 1$ we have

$$\alpha^s (|V|)_\alpha \star G_s \leq c_s(\alpha^s (|V|)_\alpha \star (\chi_{B(0,1)} g_s) + \alpha^s (|V|)_\alpha \star \tau).$$

It follows from the definition of g_s that

$$\alpha^s \|(|V|)_\alpha \star G_s\|_\infty \leq c_s(\| |V| \star (g_s \chi_{B(0,\alpha)}) \|_\infty + \alpha^s \|(|V|)_\alpha \star \tau)\|_\infty. \quad (24)$$

Let C denote the cube in R^n given by

$$C = \{x = (x_1, \cdots, x_n) \in R^n : 0 \leq x_i \leq 1, \ 1 \leq i \leq n\},$$

and let $C_k = C + k$ for all $k \in Z^n$. Then

$$\alpha^s \|(|V|)_\alpha \star \tau\|_\infty \leq \alpha^s \sup_x \{ \sum_{k \in Z^n} \sup_{y \in C_k} \tau(y)\} \int_{C_k} |V(x - \alpha y)| dy$$

$$\leq c_s \alpha^{s-n} \sup_x \int_{|x-y| \leq \alpha} |V(y)| dy \sum_{k \in Z^n} \sup_{y \in C_k} \tau(y) \quad (25)$$

Since $V \in K_{s,n}$, the last expression in (25) tends to 0 as $\alpha \to 0+$. Now (25) implies the validity of condition (ii) in Theorem 1 for $0 < s < n$.

If $s = n$, then (23) gives

$$\alpha^n \|(|V|)_\alpha \star G_n\|_\infty \leq c_n(\| |V| \star ((g_n)_{\frac{1}{\alpha}} \chi_{B(0,\alpha)})\|_\infty$$
$$+ \alpha^n \|(|V|)_\alpha \star \tau)\|_\infty \leq c_n(\| |V| \star (g_n \chi_{B(0,\alpha)})\|_\infty$$
$$+ \alpha^n \|(|V|)_\alpha \star \tau\|_\infty. \quad (26)$$

Reasoning as in the proof of (25), we see that the last term in (26) can be estimated by

$$M \sup_x \int_{|x-y|\leq \alpha} |V(y)|dy. \qquad (27)$$

Since $V \in K_{n,n}$, the expression in (27) tends to 0 as $\alpha \to 0+$. Now (26) implies condition (ii) in Theorem 1 in the case $s = n$.

If $s > n$, then $K_{s,n} = L^1_{loc,u}$, and we have

$$\alpha^s \|(|V|)_\alpha \star G_s\|_\infty \leq c_s(\alpha^{s-n}\| |V| \star \chi_{B(0,1)}\|_\infty + \alpha^s\|(|V|)_\alpha \star \tau\|_\infty).$$

Now reasoning as in (25), we get condition (ii) in Theorem 1 for $s > n$.

This completes the proof of Theorem 1.

The next theorem is one of the main results in the present paper. It gives a complete description of the classes $\tilde{K}_{s,n}$ in terms of the Bessel potentials in the case $s = 2m$ where m is a natural number.

Theorem 3 *Let $V \in S'$ and let $s = 2m$ where m is a natural number. Then condition*

$$V \in \tilde{K}_{2m,n}$$

is equivalent to conditions (i) and (ii) in Theorem 2 for $s = 2m$.

Remark 4 Theorem 3 is a generalization of Theorem 1 for the Kato classes of distributions. However, we do not know whether Theorem 3 holds for $s \neq 2m$.

Proof of Theorem 3. We will show that condition $V \in \tilde{K}_{2m,n}$ implies condition (ii) in Theorem 2, and that condition (i) in Theorem 2 implies condition $V \in \tilde{K}_{2m,n}$.

Let $V \in \tilde{K}_{2m,n}$. Then for every $\lambda \in \Lambda$ we have

$$V \star (\lambda g_{2m}) \in L^\infty. \qquad (28)$$

Recall that we denoted by r_{2m} the constant for which $(-\Delta)^m(r_{2m}g_{2m}) = \delta$. We have

$$V \star (\lambda g_{2m}) = V \star G_{2m} \star (I - \Delta)^m(\lambda g_{2m})$$

$$= \sum_{j=0}^{m-1} b_j V \star ((-\Delta)^j G_{2m}) \star (\lambda g_{2m}) + V \star G_{2m} \star (-\Delta)^m(\lambda g_{2m}) \qquad (29)$$

where b_j are the binomial coefficients. Equality (29) holds in the sense of distributions. Using (12), it is not difficult to prove that for every $0 \leq j \leq m-1$ the function $(-\Delta)^j G_{2m}$ is in L^1. Hence, we have

$$h = \sum_{j=0}^{m-1} b_j (-\Delta)^j G_{2m} \in L^1. \qquad (30)$$

Since the function λ is equal to 1 in a neighborhood of 0, we have

$$(-\Delta)^m(\lambda g_{2m}) = r_{2m}^{-1}\delta + \tau \tag{31}$$

where $\tau \in C_0^\infty$ is a function which is equal to zero near 0. It follows from (29), (30), and (31) that

$$r_{2m}^{-1} J^{-2m} V = V \star (\lambda g_{2m}) - V \star (\lambda g_{2m}) \star h - V \star G_{2m} \star \tau. \tag{32}$$

For any function τ as above there exist functions λ_1 and λ_2 such that

$$\frac{\lambda_1}{\|\lambda_1\|_\infty}, \frac{\lambda_2}{\|\lambda_2\|_\infty} \in \Lambda$$

and

$$\tau = \lambda_1 - \lambda_2. \tag{33}$$

Indeed, we may put $\lambda_1 = \tau + \rho + a$ and $\lambda_2 = \rho + a$, where $\rho \in C_0^\infty$ is a nonnegative function which is equal to zero in a neighborhood of 0 and for which $\tau + \rho \geq 0$. The function $a \in C_0^\infty$ is nonnegative and supported in a neighborhood of 0 where the functions ρ and τ are equal to 0.

It follows from (33) and from the definition of the class $\tilde{K}_{2m,n}$ that for every function $\tau \in C_0^\infty$ which is equal to zero in a neighborhood of 0, we have $V \star \tau \in L^\infty$. Hence,

$$V \star G_{2m} \star \tau \in L^\infty, \tag{34}$$

and we get from (30), (32), (34), and from the definition of the class $\tilde{K}_{2m,n}$ that

$$J^{-2m} V \in L^\infty. \tag{35}$$

Let α be such that $0 < \alpha < 1$ and let $2m - n \neq 0, 2, 4, \cdots$. Then (32) implies that for every $\lambda \in \Lambda$ there exists a function $\tau_\alpha \in C_0^\infty$, depending on λ, equal to zero near 0, and such that

$$r_{2m}^{-1} J^{-2m} V + (J^{-2m} V) \star \tau_\alpha = V \star ((\lambda)_{\frac{1}{\alpha}} g_{2m}) - V \star ((\lambda)_{\frac{1}{\alpha}} g_{2m}) \star h. \tag{36}$$

Since $h \in L^1$, it follows from the definition of the class $\tilde{K}_{2m,n}$ that the expression on the right-hand side of (36) tends to 0 in L^∞ as $\alpha \to 0+$. Now we get from (35) that $J^{-2m} V \star \tau_\alpha \in BUC$ for every $0 < \alpha < 1$. Since the space BUC is a closed subspace of L^∞, we get from (36) that

$$J^{-2m} V \in BUC.$$

This proves that for $2m - n \neq 0, 2, 4, \cdots$ condition $V \in \tilde{K}_{2m,n}$ implies condition (ii) in Theorem 2.

Let $2m - n = 0, 2, 4, \cdots$. Then

$$c_{2m}^{-1} J^{-2m} V + (J^{-2m} V) \star \rho_\alpha = V \star ((\lambda)_{\frac{1}{\alpha}} g_{2m,\alpha}) - V \star ((\lambda)_{\frac{1}{\alpha}} g_{2m,\alpha}) \star h$$

where $h \in L^1$ is given by (30), and $\rho_\alpha \in C_0^\infty$ is a function, depending on λ and satisfying $\rho_\alpha(x) = 0$ for all x in a neighborhood of 0. Next, reasoning as in the previous part of the proof, we see that condition $V \in \tilde{K}_{2m,n}$ implies condition (ii) in Theorem 2 for $2m - n = 0, 2, 4, \cdots$

Now suppose that condition (i) in Theorem 2 holds. Let $2m - n \neq 0, 2, 4, \cdots$ Then we have

$$\|V \star ((\lambda)_{\frac{1}{\alpha}} g_{2m})\|_\infty = \alpha^n \|(V)_\alpha \star ((g_{2m})_\alpha \lambda)\|_\infty$$
$$= \alpha^{2m} \|(V)_\alpha \star (g_{2m} \lambda)\|_\infty = \alpha^{2m} \|J^{-2m}(V)_\alpha \star Y\|_\infty \qquad (37)$$

where $Y = (I - \Delta)^m (\lambda g_{2m})$. It follows that

$$Y = r_{2m}^{-1} \delta + \sum_{j=0}^{m-1} b_j (-\Delta)^j (\lambda g_{2m}) + \tau \qquad (38)$$

where $\tau \in C_0^\infty$. Since

$$\sum_{j=0}^{m-1} b_j (-\Delta)^j (\lambda g_{2m}) \in L^1,$$

we have from (37), (38), and from condition (i) in Theorem 2 that

$$\lim_{\alpha \to 0+} \|V \star ((\lambda)_{\frac{1}{\alpha}} g_{2m})\|_\infty = 0.$$

Therefore, we have $V \in \tilde{K}_{2m,n}$.

If $2m - n = 0, 2, 4, \cdots$, then

$$\|V \star (\lambda)_{\frac{1}{\alpha}} g_{2m,\alpha}\|_\infty = \alpha^{2m} \|J^{-2m}(V)_\alpha \star Y\|_\infty,$$

and the proof proceeds exactly as in the case $2m - n \neq 0, 2, 4, \cdots$
This completes the proof of Theorem 3.

3 The BMO-classes

In this section we introduce the BMO-scale of distributions on R^n. First we give well-known definitions of the spaces BMO and VMO.

Let $V \in L^1_{loc}$. Then it is said that V belongs to the space BMO iff for all balls $B(x,r)$ in R^n

$$\frac{1}{m(B(x,r))} \int_{B(x,r)} |f(y) - f_{B(x,r)}| dy \leq M, \qquad (39)$$

where the constant $M > 0$ does not depend on $x \in R^n$ and $r > 0$ and where

$$f_{B(x,r)} = \frac{1}{m(B(x,r))} \int_{B(x,r)} f(u) du$$

denotes the mean value of the function V over the ball $B(x,r)$. The functions in BMO are defined up to an additive constant. The norm of a function $V \in BMO$ is given by

$$\|V\|_{BMO} = \inf\{M\}$$

where the infimum is taken with respect to all constants M satisfying (39). It is clear that $L^\infty \subset BMO$. The opposite inclusion does not hold. Various examples of unbounded functions in BMO can be found in [22], p. 141. It is also true that $BMO \subset S'$.

A function $V \in BMO$ belongs to the space VMO iff

$$\lim_{r \to 0+} \sup_{x \in R^n} \frac{1}{m(B(x,r))} \int_{B(x,r)} |f(y) - f_{B(x,r)}| dy = 0.$$

The space VMO is a closed subspace of the space BMO. It is known that

$$V \in VMO \iff \lim_{h \to 0} \|V - V(\cdot - h)\|_{BMO} = 0 \qquad (40)$$

(see [17], Theorem 1). It is also known that $BUC \subset VMO$. We refer the reader to [22] for more information concerning the spaces BMO and VMO.

In the next definition we introduce the BMO-analogues of the classes $\tilde{K}_{s,n}$.

Definition 3 *Let Let $s > 0$ be such that $s - n \neq 0, 2, 4, \cdots$. Let $V \in S'$. Then we will say that $V \in B_{s,n}$ iff for every α with $0 < \alpha \leq 1$ and $\lambda \in \Lambda$ the distribution $V \star ((\lambda)_{\frac{1}{\alpha}} g_s)$ coincides with a function from BMO, and we have*

$$\lim_{\alpha \to 0+} \|V \star (g_s(\lambda)_{\frac{1}{\alpha}})\|_{BMO} = 0.$$

In the case $s - n = 0, 2, 4, \cdots$, we say that $V \in B_{s,n}$ iff for every α with $0 < \alpha \leq 1$ and every $\lambda \in \Lambda$ the distribution $V \star ((\lambda)_{\frac{1}{\alpha}} g_{s,\alpha})$ coincides with a function from BMO, and moreover,

$$\lim_{\alpha \to 0+} \|V \star ((\lambda)_{\frac{1}{\alpha}} g_{s,\alpha})\|_{BMO} = 0.$$

The next theorem is a BMO-version of Theorem 2.

Theorem 4 *Let $V \in S'$ and $s > 0$. Then the following are equivalent:*
(i) $J^{-s}V \in BMO$ and
$$\lim_{\alpha \to 0+} \alpha^s \|J^{-s}(V)_\alpha\|_{BMO} = 0.$$

(ii) $J^{-s}V \in VMO$.

Proof of Theorem 4. The next lemma is similar to Lemma 2. The only difference in the proof is that we should use the BMO-norm instead of the L^∞-norm.

Lemma 2 *Let $V \in S'$, $s > 0$, and $J^{-s}V \in BMO$. Then for every α with $0 < \alpha < 1$ we have $J^{-s}(V)_\alpha \in BMO$. Moreover,*
$$\sup_{\alpha:0<\alpha<1} \alpha^s \|J^{-s}(V)_\alpha\|_{BMO} \leq c_s \|J^{-s}V\|_{BMO}.$$

Let us continue the proof of Theorem 4. Suppose $f \in BMO$ and $g \in L^1$. In general, the convolution $f \star g$ is not defined as an absolutely convergent integral. However, the operator $f \to f \star g$ is defined on the space BMO as the adjoint to the corresponding convolution operator on H^1. Moreover, we have

$$\|f \star g\|_{BMO} \leq \|f\|_{BMO} \|g\|_1. \tag{41}$$

Let $V \in S'$ be such that $J^{-s}V \in BMO$ and condition (i) in Theorem 4 holds. Using the previous remark, concerning the convolution of BMO and L^1 functions, we see that equality (13) can be considered as an equality for the bounded linear functionals on H^1. Next, using (13) and reasoning as in the proof of the implication (i) $\Rightarrow$ (ii) in Theorem 2 with BMO instead of L^∞, we get

$$\lim_{\alpha \to 0+} \|J^{-s}V - J^{-s}V \star (\alpha^{-n}\Phi_s(\alpha^{-1}\cdot))\|_{BMO} = 0 \tag{42}$$

where Φ_s is defined by (12). By (40), the function $h_\alpha = J^{-s}V \star (\alpha^{-n}\Phi_s(\alpha^{-1}\cdot))$ belongs to the space VMO for every $0 < \alpha \leq 1$. Since the space VMO is a closed subspace of the space BMO, (42) implies $J^{-s}V \in VMO$. This proves the implication (i) $\Rightarrow$ (ii) in Theorem 4.

Let $V \in S'$ and $J^{-s}V \in VMO$. Using (41), we see that the last two terms on the right-hand side of (16) tend to zero in BMO as $\alpha \to 0+$. Next we will prove that the first two terms also tend to zero. The proof of this fact is more complicated than the proof of the corresponding fact in Theorem 2. Here we need to show that the convolution of the function $J^{-s}V$ with the function $\alpha^{-n}\Phi_s(\alpha^{-1}\cdot)$ is defined as an absolutely convergent integral.

We have

$$\int_{R^n} \frac{|J^{-s}V(x)|}{1+|x|^{n+\delta}}dx < \infty. \tag{43}$$

Inequality (43) with $\delta = 1$ follows from (2) on p. 141 in [22]. The case $\delta > 0$ can be obtained similarly. Now we get from (43) that the function $J^{-s}V \star g$ exists as an absolutely convergent integral for any function $g \in L^1$, satisfying the estimate

$$|g(x)| \leq \frac{c}{|x|^{n+\delta}} \tag{44}$$

for all $|x| \geq 1$ and some $\delta > 0$.

Next we will show that the function Φ_s satisfies condition (44). The following formula holds (see [21]):

$$\Phi_s(x) = -\sum_{m=1}^{\infty} A_{m,s} G_{2m}(x). \tag{45}$$

Using formula (3) for the Bessel potential kernel G_{2m}, we obtain that for $|x| \geq 1$ and $0 < \delta \leq 1$,

$$\begin{aligned}
G_{2m}(x) &= [(2\pi)^{n/2} 2^m (m-1)!]^{-1} \int_0^\infty t^{\frac{2m+\delta}{2}} t^{-\frac{(n+\delta)}{2}} e^{-\frac{|x|^2}{2t}} e^{-\frac{t}{2}} \frac{dt}{t} \\
&\leq [(2\pi)^{n/2} 2^m (m-1)!]^{-1} \sup_{t>0}\{t^{-\frac{(n+\delta)}{2}} e^{-\frac{|x|^2}{2t}}\} \int_0^\infty t^{\frac{2m+\delta}{2}} e^{-\frac{t}{2}} \frac{dt}{t} \\
&\leq \frac{2^{\frac{2m+\delta}{2}} \Gamma(\frac{2m+\delta}{2})}{(2\pi)^{n/2} 2^m (m-1)!} \sup_{t>0}\{t^{-\frac{(n+\delta)}{2}} e^{-\frac{|x|^2}{2t}}\} \\
&\leq c_n 2^{\frac{\delta}{2}} \frac{\Gamma(\frac{2m+\delta}{2})}{(m-1)!} |x|^{-(n+\delta)} \sup_{t>0}\{t^{-\frac{(n+\delta)}{2}} e^{-\frac{1}{2t}}\} \\
&\leq c_n \frac{\Gamma(\frac{2m+\delta}{2})}{(m-1)!} |x|^{-(n+\delta)}. \tag{46}
\end{aligned}$$

Now using Stirling's formula (see [5]) for the gamma function, we get

$$\frac{\Gamma(\frac{2m+\delta}{2})}{(m-1)!} \leq cm^{\frac{\delta}{2}} \tag{47}$$

for all $m \geq m_0$. In (47), the constant c does not depend on δ and m, and the constant m_0 does not depend on δ. It follows from (46) and (47) that there exists a constant $\tilde{c}_{n,\delta}$ such that

$$G_{2m}(x) \leq \tilde{c}_{n,\delta} m^{\frac{\delta}{2}} |x|^{-(n+\delta)} \tag{48}$$

for all $m \geq 1$ and $|x| \geq 1$. Next (7), (45), and (48) show that the function Φ_s satisfies condition (44). Hence, the function given by

$$\int_{R^n} \alpha^{-n} |J^{-s}V(y)| |\Phi_s(\frac{x-y}{\alpha})| dy$$

is locally integrable. Therefore,

$$\|J^{-s}V \star (\alpha^{-n}\Phi_s(\alpha^{-1}\cdot))\|_{BMO} \leq \|J^{-s}V\|_{BMO} \|\Phi_s\|_1.$$

Since

$$\int_{R^n} \Phi_s(y) dy = F(\Phi_s)(0) = 1,$$

we have

$$J^{-s}V(x) - J^{-s}V \star (\alpha^{-n}\Phi_s(\alpha^{-1}\cdot))(x)$$
$$= \int_{R^n} [J^{-s}V(x) - J^{-s}V(x-y)] \alpha^{-n} \Phi_s(\frac{y}{\alpha}) dy. \qquad (49)$$

Using the definition of the space BMO and (49), we get

$$\|J^{-s}V - J^{-s}V \star (\alpha^{-n}\Phi_s(\alpha^{-1}\cdot))\|_{BMO}$$
$$\leq \int_{R^n} \alpha^{-n} |\Phi_s(\frac{y}{\alpha})| \|J^{-s}V(\cdot) - J^{-s}V(\cdot - y)\|_{BMO} dy. \qquad (50)$$

Since $J^{-s}V \in VMO$ and

$$\lim_{\alpha \to 0+} \alpha^n \int_{|y|>\epsilon} |\Phi_s(\frac{y}{\alpha})| dy$$

for every $\epsilon > 0$, we get from (40) and (50) that

$$\lim_{\alpha \to 0+} \|J^{-s}V - J^{-s}V \star (\alpha^{-n}\Phi_s(\alpha^{-1}\cdot))\|_{BMO} = 0.$$

It follows that the first two terms on the right-hand side of (16) tend to zero in BMO as $\alpha \to 0+$. This shows that the left-hand side of (16) tends to zero in BMO as $\alpha \to 0+$. This proves the implication $(ii) \Rightarrow (i)$ in Theorem 4.

The proof of Theorem 4 is thus completed.

The next theorem is the BMO-version of Theorem 3. We do not know if Theorem 5 holds in the case $s \neq 2m$.

Theorem 5 *Let $V \in S'$ and let m be a natural number. Then condition*

$$V \in B_{2m,n}$$

is equivalent to conditions (i) and (ii) in Theorem 3 for $s = 2m$.

Proof of Theorem 5. We will show that condition $V \in B_{2m,n}$ implies condition (ii) in Theorem 3, and that condition (i) in Theorem 3 implies condition $V \in B_{2m,n}$. The proof will be similar to that of Theorem 3.

Let $V \in B_{2m,n}$. Then for every $\lambda \in \Lambda$ we have
$$V \star (\lambda g_{2m}) \in BMO.$$

Using (32) and the definition of the class $B_{2m,n}$, we get that for every function $\tau \in C_0^\infty$ which is equal to zero near 0, we have $V \star \tau \in BMO$. Hence, $V \star G_{2m} \star \tau \in BMO$, and (32) gives
$$J^{-2m}V \in BMO. \tag{51}$$

Next we will prove that the expression on the right-hand side of (36) tends to 0 in BMO as $\alpha \to 0+$. Since (51) implies that $J^{-m} \star \tau_\alpha \in VMO$, and we know that VMO is a closed subspace of BMO, we obtain from (36) that $J^{-2m}V \in VMO$.

Now suppose that condition (i) in Theorem 3 holds. Let $2m - n \neq 0, 2, 4, \cdots$. Then we have
$$\|V \star ((\lambda)_{\frac{1}{\alpha}} g_{2m})\|_{BMO} = \alpha^n \|(V)_\alpha \star ((g_{2m})_\alpha \lambda)\|_{BMO}$$
$$= \alpha^{2m} \|(V)_\alpha \star (g_{2m}\lambda)\|_{BMO} = \alpha^{2m} \|(V)_\alpha \star Y\|_{BMO}$$

where $Y = (I-\Delta)^m (\lambda g_{2m})$. Then, reasoning as in the proof of Theorem 3, we get that
$$\lim_\alpha \|V \star ((\lambda)_{\frac{1}{\alpha}} g_{2m})\|_{BMO} = 0.$$

For $2m - n = 0, 2, 4, \cdots$ the proof is similar.

References

[1] M. Aizenman and B. Simon, Brownian motion and Harnack's inequality for Schrödinger operators, Comm. Pure Appl. Math. 35 (1982), 209-271.

[2] J. B. Conway, Functions of One Complex Variable, Springer, New York, 1978.

[3] E. B. Davies and A. Hinz, Kato class potentials for higher order elliptic operators, J. London. Math. Soc. (2) 58 (1998), 669-678.

[4] M. Demuth and J. A. van Casteren, Stochastic Spectral Theory for Selfadjoint Feller Operators: A functional integration approach, Birkhäuser Verlag, Basel, 2000.

[5] M. D. Greenberg, Foundations of Applied Mathematics, Prentice-Hall, Inc., Englewood Cliffs, New Jersey, 1978.

[6] A. Gulisashvili, Sharp estimates in smoothing theorems for Schrödinger semigroups, J. Functional Analysis 170 (2000), 161-187.

[7] A. Gulisashvili, On the heat equation with a time-dependent singular potential, submitted for publication.

[8] A. Gulisashvili and M. A. Kon, Smoothness of Schrödinger semigroups and eigenfunctions, Int. Math. Res. Notices 5 (1994), 193-199.

[9] A. Gulisashvili and M. A. Kon, Exact smoothing properties of Schrödinger semigroups, Amer. J. Math. 118 (1996), 1215-1248.

[10] J. W. Johnson and M. L. Lapidus, The Feynman Integral and Feynman's Operational Calculus, Oxford University Press, Oxford, 2000.

[11] T. Kato, Schrödinger operators with singular potentials, Israel J. Math. 13 (1973), 135-148.

[12] V. G. Maz'ya, Sobolev Spaces, Springer-Verlag, Berlin, 1985.

[13] Qi Zhang, On a parabolic equation with a singular lower order term, Transactions Amer. Math. Soc. 348 (1996), 2811-2844.

[14] Qi Zhang, On a parabolic equation with a singular lower order term, Part 2: The Gaussian bounds, Indiana Univ. Math. J. 46 (1997), 989-1020.

[15] F. Räbiger, A. Rhandi, R. Schnaubelt, and J. Voigt, Non-autonomous Miyadera perturbation, Differential Integral Equations 13 (2000), 341-368.

[16] S. G. Samko, A. A. Kilbas, and O. I. Marichev, Fractional Integrals and Derivatives: Theorey and Applications, Gordon and Breach Science Publishers, Amsterdam, 1993.

[17] D. Sarason, Functions of vanishing mean oscillation, Trans. Amer. Math. Soc. 207 (1975), 391-405.

[18] M. Schechter, Spectra of Partial Differential Equations, North Holland, Amsterdam, 1986.

[19] R. Schnaubelt and J. Voigt, The non-autonomous Kato class, Arch. Math. 72 (1999), 454-460.

[20] B. Simon, Schrödinger semigroups, Bull. Amer. Math. Soc. 7 (1982), 445-526.

[21] E. M. Stein, Singular Integrals and Differentiability Properties of Functions, Princeton University Press, Princeton, NJ, 1970.

[22] E. M. Stein, Harmonic Analysis: Real-Variable Methods, Orthogonality, and Oscillatory Integrals, Princeton University Press, Princeton, 1993.

[23] F. Stummel, Singulare elliptische differentialoperatoren in Hilbertschen Räumen, Math. Ann. 132 (1956), 150-176.

The Global Solution Set for a Class of Semilinear Problems

Philip Korman
Institute for Dynamics and
Department of Mathematical Sciences
University of Cincinnati
Cincinnati Ohio 45221-0025

Abstract

For a class of semilinear Dirichlet problems we present an exact multiplicity result. Our proof simplifies the previous one in T. Ouyang and J. Shi [11]. By an indirect argument we sidestep the necessity of proving positivity for linearized equation, which was the most difficult step in [11], as well as in the earlier paper of P. Korman, Y. Li and T. Ouyang [6].

1 Introduction

We consider a class of semilinear Dirichlet problems

(1.1) $\quad \Delta u + \lambda f(u) = 0 \quad \text{for } |x| < R,\ u = 0 \quad \text{for } |x| = R,$

on a ball of radius R in R^n. Here λ is a positive parameter, and the nonlinearity $f(u)$ is assumed to generalize a model case $f(u) = u(u-b)(c-u)$, with positive constants b and c, and $c > 2b$ (in case $c \leq 2b$ the problem (1.1) has no nontrivial solutions, see e.g., [6]).

We now list our assumptions on the nonlinearity $f(u)$. We assume that $f(u) \in C^2(\bar{R}_+)$, and it has the following properties

(1.2) $\quad f(0) = f(b) = f(c) = 0 \quad \text{for some constants } 0 < b < c,$

(1.3) $$f(u) < 0 \text{ for } u \in (0,b) \cup (c,\infty),$$
$$f(u) > 0 \text{ for } u \in (-\infty,0) \cup (b,c),$$

(1.4) $$\int_0^c f(u)\,du > 0,$$

(1.5) There exists an $\alpha \in (0,c)$, such that
$$f''(u) > 0 \text{ for } u \in (0,\alpha) \text{ and } f''(u) < 0 \text{ for } u \in (\alpha,c).$$

We define θ to be the smallest positive number, such that $\int_0^\theta f(s)\,ds = 0$. Clearly, $\theta \in (b,c)$. After T. Ouyang and J. Shi [11], we set $\rho = \alpha - \frac{f(\alpha)}{f'(\alpha)}$ (i.e. ρ is the first Newton iterate when solving $f(u) = 0$ with the initial guess α). We define $K(u) = \frac{uf'(u)}{f(u)}$. Our final assumption is the following. If $\theta < \rho$ we assume that

(1.6) $$K(u) > K(\theta) \quad \text{on } (b,\theta)$$
$$K(u) \text{ is nonincreasing on } (\theta,\rho)$$
$$K(u) < K(\rho) \quad \text{on } (\rho,\alpha).$$

(If $\theta \geq \rho$ this assumption is empty.)

We are now ready to state the main result.

Theorem 1.1 *Assume that $f(u)$ satisfies the conditions listed above. For the problem (1.1) there is a critical $\lambda_0 > 0$ such that the problem (1.1) has exactly 0, 1 or 2 nontrivial solutions, depending on whether $\lambda < \lambda_0$, $\lambda = \lambda_0$ or $\lambda > \lambda_0$. Moreover, all solutions lie on a single smooth solution curve, which for $\lambda > \lambda_0$ has two branches denoted by $0 < u^-(r,\lambda) < u^+(r,\lambda)$, with $u^+(r,\lambda)$ strictly monotone increasing in λ and $\lim_{\lambda \to \infty} u^+(r,\lambda) = c$ for $r \in [0,1)$. For the lower branch $\lim_{\lambda \to \infty} u^-(r,\lambda) = 0$ for $r \neq 0$, while $u^-(0,\lambda) > b$ for all $\lambda > \lambda_0$.*

In present generality this theorem was proved first by T. Ouyang and J. Shi [11]. In two dimensions (with some extra assumptions on $f(u)$) this theorem was proved in P. Korman, Y. Li and T. Ouyang [6], where the general scheme for proving such results was developed. One of the crucial things in that approach was proving positivity of any non-trivial solution of the linearized problem

(1.7) $$\Delta w + \lambda f'(u)w = 0 \quad \text{for } |x| < R, \ w = 0 \quad \text{for } |x| = R.$$

This turned out to be a difficult task, and it was the only reason the paper [6] was restricted to two dimensions. Later, T. Ouyang and J. Shi [11] were able to prove that $w(r) > 0$ by using Pohozhaev type identity. Their proof is rather involved. We also mention that one-dimensional version of this result was proved in P. Korman, Y. Li and T. Ouyang [7], where more general nonlinearities of the type $f(u) = (u-a)(u-b)(c-u)$ were considered, and where references to earlier work in case $n = 1$ by J. Smoller and A. Wasserman, and S.-H. Wang can be found.

In this work by using an indirect argument, we are able to avoid having to prove that $w(r) > 0$, which considerably simplifies the proof, and makes it more transparent. We show that it suffices to prove that $w(r)$ cannot vanish exactly once. We show that our assumptions on $f(u)$ make the function $\frac{uf'(u)}{f(u)}$ behave almost the same way as in the important paper of M.K. Kwong and L. Zhang [8], and then our proof that $w(r)$ cannot vanish exactly once is similar to Lemma 8 of [8].

We outline our arguments next. It is known that for large λ our problem (1.1) has a positive solution. When continued for increasing λ this solution, after possibly some turns, has to tend to c as $\lambda \to \infty$. When continued for decreasing λ this solution has to turn, since no positive solutions exist for $\lambda > 0$ small. The lower end of our solution curve, after possibly some turns, has to tend to 0 as $\lambda \to \infty$. If one assumes that $w(r) > 0$ at any one of the turns, we show that the result follows. It is important on this step that $w(r)$ cannot vanish exactly once. It then remains to consider the case when condition $w(r) > 0$ is violated at all turning points. Assume for simplicity there is only one turning point on the solution curve. Since condition $w(r) > 0$ is violated, it follows by the Crandall-Rabinowitz bifurcation theorem (which is recalled below) that the lower and upper solution branches intersect near the turning point. By uniqueness for initial-value problem these branches would have to intersect for all λ. But the upper branch tends to c, while the lower one tends to zero, and hence they have to separate eventually, a contradiction. In case of more than one turning point, the argument is more involved, although the idea is similar.

Next we state a bifurcation theorem of Crandall-Rabinowitz [1].

Theorem 1.2 *[1] Let X and Y be Banach spaces. Let $(\overline{\lambda}, \overline{x}) \in \mathbf{R} \times X$ and let F be a continuously differentiable mapping of an open neighborhood of $(\overline{\lambda}, \overline{x})$ into Y. Let the null-space $N(F_x(\overline{\lambda}, \overline{x})) = \text{span } \{x_0\}$ be one-dimensional and codim $R(F_x(\overline{\lambda}, \overline{x})) = 1$. Let $F_\lambda(\overline{\lambda}, \overline{x}) \notin R(F_x(\overline{\lambda}, \overline{x}))$. If Z is a complement of* span $\{x_0\}$ in X, then the solutions of $F(\lambda, x) = F(\overline{\lambda}, \overline{x})$ near $(\overline{\lambda}, \overline{x})$ form a curve $(\lambda(s), x(s)) = (\overline{\lambda}+\tau(s), \overline{x}+sx_0+z(s))$, where $s \to (\tau(s), z(s)) \in \mathbf{R} \times Z$ is a continuously differentiable function near $s = 0$ and $\tau(0) = \tau'(0) = 0$, $z(0) = z'(0) = 0$.

Throughout the paper we consider only the classical solutions of (1.1). Without loss of generality we set $R = 1$. Also notice that by the maximum principle all non-trivial solutions of (1.1) are positive.

2 Preliminary results

We list some consequences of our conditions on $f(u)$. We define $\beta > 0$ to be the unique number where $f'(\beta) = \dfrac{f(\beta)}{\beta}$. Clearly, $\beta \in (\alpha, \gamma)$, where γ is the larger root of $f'(u) = 0$. The following lemma was proved in [6].

Lemma 2.1 We have

(2.1) $$uf'(u) - f(u) \begin{cases} > 0, & \text{for } u \in (0, \beta) \\ < 0 & \text{for } u \in (\beta, c). \end{cases}$$

Lemma 2.2 $K(0) = 1$, $K(u) < 1$ on $(0, b)$.

Proof: The first statement follows by L'Hospital rule. Notice next that for $u < b$ we have $f(u) < 0$, and also by the previous lemma $f'(u)u > f(u)$. It follows that $K(u) < 1$.

Lemma 2.3 $K'(u) < 0$ on (α, β).

Proof: Compute

$$K'(u) = \frac{uf''f + f'(f - uf')}{f^2}.$$

The first term in the numerator is negative for $u > \alpha$, and the second one is negative by Lemma 2.1 (notice that $f'(u) > 0$ on (α, β)).

Lemma 2.4 $K(u) < 1$ on (β, c).

Global Solution Set for a Class of Semilinear Problems

Proof: On (β, c) we have $f(u) > 0$, and $f'(u)u < f(u)$ by Lemma 2.1, and the proof follows.

Lemma 2.5 *If $\theta < \rho$ then $K(\rho) > 1$.*

Proof: By the definition of ρ
$$f'(\alpha)(\alpha - \rho) - f(\alpha) = 0.$$
Using this and our last condition in (1.6),
$$K(\rho) > K(\alpha) = 1 + \frac{\rho f'(\alpha)}{f(\alpha)} > 1,$$
proving the lemma.

The lemmas above imply the following result.

Theorem 2.1 *Assume $\theta < \rho$. For any $u_0 \in (\theta, \rho)$ we define $\gamma = K(u_0)$. Then $\gamma > 1$, and*

$$(2.2) \quad uf'(u) - \gamma f(u) \begin{cases} > 0, & \text{for } u \in (0, u_0) \\ < 0 & \text{for } u \in (u_0, c). \end{cases}$$

Proof: The above lemmas imply that the horizontal line $y = \gamma$ intersects the graph of $y = K(u)$ exactly once, and the graph of $K(u)$ lies above the line $y = \gamma$ in the region where $f(u) > 0$. This proves the first inequality in (2.2), and second one follows similarly.

We study multiplicity of positive solutions of the Dirichlet problem, depending on a positive parameter λ

$$(2.3) \quad \Delta u + \lambda f(u) = 0 \quad \text{for } |x| < 1, \quad u = 0 \text{ on } |x| = 1,$$

with nonlinearity $f(u)$ satisfying all of our assumptions. By the classical theorem of B. Gidas, W.-M. Ni and L. Nirenberg [3] positive solutions of (2.3) are radially symmetric, which reduces (2.3) to

$$(2.4) \quad u'' + \frac{n-1}{r}u' + \lambda f(u) = 0 \quad \text{for } 0 < r < 1, \quad u'(0) = u(1) = 0.$$

We shall also need the corresponding linearized equation

$$(2.5) \quad w'' + \frac{n-1}{r}w' + \lambda f'(u)w = 0 \quad \text{for } 0 < r < 1, \quad w'(0) = w(1) = 0.$$

The following lemma was proved in [4].

Lemma 2.6 *Assume that the function $f(u) \in C^2(\bar{R}_+)$, and the problem (2.5) has a nontrivial solution w at some λ. Then*

$$(2.6) \qquad \int_0^1 f(u) w r^{n-1} \, dr = \frac{1}{2\lambda} u'(1) w'(1).$$

We recall that solution of (2.4) is called singular provided the corresponding linearized problem (2.5) has a nontrivial solution. The following lemma follows immediately from the equations (2.4) and (2.5).

Lemma 2.7 *Let (λ, u) be a singular solution of (2.4). Then*

$$(2.7) \qquad \int_0^1 (f(u) - f'(u)u) \, w r^{n-1} \, dr = 0.$$

The following lemma is a consequence of the previous two.

Lemma 2.8 *Let (λ, u) be a singular solution of (2.4). Then for any real γ*

$$(2.8) \qquad \int_0^1 (\gamma f(u) - f'(u)u) \, w r^{n-1} \, dr = \frac{\gamma - 1}{2\lambda} u'(1) w'(1).$$

Proof. Multiplying (2.6) by $\gamma - 1$, and adding (2.7), we obtain (2.8).

The following lemma is known, see e.g. E.N. Dancer [2]. We present its proof for completeness.

Lemma 2.9 *Positive solutions of the problem (2.4) are globally parameterized by their maximum values $u(0, \lambda)$. I.e., for every $p > 0$ there is at most one $\lambda > 0$, for which $u(0, \lambda) = p$.*

Proof. If $u(r, \lambda)$ is a solution of (2.4) with $u(0, \lambda) = p$, then $v \equiv u(\frac{1}{\sqrt{\lambda}} r)$ solves

$$(2.9) \qquad v'' + \frac{n-1}{r} v' + f(v) = 0, \quad v(0) = p, \, v'(0) = 0.$$

If $u(0, \mu) = p$ for some $\mu \neq \lambda$, then $u(\frac{1}{\sqrt{\mu}} r)$ is another solution of the same problem. This is a contradiction, in view of the uniqueness of solutions for initial value problems of the type (2.21), see [12].

The following lemma restricts the region where $w(r)$, solution of the linearized problem (2.5), may vanish. Its first part is due to T. Ouyang and J. Shi [11], see also J. Wei [13], and its second part is due to M.K. Kwong and L. Zhang [8].

Lemma 2.10 *Any nontrivial solution of (2.5) cannot vanish in either interval where $0 < u < \theta$, and where $\rho < u < 1$.*

In case $\theta \geq \rho$ it follows that any nontrivial solution of (2.5) is positive, and the main result of the present paper then follows similarly to [6].

The following lemma follows the idea of Lemma 8 of M.K. Kwong and L. Zhang [8].

Lemma 2.11 *Under our conditions on $f(u)$ any non-trivial solution of (2.5) $w(r)$ cannot have exactly one zero on $(0, 1)$.*

Proof. Since $w(0) \neq 0$, see [12] for the appropriate uniqueness result (if $w(0) = w'(0) = 0$ then $w \equiv 0$), we may assume that $w(0) > 0$. Assume that on the contrary $w(r)$ has exactly one root at some $r = r_0$, i.e.

$$(2.10) \qquad w(r) > 0 \text{ on } (0, r_0), \quad w(r) < 0 \text{ on } (r_0, 1).$$

By Lemma 2.10 $u(r_0) \in (\theta, \rho)$. Setting $\gamma = K(u(r_0))$, we see by Theorem 2.1 that

$$(2.11) \qquad \gamma f(u) - u f'(u) \begin{cases} < 0 & \text{for all} \quad u < u(r_0) \\ > 0 & \text{for all} \quad u > u(r_0). \end{cases}$$

Since $\gamma > 1$, we obtain by Lemma 2.8 (notice that $w'(1) > 0$ by (2.10))

$$(2.12) \qquad \int_0^1 [\gamma f(u) - u f'(u)] w(r) r^{n-1} \, dr < 0.$$

In view of (2.10) and (2.11) the quantity on the left is positive, and we have a contradiction in (2.12).

Lemma 2.12 *Let $u(r, \lambda)$ and $v(r, \lambda)$ be two solution curves of (2.4), which are continuous in λ, when the parameter λ varies in some interval I. Assume that for some $\lambda_0 \in I$ solutions $u(r, \lambda_0)$ and $v(r, \lambda_0)$ intersect. Then $u(r, \lambda)$ and $v(r, \lambda)$ intersect for all $\lambda \in I$.*

Proof: In order for the solution curves to separate, there must exist λ_1 (the last λ at which they intersect) and a point $r_1 \in [0, 1]$ at which $u(r_1, \lambda_1) = v(r_1, \lambda_1)$ and $u_r(r_1, \lambda_1) = v_r(r_1, \lambda_1)$. But this contradicts uniqueness for initial value problems.

Next we study the linearized eigenvalue problem corresponding to any solution of (2.4):

$$(2.13) \quad \varphi'' + \frac{n-1}{r}\varphi' + \lambda f'(u)\varphi + \mu\varphi = 0 \text{ on } (0,1), \quad \varphi'(0) = \varphi(1) = 0.$$

Comparing this to (2.5), we see that at any singular solution of (2.4) $\mu = 0$ is an eigenvalue, corresponding to an eigenfunction $\varphi = w$.

We shall need the following generalization of Lemma 2.6.

Lemma 2.13 *Let $\varphi > 0$ be a solution of (2.13) with $\mu \leq 0$. (I.e. φ is a principal eigenfunction of (2.13).) Then*

$$(2.14) \quad \int_0^1 f(u)\varphi r^{n-1} dr \geq \frac{1}{2\lambda} u'(1)\varphi'(1).$$

Proof. The function $v = r u_r - u_r(1)$ satisfies

$$(2.15) \quad \Delta v + \lambda f'(u)v + \mu v = \mu v - 2\lambda f(u) - \lambda f'(u)u'(1) \text{ for } |x| < 1,$$
$$v = 0 \text{ on } |x| = 1.$$

Comparing (2.15) with (2.13) we conclude by the Fredholm alternative

$$(2.16) \quad \mu \int_0^1 v\varphi r^{n-1} dr - 2\lambda \int_0^1 f(u)\varphi r^{n-1} dr - \lambda u'(1) \int_0^1 f'(u)\varphi r^{n-1} dr = 0.$$

Integrating (2.13)

$$-\lambda \int_0^1 f'(u)\varphi r^{n-1} dr = \varphi'(1) + \mu \int_0^1 \varphi r^{n-1} dr.$$

Using this in (2.16), we have

$$2\lambda \int_0^1 f(u)\varphi r^{n-1} dr = \mu \int_0^1 ru_r \varphi r^{n-1} dr + u'(1)\varphi'(1) + \mu u'(1) \int_0^1 \varphi r^{n-1} dr,$$

and the proof follows.

We now define Morse index of any solution of (2.4) to be the number of negative eigenvalues of (2.13). The following lemma is based on K. Nagasaki and T. Suzuki [10].

Global Solution Set for a Class of Semilinear Problems

Lemma 2.14 *Assume that (λ, u) is a singular solution of (2.4) such that $w'(1) < 0$ and*

$$(2.17) \qquad \int_0^1 f''(u) w^3 r^{n-1} dr < 0.$$

Then at (λ, u) a turn to "the right" in (λ, u) "plane" occurs, and as we follow the solution curve in the direction of decreasing $u(0, \lambda)$, the Morse index is increased by one.

Proof. To see that the turn is to the right, we observe that the function $\tau(s)$, defined in Crandall-Rabinowitz theorem, satisfies $\tau(0) = \tau'(0) = 0$ and

$$(2.18) \qquad \tau''(0) = -\frac{\lambda \int_0^1 f''(u) w^3 r^{n-1} dr}{\int_0^1 f(u) w r^{n-1} dr},$$

see [6] for more details. By our assumption the numerator in (2.18) is negative, while by Lemma 2.2 the denominator is positive. It follows that $\tau''(0) > 0$, and hence $\tau(s)$ is positive for s close to 0, which means that the turn is to the right.

At a turning point one of the eigenvalues of (2.13) is zero. Assume it is the ℓ-th one, and denote $\mu = \mu_e$. Here $\mu = \mu(s)$, and $\mu(0) = 0$. We now write (2.13) in the corresponding PDE form and differentiate this equation in s

$$(2.19) \quad \Delta\varphi_s + \lambda f'(u)\varphi_s + \lambda' f'(u)\varphi + \lambda f''(u) u_s \varphi + \mu' \varphi + \mu \varphi_s = 0$$
$$\text{for } |x| < 1, \quad \varphi_s = 0 \text{ on } |x| = 1.$$

At (λ, u) the Crandall-Rabinowitz theorem applies, and hence we have: $\mu(0) = 0$, $\varphi(0) = w$, $\lambda'(0) = 0$, and $u_s(0) = -w$ (considering the chosen parameterization). Here w is a solution of the linearized equation (2.5). The equation (2.19) becomes

$$(2.20) \qquad \Delta\varphi_s - \lambda f''(u) w^2 + \lambda f'(u)\varphi_s + \mu'(0) w = 0.$$

Multiplying (2.5) by φ_s, (2.20) by w, subtracting and integrating, we have

$$\mu'(0) = \frac{\lambda \int_0^1 f''(u) w^3 r^{n-1} dr}{\int_0^1 w^2 r^{n-1} dr} < 0.$$

It follows that across the turning point one of the positive eigenvalues crosses into the negative region, increasing the Morse index by one.

Lemma 2.15 *Assume that (λ_0, u_0) is a singular solution of (2.4), i.e. the problem (2.5) has a nontrivial solution $w(r)$. Then*

(2.21) $$w(r) > 0 \quad \text{for all } r \in [0, 1)$$

if and only if for λ close to λ_0 any two solutions on the solution curve passing through (λ_0, u_0) do not intersect.

Proof: In view of Lemma 2.6 the Crandall-Rabinowitz theorem applies at (λ_0, u_0) (see [6] for more details). According to that theorem near the point (λ_0, u_0) solutions differ asymptotically by a factor of $w(r)$, which implies the lemma.

The following lemma was proved in [6], see also [11] and [13].

Lemma 2.16 *Assume that (λ_0, u_0) is a singular solution of (2.4), such that (2.21) holds. Then the inequality (2.17) holds, and the conclusions of the Lemma 2.14 apply.*

Next we study eigenvalues and eigenfunctions of radial solutions of Laplace equation on a ball. Since singularity at $r = 0$ is introduced by the polar coordinates, and is not present in the original equation, it is natural to expect spectral properties similar to that of regular Sturm-Liouville problems. Surprisingly, we were not able to find any references.

Lemma 2.17 *Consider an eigenvalue problem*

(2.22) $$y'' + \frac{a}{r}y' + b(r)y + \lambda y = 0, \quad \text{for } 0 < r < 1, \; y'(0) = y(1) = 0,$$

with a constant $a > 1$, and $b(r) \in C^2[0, 1]$. Assume that $\lambda = 0$ is an eigenvalue of (2.22), and let $y_0(r)$ be the corresponding eigenfunction. Then the problem (2.22) has an infinite sequence of eigenvalues $\lambda_1 < \lambda_2 < \ldots$, with $\lambda_n \to \infty$ as $n \to \infty$, and the n-th eigenfunction has precisely $n - 1$ roots on $(0, 1)$ for all $n \geq 1$. (One of λ_k's is equal to zero.)

Proof: We convert the problem (2.22) into an integral equation, using the modified Green's function. We claim that any solution of the equation

(2.23) $$y'' + \frac{a}{r}y' + b(r)y = 0,$$

that is bounded at $r = 0$ must be a multiple of the first eigenfunction $y_0(r)$. Indeed, writing the first two terms of the Taylor's series of the solution with a remainder term, we easily conclude that $y'(0) = 0$ for any bounded solution. If we now fix a constant α so that $y(0) = \alpha y_0(0)$, then we shall have $y(r) = \alpha y_0(r)$ for all $r > 0$, in view of uniqueness for initial value problems of the type (2.23), see [12]. Let $y_2(r)$ be a solution of (2.23) with $y_2(1) = 1$. Since y_2 is not a multiple of y_0, it follows that $y_2(r) \to \infty$ as $r \to +0$. A formal use of Frobenius method at $r = 0$ shows that $y_2 \sim \beta r^{-a+1}$ as $r \to 0$, with some constant β. Setting $y(r) = r^{-a+1}z(r)$, we see that the resulting equation for $z(r)$ has all solutions bounded near $r = 0$, which justifies the asymptotic formula for $y_2(r)$ near $r = 0$.

Notice that the problem (2.23) can be put into an equivalent self-adjoint form
$$(2.24) \qquad (r^a y')' + r^a b(r) y = 0.$$

The modified Green's function for (2.24) subject to the boundary conditions $y'(0) = y(1) = 0$ has the form

$$(2.25) \qquad G(r, \xi) = \begin{cases} \frac{y_0(r) y_2(\xi)}{K} & \text{for } r < \xi \\ \frac{y_0(\xi) y_2(r)}{K} & \text{for } r > \xi, \end{cases}$$

where K is a constant. By the above remarks we have, with some constant $c > 0$,

$$(2.26) \qquad \begin{aligned} |G(r,\xi)| &\leq c\xi^{1-a} \text{ for } r < \xi \\ |G(r,\xi)| &\leq cr^{1-a} \text{ for } r > \xi. \end{aligned}$$

We now multiply the equation (2.22) by r^a, and convert it into an integral equation for the function $z(r) = r^{\frac{a}{2}} y(r)$

$$(2.27) \qquad z(r) = \lambda \int_0^1 \bar{G}(r,\xi) z(\xi)\, d\xi,$$

with the kernel $\bar{G}(r,\xi) = G(r,\xi) r^{\frac{a}{2}} \xi^{\frac{a}{2}}$. Using (2.26) it is a standard exercise to show that $\int_0^1 \int_0^1 \bar{G}^2 \, dr d\xi < \infty$, see pages 178 and 421 in [14]. This means that (2.27) is an integral equation with a compact and symmetric kernel. It follows that its spectrum is discrete, and eigenvalues tend to infinity. Moreover, we conclude that the minimum characterization of eigenvalues applies, from which it follows that the k-th eigenfunction cannot have more than $k-1$ interior roots, see p. 173 in [14]. On the other hand, the same minimum characterization implies that y_1 is of one sign, and y_2 must vanish at least once. Also, by Sturm's comparison theorem y_{k+1} must have at least one more interior root than y_k. We then conclude that y_2, then y_3, and so on have the desired number of interior roots.

3 Proof of the main result

We are now ready to prove the Theorem 1.1. We begin by noticing that existence of positive solutions under our conditions follows by the Theorem 1.5 in P.L. Lions [9], see also [11]. Indeed the result in [9] implies existence of a critical $\bar{\lambda}$, so that for $\lambda \geq \bar{\lambda}$ there exists a maximal positive solution of (2.3), while for $\lambda > \bar{\lambda}$ there exists at least two positive solutions. Since positive solutions of our problem (2.3) are radial, we consider its ODE version (2.4). We now continue the curve of maximal solutions for decreasing λ. It was shown in [6] that this curve cannot be continued for all $\lambda > 0$, and hence a critical point (λ_0, u_0) must be reached, at which the curve will turn. By the definition of a critical point, the linearized equation (2.5) has a nontrivial solution $w(r)$. We claim that the theorem follows provided that

(3.1) $$w(r) > 0 \quad \text{for all } r \in [0,1).$$

By the Crandall-Rabinowitz Theorem near the turning point (λ_0, u_0) the solution set has two branches $u^-(r,\lambda) < u^+(r,\lambda)$, for $r \in [0,1)$, $\lambda > \lambda_0$. By the Crandall-Rabinowitz Theorem we also conclude

(3.2) $$u_\lambda^+(r,\lambda) > 0 \text{ for } \lambda \text{ close to } \lambda_0 \text{ (for all } r \in [0,1)).$$

Arguing like in P. Korman, Y. Li and T. Ouyang [6], we show that the same inequality holds for all $\lambda > \lambda_0$ (until a possible turn), see also T. Ouyang and J. Shi [11] and J. Wei [13]. We claim next that solutions $u^+(r,\lambda)$ are stable, i.e. all eigenvalues of (2.13) are positive. Indeed, let on the contrary $\mu \leq 0$ be the principal eigenvalue of (2.13), and $\varphi > 0$ the corresponding eigenvector. The equation for u_λ is

(3.3) $$u_\lambda'' + \tfrac{n-1}{r} u_\lambda' + \lambda f'(u) u_\lambda + f(u) = 0 \text{ for } r \in (0,1],$$
$$u_\lambda'(0) = u_\lambda(1) = 0.$$

From the equations (2.13) and (3.3) we obtain

(3.4) $$r^{n-1}(\varphi' u_\lambda - u_\lambda' \varphi)|_0^1 = -\mu \int_0^1 \varphi u_\lambda r^{n-1} dr + \int_0^1 f(u) \varphi r^{n-1} dr.$$

Global Solution Set for a Class of Semilinear Problems

The right hand side in (3.4) is positive by our assumptions, inequality (3.2), and Lemma 2.13, while the quantity on the left is zero, a contradiction.

We show next that for $\lambda > \lambda_0$ both branches $u^+(r,\lambda)$ and $u^-(r,\lambda)$ have no critical points. Indeed, if we had a critical point on the upper branch $u^+(r,\lambda)$ at some $\overline{\lambda} > \lambda_0$, then by the Crandall-Rabinowitz Theorem solution of the linearized equation would be positive at $\lambda = \overline{\lambda}$ (since $u_\lambda > 0$ as we enter the critical point). But then by Lemma 2.14 we know precisely the structure of solution set near $(\overline{\lambda}, u^+(r,\overline{\lambda}))$, namely it is a parabola-like curve with a turn to the right. This is impossible, since the solution curve has arrived at this point from the left. Turning to the lower branch $u^-(r,\lambda)$, we know by Lemma 2.14 that each solution on this branch has Morse index of one, until a possible critical point. At the next possible turning point one of the eigenvalues becomes zero, which means that the Morse index of the turning point is either zero or one. If Morse index is zero, it means that zero is a principal eigenvalue, and so solutions of the corresponding linearized equation are of one sign, and then we obtain a contradiction the same way as on the upper branch. If Morse index = 1, it means that zero is a second eigenvalue, i.e. by Lemma 2.17 $w(r)$ changes sign exactly once, but that is impossible by Lemma 2.11. It follows that if condition (3.1) is satisfied at the first turning point, then our Theorem 1.1. follows. But exactly the same arguments show that having $w(r) > 0$ at any turning point will imply Theorem 1.1.

It remains to rule out the possibility that condition (3.1) fails at all turning points. By Lemma 2.15 this means that the branches $u^-(r,\lambda)$ and $u^+(r,\lambda)$ intersect near any turning point (λ_0, u_0). When we continue the upper branch $u^+(r,\lambda)$ for increasing λ, then, after possibly some more turns, $u^+(r,\lambda) \to c$ as $\lambda \to \infty$ for all $r \in [0,1)$, see [6] for more details. Similarly, for the lower branch we have $u^-(r,\lambda) \to 0$ as $\lambda \to \infty$ for all $r \in (0,1)$, after possibly some additional turns, see [6]. (Notice that $u^-(0,\lambda) > \theta$.) It follows that for λ sufficiently large

$$(3.5) \qquad u^-(r,\lambda) < u^+(r,\lambda) \quad \text{for all } r \in [0,1).$$

We now pick the leftmost turning point on our curve (i.e. the turning point with smallest λ; if there is more than one such point, take any one of them). In Figure 1 this is the point A. By above, condition (3.1) is violated at this point, and hence solution branches contain intersecting solutions near A. As we increase λ solutions on both branches continue to intersect by Lemma 2.12, until a possible turning point. If both branches have no more

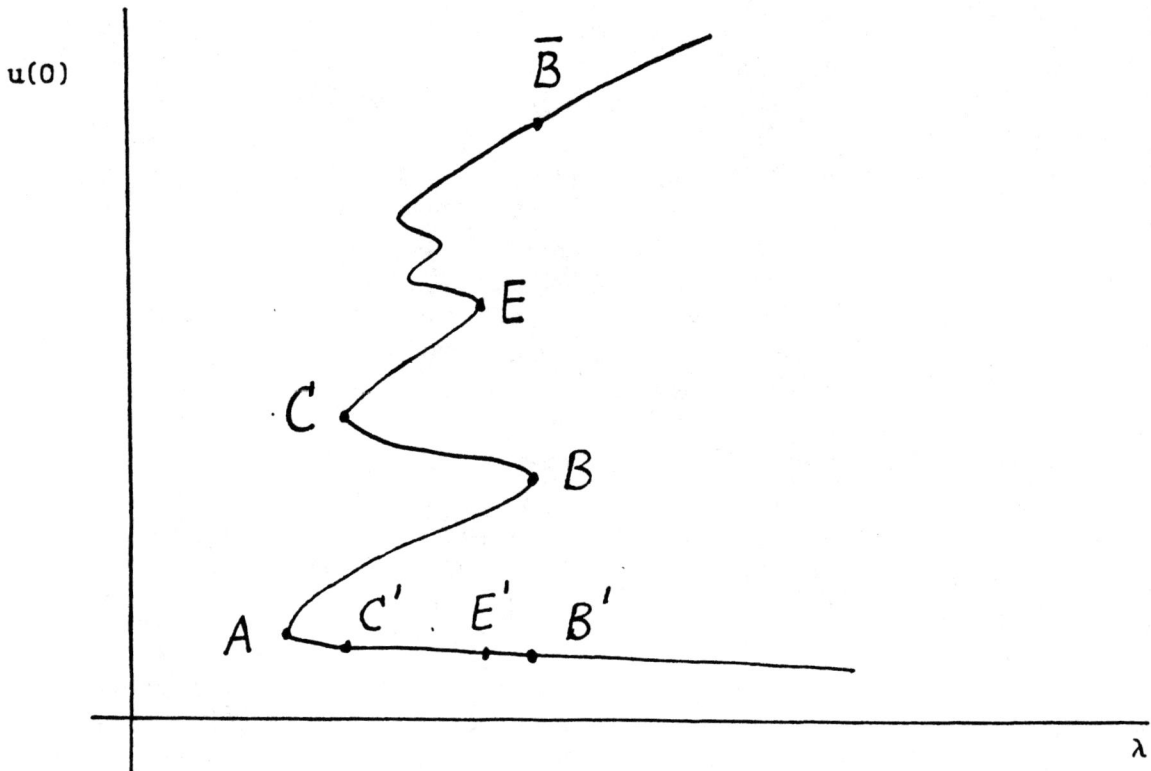

Figure. 1. Solution curve with several turning points.

turning points, solutions will intersect for all λ, contradicting (3.5). Assume that as we continue both branches $u^-(r,\lambda)$ and $u^-(r,\lambda)$ for increasing λ the first turning point happens, say, at the upper branch at a point B. By B' we denote the point on the lower branch, which has the same λ coordinate as B. By Lemma 2.12 solutions at B and B' intersect. We now continue the upper branch for decreasing λ until the next turning point, which we call C. By C' we denote the point on the lower branch, which has the same λ coordinate as C. Moving leftwards on both branches, we conclude by Lemma 2.12 that solutions at C and C' intersect. We denote by E the next turning point on the upper branch (if it exists), and by E' the corresponding point under it on the lower branch. By moving to the right on both branches and using Lemma 2.12, we conclude that solutions at E and E' intersect. We continue the process until the upper branch passes over B for the last time at a point $\bar{B}$. We conclude that solutions at B' and $\bar{B}$ intersect. We now resume moving forward in λ on both lower and upper branches. If another turning point is encountered, we repeat the above procedure. We conclude that solutions on upper and lower branches corresponding to the same λ intersect for all λ. This contradicts (3.5). We conclude that $w(r) > 0$ at any turning point, and the theorem follows.

Remark. After completing the proof, we conclude that the solution curve has exactly one turn, and $w(r) > 0$ there. This is simpler than the previous strategy (of [6], [11] and [13]) of directly proving that $w(r) > 0$.

Acknowledgement. It is a pleasure to thank S. Aizicovici and N. Pavel for a very well organized and stimulating workshop, and Y. Li and T. Ouyang for useful comments.

References

[1] M.G. Crandall and P.H. Rabinowitz, Bifurcation, perturbation of simple eigenvalues and linearized stability, *Arch. Rational Mech. Anal.* **52**, 161-180 (1973).

[2] E.N. Dancer, On the structure of solutions of an equation in catalysis theory when a parameter is large, *J. Differential Equations* **37**, 404-437 (1980).

[3] B. Gidas, W.-M. Ni and L. Nirenberg, Symmetry and related properties via the maximum principle, *Commun. Math. Phys.* **68**, 209-243 (1979).

[4] P. Korman, Solution curves for semilinear equations on a ball, *Proc. Amer. Math. Soc.* **125(7)**, 1997-2006 (1997).

[5] P. Korman, Exact multiplicity of positive solutions for a class of semilinear equations on a ball, Preprint.

[6] P. Korman, Y. Li and T. Ouyang, An exact multiplicity result for a class of semilinear equations, *Commun. PDE.* **22** (3&4), 661-684 (1997).

[7] P. Korman, Y. Li and T. Ouyang, Exact multiplicity results for boundary value problems with nonlinearities generalising cubic, *Proc. Royal Soc. Edinburgh* **126A**, 599-616 (1996).

[8] M.K. Kwong and L. Zhang, Uniqueness of the positive solution of $\Delta u + f(u) = 0$ in an annulus, *Differential and Integral Equations* **4**, 582-599 (1991).

[9] P.L. Lions, On the existence of positive solutions of semilinear elliptic equations, *SIAM Review* **24**, 441-467 (1982).

[10] K. Nagasaki and T. Suzuki, Spectral and related properties about the Emden-Fowler equation $-\Delta u = \lambda e^u$ on circular domains, *Math. Ann* **299**,1-15 (1994).

[11] T. Ouyang and J. Shi, Exact multiplicity of positive solutions for a class of semilinear problems, *J. Differential Equations* **146**, 121-156 (1998).

[12] L.A. Peletier and J. Serrin, Uniqueness of positive solutions of semilinear equations in R^n, *Arch. Rat. Mech. Anal.* **81**, 181-197 (1983).

[13] J. Wei, Exact multiplicity for some nonlinear elliptic equations in balls, *Proc. Amer. Math. Soc.* **125**, 3235-3242 (1997).

[14] H.F. Weinberger, A First Course in Partial Differential Equations, John Wiley & Sons (1965).

Optimal Control and Algebraic Riccati Equations under Singular Estimates for $e^{At}B$ in the Absence of Analyticity. Part I: The Stable Case

Irena Lasiecka and Roberto Triggiani*
Department of Mathematics
Kerchof Hall
University of Virginia
Charlottesville, VA 22904

Abstract

We study the quadratic optimal control problem over an infinite time horizon in the case where the free dynamics operator A and the control operator B yield a singular estimate for $e^{At}B$. Here, e^{At} is the corresponding s.c. semigroup which, by assumption, is *not* analytic. In this Part I, e^{At} is assumed (exponentially) stable. The resulting abstract model covers systems of coupled Partial Differential Equations, which possess an analytic *component*, but which are *not themselves analytic*. Two applications are given to hyperbolic/parabolic structural acoustic problems. Here a hyperbolic PDE (a wave equation within an acoustic chamber) is coupled with a parabolic PDE (the flexible wall which is either modeled by an elastic equation with structural damping [A-L.1] or else by a thermoelastic equation with no rotational inertia [L-T.4-6]).

1 Mathematical Setting and Formulation of the Control Problem

Dynamical Model. Let U (control), Y (state) be separable Hilbert spaces. In this paper, we consider the following abstract state equation

$$\dot{y}(t) = Ay(t) + Bu(t) + w(t) \text{ on, say, } [\mathcal{D}(A^*)]'; \; y(0) = y_0 \in Y, \tag{1.1}$$

subject to the following assumptions, to be maintained throughout the paper:

---

*Research partially supported by the National Science Foundation under Grant DMS-9804056.

(H.1) $A : Y \supset \mathcal{D}(A) \to Y$ is the infinitesimal generator of a strongly continuous (s.c.) semigroup e^{At} on Y. Moreover, e^{At} is (exponentially) uniformly stable: that is, there exist constants $M \geq 1, \omega > 0$, such that

$$\|e^{At}\|_{\mathcal{L}(Y)} \leq M e^{-\omega t}, \quad t \geq 0. \tag{1.2}$$

(H.2) B is a linear operator $U = \mathcal{D}(B) \to [\mathcal{D}(A^*)]'$, the dual space of the domain $\mathcal{D}(A^*)$, with respect to the pivot space Y. Here A^* is the adjoint of A in Y. Thus, e^{At} can be extended as a s.c. semigroup on $[\mathcal{D}(A^*)]'$ as well.

(H.3) There exist constants $0 \leq \gamma < 1$ and $T > 0$, such that the following singular estimate holds true:

$$\|e^{At}B\|_{\mathcal{L}(U;Y)} = \|B^* e^{A^*t}\|_{\mathcal{L}(Y;U)} \leq \frac{c_T}{t^\gamma}, \quad \forall\, 0 < t \leq T, \tag{1.3}$$

where $(Bu, v)_Y = (u, B^*v)_U$, $u \in U$, $v \in \mathcal{D}(B^*) \supset \mathcal{D}(A^*)$.

(H.4) The function w is a deterministic disturbance, satisfying

$$w \in L_2(0, \infty; Y) \tag{1.4}$$

to be kept fixed throughout.

Optimal Control Problem. With the dynamics (1.1), we associate the following quadratic cost functional over an infinite time horizon:

$$J_w(u, y) = \int_0^\infty [\|Ry(t)\|_Z^2 + \|u(t)\|_U^2] dt, \tag{1.5}$$

where $y(t) = y(t; y_0)$ is the solution of (1.1) due to $u(t)$, for fixed w, and, moreover,

(H.5)
$$R \in \mathcal{L}(Y; Z), \tag{1.6}$$

where Z is another Hilbert space. The corresponding Optimal Control Problem is:

> For fixed w as in (1.4), minimize $J_w(u, y)$ over all $u \in L_2(0, \infty; U)$, where y is the solution of (1.1) due to u (and w). (1.7)

Remark 1.1. What makes the above optimal control problem different from those studied in the literature [B-D-D-M], [L-T.1], [L-T.2], is, of course, the new set of assumptions imposed on model (1.1); in particular, the presence of the singular estimate of hypothesis (H.3) = (1.3), while, however, the semigroup e^{At} is only assumed to be strongly continuous. Explicitly, e^{At} is *not* assumed to be analytic. In [L-T.1, Chapters 1, 2, 6], the singular estimate (H.3) = (1.3) was also available in the treatment of those chapters; however, it was

Optimal Control and Algebraic Riccati Equations

so *a-posteriori*, as a *consequence* of the original assumption that the s.c. semigroup e^{At} be, moreover, analytic. Thus, writing $e^{At}B = (-A)^\gamma e^{At}(-A)^{-\gamma}B$ would at once yield estimate (1.3) under the two assumptions in those aforementioned chapters: (i) analyticity of the semigroup of e^{At} on Y, and (ii) the property $(-A)^{-\gamma}B \in \mathcal{L}(U;Y)$. In the present paper, by contrast, the s.c. semigroup e^{At} is *not* assumed to be analytic. The above set of assumptions on (1.1) are motivated by the structural acoustic problem, with hyperbolic/parabolic coupling, where these assumptions are, in fact, properties of the coupled dynamics: see the examples in Section 4 below, with both $\gamma < \frac{1}{2}$ (the easier case: see (3.1.16) below) and $\frac{1}{2} < \gamma < 1$ (the more challenging case). Similarly, in line with the structural acoustic problem, we are taken a pure distributed L_2-disturbance w (that is G = Identity in the notation of [L-T.2, Eqn. (6.1.1.1) of Chapter 6]. □

Lemma 1.1. Assume (H.1), (H.2), (H.3). Then, for any $0 < \omega_0 < \omega$, there exists a constant $k > 0$ (depending on M, c_T, ω, ω_0, T, γ, and computed below), such that

$$\left\|e^{At}B\right\|_{\mathcal{L}(U;Y)} = \left\|B^* e^{A^* t}\right\|_{\mathcal{L}(Y;U)} \leq k \frac{e^{-\omega_0 t}}{t^\gamma}, \quad \forall\, t > 0;\ 0 < \omega_0 < \omega. \tag{1.8}$$

Proof. Let $t > T$, the latter being the constant defined in (1.3). By using (1.3) and (1.2), we obtain the following estimate in the operator norm:

$$\left\|e^{At}B\right\|_{\mathcal{L}(U;Y)} = \left\|e^{A(t-T)}e^{AT}B\right\|_{\mathcal{L}(U;Y)} \leq M e^{\omega T} c_T \frac{e^{-\omega t}}{T^\gamma} \leq k \frac{e^{-\omega_0 t}}{t^\gamma}, \quad \forall\, t \geq T, \tag{1.9}$$

since $f(t) = \left(\frac{t}{T}\right)^\gamma e^{-(\omega-\omega_0)t}$ has its maximum at $t = \bar{t} = \gamma/(\omega - \omega_0)$ given by $\{\gamma/[T(\omega-\omega_0)]\}^\gamma e^{-\gamma}$ over $[0,\infty]$. For all $t \geq T$, we then have: $f(t) \leq f(\bar{t})$ if $\bar{t} > T$, and $f(t) \leq f(T)$ if $\bar{t} \leq T$. This identifies the constant k. □

Preliminaries. The solution to problem (1.1) is given by

$$y(t) = e^{At}y_0 + (Lu)(t) + (Ww)(t); \tag{1.10}$$

$$(Lu)(t) = \int_0^t e^{A(t-\tau)}Bu(\tau)d\tau \tag{1.11a}$$

$$: \text{continuous } L_2(0,\infty;U) \to L_2(0,\infty;Y) \tag{1.11b}$$

$$: \text{continuous } C([0,\infty];U) \to C([0,\infty];Y); \tag{1.11c}$$

$$(Ww)(t) = \int_0^t e^{A(t-\tau)}w(\tau)d\tau \tag{1.12a}$$

$$: \text{continuous } L_2(0,\infty;Y) \to L_2(0,\infty;Y) \cap C([0,\infty];Y). \tag{1.12b}$$

The regularity properties noted in (1.11b-c), (1.12b), due to the Young's inequality [S.1] and (H.3), will be formalized in Proposition 3.1.2 below. The L_2-adjoints of L and W, are, respectively,

$$(L^*f)(t) = \int_t^\infty B^* e^{A^*(\tau-t)} f(\tau) d\tau \tag{1.13a}$$

$$: \text{continuous } L_2(0,\infty;Y) \to L_2(0,\infty;U) \tag{1.13b}$$

$$: \text{continuous } C([0,\infty];Y) \to C([0,\infty];U); \tag{1.13c}$$

$$(W^*v)(t) = \int_t^\infty e^{A^*(\tau-t)} v(\tau) d\tau \tag{1.14a}$$

$$: \text{continuous } L_2(0,\infty;Y) \to L_2(0,\infty;Y) \tag{1.14b}$$

$$: \text{continuous } C([0,\infty];Y) \to C([0,\infty];Y). \tag{1.14c}$$

2 Statement of Main Results

The main result of this paper is the following theorem.

Theorem 2.1. Assume (H.1), (H.2), (H.3) = (1.3), (H.4) = (1.4), (H.5) = (1.6). Then:
(a1) For each $y_0 \in Y$ and $w \in L_2(0,\infty;Y)$ fixed, there exists a unique optimal pair $\{u_w^0(t;y_0), y_w^0(t;y_0)\}$ of the optimal control problem (1.5), (1.7) for the dynamics (1.1), or (1.10), which satisfies the following properties:

$$u_w^0(\,\cdot\,;y_0) \in L_2(0,\infty;U) \cap C([0,\infty];U); \tag{2.1}$$

$$y_w^0(\,\cdot\,;y_0) \in L_2(0,\infty;Y) \cap C([0,\infty];Y). \tag{2.2}$$

(a2) The operator $\Phi(t) \in \mathcal{L}(Y)$ defined by

$$\Phi(t)y_0 \equiv y_{w=0}^0(t;y_0) \in C([0,\infty];Y) \cap L_2(0,\infty;Y), \quad y_0 \in Y, \tag{2.3}$$

describes a s.c. semigroup on Y, which, moreover, is (exponentially) uniformly stable on Y.
(a3) The bounded operator $P \in \mathcal{L}(Y)$ defined on Y by

$$Px \equiv \int_0^\infty e^{A^*t} R^* R \Phi(t) x \, dt \in \mathcal{L}(Y), \quad x \in Y \tag{2.4}$$

is non-negative, self-adjoint on Y : $P = P^* \geq 0$,

$$(Px_1, x_2)_Y = \int_0^\infty \left\{ (R y_{w=0}^0(t;x_1), R y_{w=0}^0(t;x_2))_Z + (u_{w=0}^0(t;x_1), u_{w=0}^0(t;x_2))_U \right\} dt. \tag{2.5}$$

(a4) The gain operator B^*P is bounded $Y \to U$:

$$B^*P \in \mathcal{L}(Y;U). \tag{2.6}$$

Optimal Control and Algebraic Riccati Equations

(a5) The infinitesimal generator of the s.c. semigroup $\Phi(t)$ in (a2) is the operator $A_P = A - BB^*P$, with maximal domain $\mathcal{D}(A_P)$ on Y,

$$A_P = A - BB^*P, \quad \mathcal{D}(A_P) = \{x \in Y : [I - A^{-1}B(B^*P)]x \in \mathcal{D}(A)\}, \qquad (2.7a)$$

so that the s.c. uniformly stable semigroup $\Phi(t)$ may be also denoted as

$$\Phi(t) = e^{A_P t} = e^{(A-BB^*P)t}, \ t \geq 0; \quad \|e^{A_P t}\|_{\mathcal{L}(Y)} \leq M_P e^{-\omega_P t}, \ t \geq 0 \qquad (2.7b)$$

for some constants $M_P \geq 1, \omega_P > 0$.

(a6) The following singular estimate holds true: for any $0 < \omega_1 < \omega_P$ (where ω_P is defined in (2.7b)), there exists a constant k_P (depending on $M_P, \omega_P, \gamma, T, \omega_1$) such that

$$\|\Phi(t)B\|_{\mathcal{L}(U;Y)} = \|e^{A_P t}B\|_{\mathcal{L}(U;Y)} \leq k_P \frac{e^{-\omega_1 t}}{t^\gamma}, \ \forall\, t > 0. \qquad (2.8)$$

(a7) The operator P possesses the following additional regularity properties:

$$P \in \mathcal{L}(\mathcal{D}(A_P); \mathcal{D}(A^*)), \text{ or } A^*P \in \mathcal{L}(\mathcal{D}(A_P); Y); \qquad (2.9)$$

$$P \in \mathcal{L}(\mathcal{D}(A); \mathcal{D}(A_P^*)), \text{ or } A_P^*P \in \mathcal{L}(\mathcal{D}(A); Y). \qquad (2.10)$$

(a8) The operator P in (2.4) satisfies the following Algebraic Riccati Equation,

$$(A^*Px, z)_Y + (PAx, z)_Y + (Rx, Rz)_Z = (B^*Px, B^*Pz)_U$$

$$\text{either } \forall\, x, y \in \mathcal{D}(A); \text{ or } \forall\, x, y \in \mathcal{D}(A_P). \qquad (2.11)$$

(a9) Moreover, the operator P in (2.4) is the unique operator satisfying the ARE (2.11) within the class of non-negative, self-adjoint operators $\bar{P} = \bar{P}^* \in \mathcal{L}(Y)$ such that $B^*\bar{P} \in \mathcal{L}(Y;U)$ [a property enjoyed by P by (a4)].

(a10) The optimal pair $\{u_w^0(t; y_0), y_w^0(t; y_0)\}$ satisfies the following feedback relation:

$$u_w^0(t; y_0) = -B^*p_w(t; y_0) = -B^*[Py_w^0(t; y_0) + r_w(t)] \qquad (2.12)$$

$$= -B^*Py_w^0(t; y_0) - B^*r_w(t) \in L_2(0, \infty; U) \cap C([0, \infty]; U). \qquad (2.13)$$

$$B^*Py_w^0(\,\cdot\,; y_0) \text{ and } B^*r_w(t) \in L_2(0, \infty; U) \cap C([0, \infty]; U). \qquad (2.14)$$

Here, $r_w(t)$ is defined by

$$r_w(t) \equiv p_w(t; y_0 = 0) - Py_w^0(t; y_0 = 0) \in L_2(0, \infty; Y) \cap C([0, \infty]; Y), \qquad (2.15)$$

with

$$p_w(t; y_0) \equiv \int_t^\infty e^{A^*(\tau-t)} R^* R y_w^0(\tau; y_0)\, d\tau \in L_2(0, \infty; Y) \cap C([0, \infty]; Y),\ y_0 \in Y. \qquad (2.16)$$

(a11) Moreover, $r_w(t)$ is likewise given by

$$r_w(t) = \int_t^\infty e^{A_P^*(\tau-t)} Pw(\tau)d\tau, \text{ so that } B^* r_w(t) = \int_t^\infty B^* e^{A_P^*(\tau-t)} Pw(\tau)d\tau. \tag{2.17}$$

Thus $p_w(t; y_0)$ is the unique solution of the problem

$$\begin{cases} \dot{p}_w(t; y_0) = -A^* p_w(t; y_0) - R^* R y_w^0(t; y_0), \ t > 0, \ y_0 \in Y; & (2.18\text{a}) \\ \lim_{T \to \infty} p_w(T; y_0) = 0, & (2.18\text{b}) \end{cases}$$

while $r_w(t)$ is the unique solution of the problem

$$\begin{cases} \dot{r}_w(t) = -A_P^* r_w(t) - Pw(t), \ t > 0, \ \text{in } [\mathcal{D}(A)]'; & (2.19\text{a}) \\ \lim_{T \to \infty} r_w(T) = 0, & (2.19\text{b}) \end{cases}$$

(a12) Finally, the optimal dynamics may be rewritten as

$$\dot{y}_w^0(t; y_0) = A_P y_w^0(t; y_0) - BB^* r_w(t) + w \text{ in } [\mathcal{D}(A^*)]'; \tag{2.20}$$

$$y_w^0(t; y_0) = e^{A_P t} y_0 - \int_0^t e^{A_P(t-\tau)} BB^* r_w(\tau) d\tau + \int_0^t e^{A_P(t-\tau)} w(\tau) d\tau$$

$$\in L_2(0, \infty; Y) \cap C([0, \infty]; Y). \quad \square \tag{2.21}$$

Additional results are given in the sections below.

3 Proof of Theorem 2.1

3.1 Existence of a Unique Optimal Pair, Characterization, and Regularity Properties: Proof of (a1)

Proof of (a1). First, as already noted in (1.11b) and (1.12b), assumptions (H.1), (H.2), (H.3) guarantee the regularity properties $Lu \in L_2(0, \infty; Y)$ and $Ww \in L_2(0, \infty; Y)$, thus fulfilling the notion of well-posedness of the dynamics (1.1), needed in the cost functional (1.5). Next, the stability assumption (1.2) implies that both the Finite Cost Condition and the Detectability Condition (see Equations (2.1.12) and (2.1.13) of [L-T.2, Chapter 2]) are automatically satisfied. Then, the usual argument of, say, [L-T.2, Theorem 1.2.1.1 of Chapter 1, p. 14; Theorem 6.2.1.1 of Chapter 6, p. 563, etc.] yields that: there exists a unique optimal pair $\{u_w^0(t; y_0), y_w^0(t; y_0)\}$ in $L_2(0, \infty; U) \times L_2(0, \infty; Y)$ of the optimal control problem, satisfying the optimality condition. We obtain

Proposition 3.1.1. Assume (H.1) through (H.5). Then:

Optimal Control and Algebraic Riccati Equations

(i) for any $w \in L_2(0, \infty; Y)$ fixed, the optimal control problem (1.5), (1.7) for the dynamics (1.1), or (1.10), admits a unique optimal pair $\{u_w^0(\,\cdot\,; y_0), y_w^0(\,\cdot\,; y_0)\}$ satisfying the optimality condition

$$u_w^0(\,\cdot\,; y_0) = -L^* R^* R y_w^0(\,\cdot\,; y_0) \in L_2(0, \infty; U). \tag{3.1.1}$$

(ii) the optimal pair is given explicitly by the following formulas:

$$u_w^0(\,\cdot\,; y_0) = -[I + L^* R^* R L]^{-1} L^* R^* R \left[e^A \cdot y_0 + Ww\right] \in L_2(0, \infty; U) \tag{3.1.2a}$$

$$= u_{w=0}^0(\,\cdot\,; y_0) + u_w^0(\,\cdot\,; y_0 = 0); \tag{3.1.2b}$$

$$y_w^0(\,\cdot\,; y_0) = [I + L L^* R^* R]^{-1} \left[e^A \cdot y_0 + Ww\right] \in L_2(0, \infty; Y) \tag{3.1.3a}$$

$$= \{I - L[I + L^* R^* R L]^{-1} L^* R^* R\} \left[e^A \cdot y_0 + Ww\right] \tag{3.1.3b}$$

$$= y_{w=0}^0(\,\cdot\,; y_0) + y_w^0(\,\cdot\,; y_0 = 0), \tag{3.1.3c}$$

$$R y_w^0(\,\cdot\,; y_0) = [I + R L L^* R^*]^{-1} R \left[e^A \cdot y_0 + Ww\right] \in L_2(0, \infty; Z) \tag{3.1.4a}$$

$$= R y_{w=0}^0(\,\cdot\,; y_0) + R y_w^0(\,\cdot\,; y_0 = 0), \tag{3.1.4b}$$

where the inverse operators in the above formulas are well-defined as bounded operators on all of $L_2(0, \infty; \,\cdot\,)$ by (1.11b), (1.12b), (1.13b), (1.14b) (for the inverse occurring in (3.1.3a), see [L-T.2, Chapter 2, Appendix 2A]). Moreover, with $y_0 \in Y$, the corresponding optimal dynamics is

$$y_w^0(t; y_0) = e^{At} y_0 + \{L u_w^0(\,\cdot\,; y_0)\}(t) + \{Ww\}(t) \in L_2(0, \infty; Y). \tag{3.1.5}$$

(iii) For $y_0 \in Y$, the optimal cost satisfies the following relations

$$J_w^0(y_0) \equiv J_w(u_w^0(\,\cdot\,; y_0), y_w^0(\,\cdot\,; y_0)) = J_{w=0}^0(y_0) + J_w^0(y_0 = 0) + \mathcal{X}_w(y_0); \tag{3.1.6}$$

$$J_{w=0}^0(y_0) = \left(Re^A \cdot y_0, [I + R L L^* R^*]^{-1} Re^A \cdot y_0\right)_{L_2(0,\infty;Z)}; \tag{3.1.7}$$

$$J_w^0(y_0 = 0) = \left(w, W^* R^* [I + R L L^* R^*]^{-1} R W w\right)_{L_2(0,\infty;Y)}; \tag{3.1.8}$$

$$\mathcal{X}_w(y_0) = 2 \left(Re^A \cdot y_0, [I + R L L^* R^*]^{-1} R W w\right)_{L_2(0,\infty;Z)}$$

$$= \text{linear in } w. \quad \square \tag{3.1.9}$$

We next prove the additional regularity $C([0, \infty]; \,\cdot\,)$ for u_w^0 and y_w^0. To this end, and generally to build further a theory, as described by Theorem 2.1, we make the preliminary observation that the abstract model of the present chapter differs critically from the two main settings of the literature [B-D-D-M], [L-T.1], [L-T.2]. Indeed, neither is the s.c. semigroup e^{At} analytic—as in [L-T.2, Chapters 1, 2, and 6]—nor does the pair $\{A, B\}$ satisfy the 'abstract trace condition' as in [L-T.2, Chapters 7, 9]. The key feature of the present new setting is assumption (H.3) = (1.3) on $0 < t \leq T$—hence, its consequence (1.8) for all $t > 0$ under the stability hypothesis (1.2). As noted in Remark 1.1, this makes the present setting a shadowy resemblance akin to the 'analytic case' of [L-T.2, Chapters 1, 2, or 6]. This observation provides the key guide to the treatment that follows.

First, all the critical smoothing properties, in the analytic or parabolic case, of the operators L, L^*—beginning with the preliminary regularity $L_2 \to L_2$ in (1.11b), (1.13b)—*continue to hold true* under assumptions (H.1), (H.2), (H.3). This is so, since it is the singular estimate (1.8) on the kernels of these operators, which plays the key role in these results in the parabolic or analytic chapters of [L-T.2]. Thus, we still have available, under the present setting (H.1), (H.2), (H.3), the following properties:

(i) The regularizing properties of the operators L and L^* on L_p-spaces ([L-T.2, Theorem 1.4.4.3 of Chapter 1, p. 38 for T finite; Theorem 2.3.5.1 of Chapter 2, p. 143 for $T = \infty$, as well as Theorem 6.9.1 and Theorem 6.23.1 of Chapter 6, p. 590 and p. 620, the latter for $T = \infty$]). This is recorded as Proposition 3.1.2 below.

(ii) The regularizing properties of the operators L and L^* on the space ${}_\gamma C([\ ,\ ]';\cdot)$ with singularity on the left ([L-T.2, Proposition 1.4.5.4 of Chapter 1, p. 49]). This is recorded in Proposition 3.3.1 below.

We collect the conclusions reached in (i) above in the next statement.

Proposition 3.1.2. Assume (H.1), (H.2), (H.3), (H.5). Then, with reference to the operators L and L^* in (1.11a), (1.13a), we have:

(i)
$$L\ :\ \text{continuous}\ L_2(0,\infty;U) \to L_2(0,\infty;Y) \qquad (3.1.10)$$
$$\text{continuous}\ C([0,\infty];U) \to C([0,\infty];Y); \qquad (3.1.11)$$

(ii)
$$L^*\ :\ \text{continuous}\ L_2(0,\infty;Y) \to L_2(0,\infty;U) \qquad (3.1.12)$$
$$\text{continuous}\ C([0,\infty];Y) \to C([0,\infty];U); \qquad (3.1.13)$$

(iii)
$$[I + L^*R^*RL]^{-1} \in \mathcal{L}(L_2(0,\infty;U)) \cap \mathcal{L}(C([0,\infty];U)); \qquad (3.1.14)$$

(iv)
$$[I + LL^*R^*R]^{-1} \in \mathcal{L}(C([0,\infty];Y)) \cap \mathcal{L}(L_2(0,\infty;Y)); \qquad (3.1.15)$$

(v)
$$L : \text{continuous}\ L_2(0,\infty;U) \to C([0,\infty];Y),\ \text{if}\ \gamma < \frac{1}{2}; \qquad (3.1.16)$$

Optimal Control and Algebraic Riccati Equations

(vi)
$$L : \text{continuous } L_p(0,\infty;U) \to C([0,\infty];Y), \text{ if } p > 1/(1-\gamma); \qquad (3.1.17)$$

(vii)
$$L : \text{continuous } L_{r_1}(0,\infty;U) \to L_{r_2}(0,\infty;Y), \qquad (3.1.18)$$

where r_1 is any positive number satisfying $r_1 < 2/(2\gamma - 1)$, where $2/(2\gamma - 1) > 2$, for $\frac{1}{2} \leq \gamma < 1$; for $0 \leq \gamma < \frac{1}{2}$ we may take $r_1 = \infty$; and r_2 is any positive number satisfying $r_2 < 2/(4\gamma - 3)$, where $2/(4\gamma - 3) > r_1$ for $\frac{3}{4} \leq \gamma < 1$; for $0 \leq \gamma < \frac{3}{4}$, we may take $r_2 = \infty$.

(viii) there exists a positive integer $n_0(\gamma)$ depending on γ, such that for all positive integers $n \geq n_0(\gamma)$, we have

$$(L^*R^*RL)^n : L_2(0,\infty;U) \to C([0,\infty];Y). \qquad \square \qquad (3.1.19)$$

Regarding properties (3.1.14) and (3.1.15) in the space C—that is, $[I + L^*R^*RL]^{-1} \in C([0,\infty];U)$ and $[I + LL^*R^*R]^{-1} \in C([0,\infty];Y)$—these are achieved by the usual boot-strap argument as in, say, [L-T.2, Corollary 2.3.5.2 in Chapter 2, p. 145, and Corollary 6.23.2 of Chapter 6, p. 621]. This plays alternatively between the optimality condition (3.1.1) and the optimal dynamics (3.1.5), starting with the $L_2(0,\infty;\,\cdot\,)$-regularity for u_w^0 and y_w^0 as in (3.1.2a), (3.1.3b), and using the smoothing properties of L and L^*, while $Ww \in C([0,\infty];Y)$ by (1.12b).

Proposition 3.1.3. Assume (H.1) through (H.5). Then, for any $y_0 \in Y$ and any $w \in L_2(0,\infty;Y)$ (fixed), the optimal pair established in Proposition 3.1.1 satisfies the additional regularity properties

$$u_w^0(\,\cdot\,;y_0) \in C([0,\infty];U); \quad y_w^0(\,\cdot\,;y_0) \in C([0,\infty];Y). \qquad (3.1.20)$$

Proof. Apply (3.1.14) and (3.1.15) in the space C to the explicit formulas (3.1.2a) for u_w^0 and (3.1.3a) for y_w^0, where $[e^{A\,\cdot}\,y_0 + Ww] \in C([0,\infty];Y)$ by (1.2) and (1.12b). Of course, (1.13c) for L^* is also used. This way, the regularity properties in (3.1.20) are achieved. $\square$

The proof of property (a1) of Theorem 2.1 is complete.

3.2 The Operator $\Phi(t)$; The Functions p_w, the Operator P: Proof of (a2), (a3), (a4), (a5)

The Operator $\Phi(t)$. This was defined in (2.3) by (recall (3.1.3a)):

$$\Phi(t)y_0 = y_{w=0}^0(t;y_0) = \left\{[I + LL^*R^*R]^{-1}\left[e^{A\,\cdot}\,y_0\right]\right\}(t)$$

$$\in L_2(0,\infty;Y) \cap C([0,\infty];Y). \qquad (3.2.1)$$

The regularity noted in (3.2.1) follows from (3.1.2a) and (3.1.3a) for L_2, as well as (3.1.20) for C.

Proposition 3.2.1. (property (a2)) Assume (H.1) through (H.5). Then the operator $\Phi(t)$ defined in (3.2.1) is a s.c. semigroup on Y, $t \geq 0$, which, moreover, is (exponentially) uniformly stable.

Proof. Because of the regularity $\Phi(t)y_0 \in C([0, \infty]; Y)$ in (3.2.1) already established, it is the semigroup property that needs to be proved next. But this can be done precisely as in past chapters, several times, as a consequence of the optimality condition: see [L-T.2, Proposition 1.4.3.1(ii), Eqn. (1.4.3.3), p. 32 (evolution property in the case $T < \infty$) of Chapter 1; Lemma 2.3.2.1, p. 132, or Theorem 2.3.6.1, p. 146, of Chapter 2; Theorem 6.10.1, p. 593, and Theorem 6.12.1, p. 596 of Chapter 6; Lemma 6.24.2, p. 622 for $T = \infty$ in the stable case, Chapter 6]. See also [L-T.2, Theorem 6.25.1 of Chapter 6]. Uniform (exponential) stability follows, as usual, by Datko's Theorem, since $\Phi(t)y_0 \in L_2(0, \infty; Y)$ for all $y_0 \in Y$. □

The Functions $p_w(t; y_0)$, $r_w(t)$, and the Operator $P \in \mathcal{L}(Y)$. For $y_0 \in Y$, we define (as in [L-T.2, (6.2.2.1) of Section 6.2.2, p. 564, of Chapter 6]) the function

$$p_w(t; y_0) = \int_t^\infty e^{A^*(\tau-t)} R^* R y_w^0(\tau; y_0) d\tau, \quad y_0 \in Y, \tag{3.2.2}$$

which is the unique solution of the problem

$$\begin{cases} \dot{p}_w(t; y_0) = -A^* p_w(t; y_0) - R^* R y_w^0(t; y_0); & (3.2.3a) \\ \lim_{T \uparrow \infty} p_w(T; y_0) = 0, & (3.2.3b) \end{cases}$$

with zero initial condition at $T = \infty$. Moreover, we introduce the operator $P \in \mathcal{L}(Y)$ (see [L-T.1], [L-T.2]),

$$P y_0 = \int_0^\infty e^{A^* t} R^* R \Phi(t) y_0 = \int_t^\infty e^{A^*(\tau-t)} R^* R \Phi(\tau - t) y_0 d\tau. \tag{3.2.4}$$

Lemma 3.2.2. (property (a4)) Assume (H.1) through (H.5). Then, with reference to (3.2.2) and (3.2.4), we have for $y_0 \in Y$:
(i)

$$p_w(t; y_0) = p_{w=0}(t; y_0) + p_w(t; y_0 = 0) \in L_2(0, \infty; Y) \cap C([0, \infty]; Y); \tag{3.2.5}$$

$$p_{w=0}(t; y_0) = \int_t^\infty e^{A^*(\tau-t)} R^* R y_{w=0}^0(\tau; y_0) d\tau \tag{3.2.6}$$

$$= \int_t^\infty e^{A^*(\tau-t)} R^* R \Phi(\tau - t) \Phi(t) y_0 d\tau$$

$$= P\Phi(t)y_0 \in L_2(0,\infty;Y) \cap C([0,\infty];Y). \tag{3.2.7}$$

(ii)

$$-u_w^0(t;y_0) = \{L^* R^* R y_w^0(\,\cdot\,;y_0)\}(t) = \int_t^\infty B^* e^{A^*(\tau-t)} R^* R y_w^0(\tau;y_0) d\tau \tag{3.2.8}$$

$$= B^* p_w(t;y_0) \in L_2(0,\infty;U) \cap C([0,\infty];U). \tag{3.2.9}$$

(iii) $P \in \mathcal{L}(Y)$ and moreover B^*P: continuous $Y \to U$:

$$\begin{cases} B^*P \in \mathcal{L}(Y;U); & (3.2.10) \\ B^*Px = \int_0^\infty B^* e^{A^*t} R^* R \Phi(t) x\, dt = B^* p_{w=0}(0;x), \ x \in Y. & (3.2.11) \end{cases}$$

(iv)

$$-u_{w=0}^0(t;y_0) = B^* p_{w=0}(t;y_0) = B^* P \Phi(t) y_0 \in L_2(0,\infty;U) \cap C([0,\infty];U). \tag{3.2.12}$$

Proof. (i) The decomposition of p_w in (3.2.5) is a consequence of using the decomposition of y_w^0 in (3.1.3c) in the definition of (3.2.2). Moreover, the regularity of p_w noted in (3.2.5) follows from using the stability assumption (1.2) and the regularity (3.1.3a) [no need of (3.1.20)] for y_w^0 in the defining integral (3.2.2).

(ii) The steps in (3.2.8), (3.2.9) are self-explanatory, once one recalls the optimality condition (3.1.1), and then (1.13a) for L^* and (3.2.2) for p_w. The regularity noted in (3.2.9) is the one established in (3.1.1) and (3.1.20) for u_w^0.

(iii) The operator P defined in (3.2.4) is plainly in $\mathcal{L}(Y)$, by the stability hypothesis (1.2) and the regularity (3.1.3a) of y_w^0. It is, however, at the level of establishing $B^*P \in \mathcal{L}(Y;U)$ from its definition (3.2.11) that we critically use the assumed singular estimate (1.3) on $0 < t \leq T$, as propagated to all $t > 0$ in (1.8): since $0 \leq \gamma < 1$ by hypothesis, then (3.2.11) yields via (1.8) and (3.2.1) in the C-space:

$$\|B^*Px\|_Y \leq \int_0^\infty \|B^* e^{A^*t}\| \|R^* R \Phi(t) x\| dt \tag{3.2.13}$$

$$\text{(by (1.8), (3.2.1))} \leq k \left(\int_0^\infty \frac{e^{-\omega_0 t}}{t^\gamma} dt \right) \|\Phi(\,\cdot\,)x\|_{C([0,\infty];Y)} \tag{3.2.14}$$

$$\leq \text{const } \|x\|_Y, \quad \forall\, x \in Y, \tag{3.2.15}$$

and (3.2.15) proves (3.2.10), as desired.

(iv) We return to $p_{w=0}(t; y_0) = P\Phi(t)y_0$ in (3.2.7), apply B^* on both sides and obtain (3.2.12), where the noted regularity follows from (3.2.10) on B^*P and (3.2.1) on $\Phi(t)y_0$. □

Property (a3). The non-negative, self-adjoint property of P is contained in identity (2.5): this is then proved in the same way as, say, [L-T.2, Proposition 1.4.4.8 of Chapter 1, p. 44; Corollary 6.26.1.2, Eqn. (6.26.1.7) of Chapter 6, p. 628].

Property (a5). One only needs to show that the infinitesimal generator of the s.c. (exponential) uniformly stable semigroup $\Phi(t)$ is, in fact, $A_P = A - BB^*P$. But this follows, as usual (see the proof of [L-T.2, Theorem 2.3.8.1, p. 150 of Chapter 2]), as a consequence of the optimal dynamics (3.1.5) for $w = 0$, via (3.2.1) and (3.2.12) for $u^0_{w=0}$. □

3.3 Singular Estimate for $e^{A_P t}B$: Proof of (a6)

Orientation. As in the proof of, say, [L-T.2, Theorem 6.26.3.1], in order to derive that the operator P defined in (3.2.4) satisfies the Algebraic Riccati Equation (2.17) on $\mathcal{D}(A)$, we need to differentiate strongly $e^{A_P t}$ on $\mathcal{D}(A)$. This, in turn, is accomplished if we can establish the same singular estimate for $e^{A_P t}B$ that holds true under (H.3) for $e^{At}B$. The resulting Theorem 3.3.2 below is a delicate new point of the present development, which was not explicitly needed in our treatment of the abstract analytic or parabolic case of [L-T.2, Chapters 1, 2, and 6].

We begin by collecting the conclusions reached in point (1ii) of Section 3.1.

Proposition 3.3.1. Assume (H.1) through (H.5). With reference to the operators L and L^* in (1.11) and (1.13), and recalling that $0 \leq \gamma < 1$ as postulated in assumption (H.3), we have:

(i) Let $0 < r < 1$. Then for any $0 < T \leq \infty$,

$$L : \text{continuous } {}_rC([0,T];U) \to_{(r+\gamma-1)} C([0,T];Y). \tag{3.3.1}$$

(ii) Let $r > 0$, and $\epsilon > 0$ arbitrary,

$$L^* : \text{continuous } {}_rC([0,T];Y) \to_{(r+\gamma-1+\epsilon)} C([0,T];U). \tag{3.3.2}$$

(iii) Let $0 < r < 1$. Then, there exists a positive integer $m = m(r)$ such that

$$(L^*R^*RL)^m : \text{continuous } {}_rC([0,T];U) \to C([0,T];U). \tag{3.3.3}$$

(iv)
$$[I + L^*R^*RL]^{-1} \in \mathcal{L}({}_\gamma C([0,T];Y)). \tag{3.3.4}$$

(v)
$$[I + LL^*R^*R]^{-1} = I - L[I + L^*R^*RL]^{-1}L^*R^*R \in \mathcal{L}({}_\gamma C([0,T];Y)), \tag{3.3.5}$$

Optimal Control and Algebraic Riccati Equations

where we recall that, if X is a Banach space, then

$$_rC([0,T];X) \equiv \left\{ f(t) \in C((0,T];X) : \|f\|_{rC([0,T];X)} = \sup_{0<t\leq T} t^r\|f(t)\|_X < \infty \right\}, \quad (3.3.6)$$

see [L-T.2, p. 3, p. 46 of Chapter 1].

Proof. (i), (ii), (iii). Properties (3.3.1), (3.3.2), (3.3.3) are precisely the properties established in [L-T.2, Proposition 1.4.5.4, p. 49, of Chapter 1], in the analytic case. As argued in point (1ii) of Section 3.1, the proofs given there continue to hold true on any $[0,T]$, $0 < T < \infty$, under assumption (H.3): this is so since the key role in these proofs in precisely the singular estimate (1.3) for the kernels of the operators L and L^*. Moreover, in view of the stability assumption (1.2), we may also take the case $T = \infty$ in these three results (i), (ii), (iii).

(iv) The proof of property (3.3.4) on the space $_rC([0,T]; \cdot)$ with singularity on the left is nothing but the exact counterpart of the property

$$[I + L^*R^*RL]^{-1} \in \mathcal{L}(C_\gamma([0,T];U), \quad (3.3.7)$$

which is [L-T.2, Theorem 1.4.4.4, p. 40, of Chapter 1] on the space $C_\gamma([0,T];U)$ with singularity on the right. Indeed, property (3.3.7) was proved there as a consequence of two main ingredients: that given $0 < \gamma < 1$, there is a corresponding positive integer n (depending on γ) such that

$$(L^*R^*RL)^n \; : \; \text{continuous } C_\gamma([0,T];U) \to C([0,T];U); \quad (3.3.8)$$

$$\; : \; \text{continuous } L_2(0,T;U) \to C([0,T];U) \quad (3.3.9)$$

(see [L-T.2, Corollary 1.4.4.2, p. 38, and Theorem 1.4.4.3(v), Eqn. (1.4.4.19), p. 39]). Then, (3.3.8), (3.3.9) $\Rightarrow$ (3.3.7) in the proof of [L-T.2, Theorem 1.4.4.4].

But the counterpart of property (3.3.8) this time on $_rC([0,T];U)$ with singularity on the left continues to hold true in the present setting under assumption (H.3), by (3.3.3). The same holds true for property (3.3.9), which was noted in (3.1.19). Then, the same argument given in the proof of [L-T.2, Theorem 1.4.4.4] yields (3.3.4).

(v) This follows from property (iv). □

Theorem 3.3.2. Assume (H.1) through (H.5). Then, with reference to the semigroup $\Phi(t) = e^{A_P t}$ (feedback semigroup when $w \equiv 0$), we have:

(i)

$$\left\|e^{A_P t}B\right\|_{\mathcal{L}(U;Y)} = \left\|B^* e^{A_P^* t}\right\|_{\mathcal{L}(Y;U)} \leq c_T \frac{1}{t^\gamma}, \; 0 < t \leq T; \quad (3.3.10)$$

(ii) for any $0 < \omega_1 < \omega_P$, there exists a constant $k_1 > 0$ (depending on M_P, ω_P, ω_1, T, γ) such that

$$\left\|e^{A_P t}B\right\|_{\mathcal{L}(U;Y)} = \left\|B^* e^{A_P^* t}\right\|_{\mathcal{L}(Y;U)} \leq k_1 \frac{e^{-\omega_1 t}}{t^\gamma}, \; \forall \, t > 0, \; 0 < \omega_1 < \omega_P. \quad (3.3.11)$$

Proof. (i) We return to the explicit formula (3.2.1) for $w = 0$, which we apply with Bx, $x \in Y$, in place of y_0, thus obtaining

$$\Phi(t)Bx \equiv e^{A_P t}Bx = [I + LL^*R^*R]^{-1}[e^{A\cdot}Bx], \quad x \in Y. \tag{3.3.12}$$

By assumption (H.3) = (1.3), we have $e^{A\cdot}Bx \in_\gamma C([0,T];Y)$ continuously on $x \in Y$. We then invoke Proposition 3.3.1(v), Eqn. (3.3.5) on (3.3.12), and then conclude that

$$e^{A_P t}Bx \in {_\gamma}C([0,T];Y)$$

continuously on $x \in Y$. But this, in view of (3.3.6), means precisely estimate (3.3.10).

(ii) Part (i), Eqn. (3.3.10), implies Part (ii), Eqn. (3.3.11) as in the proof of Lemma 1.1. □

As a consequence of Theorem 3.3.2, we obtain

Proposition 3.3.3. Assume (H.1) through (H.5). Then, the s.c. semigroup $\Phi(t) = e^{A_P t}$ in (3.2.1) is strongly differentiable on $\mathcal{D}(A)$: that is, more precisely, if $x \in \mathcal{D}(A)$, then for $t > 0$:

$$\frac{d}{dt}e^{A_P t}x = e^{A_P t}A_P x = e^{A_P t}(A - BB^*P)x \tag{3.3.13}$$

$$= e^{A_P t}Ax - e^{A_P t}B(B^*Px) \in Y, \; x \in \mathcal{D}(A), \; t > 0, \tag{3.3.14}$$

and in fact,

$$\left\|\frac{d}{dt}e^{A_P t}x\right\|_Y \leq M_P e^{-\omega_P t}\|Ax\| + k_1 \frac{e^{-\omega_1 t}}{t^\gamma}\|B^*P\|_{\mathcal{L}(Y;U)}\|x\|_Y, \; t > 0. \tag{3.3.15}$$

Proof. The steps in (3.3.13), (3.3.14) are self-explanatory. Of course, (3.3.13) makes sense, at least, in $[\mathcal{D}(A_P^*)]'$. The point is that, under present assumptions, (3.3.14) makes sense actually in Y. Estimate (3.3.15) follows from (3.3.14) by invoking (2.7b) (already proved in Section 3.2) and (3.3.11) of Theorem 3.3.2 on $e^{A_P t}$ and $e^{A_P t}B$, respectively, as well as (3.2.10) of Lemma 3.2.2(iii) on B^*P. □

3.4 Additional Regularity Properties of P: Proof of (a7)

We shall recall [from [F-L-T.1], [L-T.3], [L-T.2, Chapter 11, vol. 3]] more precise versions of the regularity properties (2.9) and (2.10) of Theorem 2.1, (a7). For these, assumption (H.3) is not really needed.

Proposition 3.4.1. Assume (H.1), (H.2), (H.5). Then the following identities hold true:

Optimal Control and Algebraic Riccati Equations

(i)
$$A^*Px = -R^*Rx - PA_Px \in Y, \ \forall\, x \in \mathcal{D}(A_P), \qquad (3.4.1a)$$

and so
$$A^*P : \text{ continuous } \mathcal{D}(A_P) \to Y. \qquad (3.4.1b)$$

(ii)
$$A_P^*Px = -R^*Rx - PAx \in Y, \ \forall\, x \in \mathcal{D}(A), \qquad (3.4.2a)$$

and so
$$A_P^*P : \text{ continuous } \mathcal{D}(A) \to Y. \qquad (3.4.2b)$$

Proof. This result was established already in [F-L-T.1] without assuming the stability hypothesis (H.3). Here we provide the counterpart proof under (H.3).

(i) Let $x \in \mathcal{D}(A_P)$. Then, recalling (3.2.4), we integrate by parts and obtain

$$A^*Px \ = \ \int_0^\infty A^* e^{A^*t} R^* R e^{A_P t} x \, dt \qquad (3.4.3)$$

$$= \ \left[e^{A^*t} R^* R e^{A_P t} x \right]_{t=0}^{t=\infty} - \int_0^\infty e^{A^*t} R^* R e^{A_P t} A_P x \, dt \qquad (3.4.4)$$

(by (3.2.4)) $\quad = \ -R^*Rx - PA_Px \in Y, \qquad (3.4.5)$

recalling the exponential stability in (H.3) = (1.2) and in (2.7) in (a4) (already proved in Section 3.2) at $t = \infty$. Then (3.4.5) establishes (3.4.1a) from which (3.4.1b) follows by the closed graph theorem: indeed, the bounded operator P is acted upon by the closed operator A^* which is boundedly invertible, so that A^*P is closed [K.1, p. 167].

(ii) From (3.2.4) we obtain for $x, y \in Y$:

$$(Px, y)_Y = \left(\int_0^\infty e^{A^*t} R^* R e^{A_P t} x \, dt, y \right)_Y = \left(x, \int_0^\infty e^{A_P^* t} R^* R e^{At} y \, dt \right)_Y, \qquad (3.4.6)$$

and so
$$P^*y = \int_0^\infty e^{A_P^* t} R^* R e^{At} y \, dt, \ y \in Y. \qquad (3.4.7)$$

But $P = P^*$ by part (a3) (already proved in Section 3.2). Hence

$$Px = \int_0^\infty e^{A_P^* t} R^* R e^{At} x \, dt, \ x \in Y, \qquad (3.4.8)$$

which provides an alternative expression for P over (3.2.4). Next, we apply to (3.4.8) the counterpart argument to that employed in (i). For $x \in \mathcal{D}(A)$, we integrate by parts and obtain

$$A_P^* Px = \int_0^\infty A_P^* e^{A_P^* t} R^* R e^{At} x \, dt \qquad (3.4.9)$$

$$= \left[e^{A_P^* t} R^* R e^{At} x \right]_{t=0}^{t=\infty} - \int_0^\infty e^{A_P^* t} R^* R e^{At} A x \, dt \qquad (3.4.10)$$

$$\text{(by (3.4.8))} \quad = -R^* R x - P A x \in Y. \qquad (3.4.11)$$

Thus, (3.4.11) establishes (3.4.2a) from which (3.4.2b) follows by the closed graph theorem. □

3.5 The Operator P Satisfies the ARE on $\mathcal{D}(A)$: Proof of (a8)

We proceed as, say, in the proof of [L-T.2, Theorem 2.3.9.1 of Chapter 2, Theorem 6.15.1 and Theorem 6.26.3.1 of Chapter 6] in the present circumstances, by critically using Proposition 3.3.3.

Theorem 3.4.1. Assume (H.1) through (H.5). Then, the operator $P \in \mathcal{L}(Y)$ defined in (3.2.4)—which was noted in property (a3) to be non-negative, self-adjoint: $P = P^* \geq 0$—satisfies the following Algebraic Riccati Equation on $\mathcal{D}(A)$, or on $\mathcal{D}(A_P)$: that is

$$(A^* Px, z)_Y + (PAx, z)_Y + (Rx, Rz)_Z = (B^* Px, B^* Pz)_U$$

$$\forall \, x, z \in \mathcal{D}(A); \text{ or else } \forall \, x, z \in \mathcal{D}(A_P). \qquad (3.5.1)$$

Proof. Let, at first, $x, z \in Y$. From the definition (3.2.4) with $\Phi(t) = e^{A_P t}$, we have

$$(Px, z)_Y = \int_0^\infty \left(R e^{A_P t} x, R e^{At} z \right)_Z dt = \int_t^\infty \left(R e^{A_P(\tau-t)} x, R e^{A(\tau-t)} z \right)_Z d\tau, \; x, y \in Y. \qquad (3.5.2)$$

We next specialize to $x, z \in \mathcal{D}(A)$ and differentiate (3.5.2) in t. We obtain, recalling Proposition 3.3.3,

$$0 = -\int_t^\infty \left(R e^{A_P(\tau-t)} A_P x, R e^{A(\tau-t)} z \right)_Z dt$$

$$- \int_t^\infty \left(R e^{A_P(\tau-t)} x, R e^{A(\tau-t)} A z \right)_Z dt - (Rx, Rz)_Z, \; x, z \in \mathcal{D}(A). \qquad (3.5.3)$$

At first, we may consider the first integral in (3.5.3) as an improper integral, since (3.3.14) applies for $\tau > t$. However, invoking (3.5.2) and $A_P = A - BB^* P$, we rewrite (3.5.3) first as

$$(PA_P x, z)_Y + (Px, Az)_Y + (Rx, Rz)_Z = 0, \quad x, z \in \mathcal{D}(A). \qquad (3.5.4)$$

Optimal Control and Algebraic Riccati Equations

which is well-defined, by recalling (3.4.2a) of Proposition 3.4.1, so that $A_P^* P z \in Y$, and then, since $A_P = A - BB^*P$, as

$$(P(A - BB^*P)x, z)_Y + (Px, Az)_Y + (Rx, Rz)_Z = 0, \quad x, z \in \mathcal{D}(A). \tag{3.5.5}$$

Finally, recalling that $B^*P \in \mathcal{L}(Y;U)$ by (3.2.10), we obtain from (3.5.4),

$$(PAx, z)_Y + (A^*Px, z)_Y + (Rx, Rz)_Z = (B^*Px, B^*Pz)_U, \tag{3.5.6}$$

where each term is well-defined and Theorem 3.4.1 is proved for $x, z \in \mathcal{D}(A)$.

Let now $x, z \in \mathcal{D}(A_P)$. Then, the steps through (3.5.4) hold plainly true: finally (3.5.4) again yields (3.5.5), this time by virtue of the property $A^*P \in \mathcal{L}(\mathcal{D}(A_P); Y)$ claimed in (3.4.1) of Proposition 3.4.1. □

3.6 The Function r_w. Feedback Synthesis of Optimal Control: Proof of (a10), (a11)

Proposition 3.6.1. (property (a10)) Assume (H.1) through (H.5). Then, with reference to the function $p_w(t; y_0)$ defined in (3.2.2), we have:

(i)
$$p_w(t; y_0) = P\Phi(t)y_0 + p_w(t; y_0 = 0) = Py_{w=0}^0(t; y_0) + p_w(t; y_0 = 0) \tag{3.6.1}$$

$$= Py_w^0(t; y_0) + r_w(t) \in L_2(0, \infty; Y) \cap C([0, \infty]; Y), \tag{3.6.2}$$

where the function $r_w(t)$ is defined by

$$r_w(t) = p_w(t; y_0 = 0) - Py_w^0(t; y_0 = 0) \in L_2(0, \infty; Y) \cap C([0, \infty]; Y); \tag{3.6.3}$$

(ii) for $y_0 \in Y$, the optimal control is written in feedback synthesis as

$$-u_w^0(t; y_0) = B^* p_w(t; y_0) = B^*[Py_w^0(t; y_0) + r_w(t)] \tag{3.6.4}$$

$$= B^* Py_w^0(t; y_0) + B^* r_w(t) \in L_2(0, \infty; U) \cap C([0, \infty]; U), \tag{3.6.5}$$

where

$$B^* Py_w^0(\,\cdot\,; y_0) \text{ and } B^* r_w(t) \in L_2(0, \infty; U) \cap C([0, \infty]; U); \tag{3.6.6}$$

(iii) the optimal dynamics may thus be rewritten as

$$\dot{y}_w^0(t; y_0) = A_P y_w^0(t; y_0) - BB^* r_w(t) + w \text{ in } [\mathcal{D}(A^*)]', \tag{3.6.7}$$

i.e., in the sense that

$$(\dot{y}_w^0(t; y_0), x)_Y = ([I - A^{-1}BB^*P]y_w^0(t; y_0), A^*x)_Y$$
$$- (B^* r_w(t), B^{*-1}A^*x)_U + (w, x)_Y, \quad x \in \mathcal{D}(A^*); \tag{3.6.8}$$

$$y_w^0(t; y_0) = e^{A_P t} y_0 - \int_0^t e^{A_P(t-\tau)} B B^* r_w(\tau) d\tau + \int_0^t e^{A_P(t-\tau)} w(\tau) d\tau$$

$$\in L_2(0, \infty; Y) \cap C([0, \infty]; Y). \tag{3.6.9}$$

Proof. (i) Eqn. (3.6.1) results from substituting (3.2.7) into the right side of (3.2.5), and recalling the definition (3.2.1) of $\Phi(t)y_0$. The passage from (3.6.1) to (3.6.2) uses identity (3.1.3c) on y_w^0 via the definition of r_w in (3.6.3).

(ii) The first identity in (3.6.4) was proved in (3.2.9) of Lemma 3.2.2(ii). Then, the second identity in (3.6.4) follows by substituting (3.6.2). The final breaking up as in (3.6.5) is legal, once we establish the regularity properties in (3.6.6). The regularity (3.6.6) for $B^* P y_w^0$ follows at once from $B^* P \in \mathcal{L}(Y; U)$ in (3.2.10), and from the regularity of y_w^0 in (3.1.3c) (in L_2) and in (3.1.20) (in C). Similarly, from (3.6.3),

$$B^* r_w(t) = B^* p_w(t; y_0 = 0) - B^* P y_w^0(t; y_0 = 0) \tag{3.6.10}$$

$$= -u_w^0(t; y_0 = 0) - B^* P y_w^0(t; y_0 = 0) \in L_2(0, \infty; U) \cap C([0, \infty]; U) \tag{3.6.11}$$

we establish the regularity of $B^* r_w$ in (3.6.6), since $u_w^0(\cdot; y_0 = 0)$ and $y_w^0(\cdot; y_0 = 0)$ have the required regularity, see (3.1.2a), (3.1.3a) (in L_2) and (3.1.20) (in C), and $B^* P$ is bounded as in (3.2.10).

(iii) Eqn. (3.6.7) is obtained by substituting (3.6.5) into the optimal dynamics, and recalling the definition of A_P. Then, (3.6.9) is the unique solution of (3.6.7).

We note explicitly, that the first integral term in (3.6.9) is well-defined in $L_2(0, \infty; Y) \cap C([0, \infty]; Y)$, by virtue of the singular estimate (3.3.11) on $e^{A_P t} B$, and the regularity in (3.6.6) for $B^* r_w(t)$. □

Property (a11). Equation satisfied by p_w. It remains to prove property (a11). That p_w defined by (3.2.2) satisfies the backward initial value problem (2.18) follows by direct differentiation using the regularity (3.1.20) for y_w^0. The latter, plus the stability (1.2), justify that $p_w(\infty; y_0) = 0$, as stated in (2.18b), again by (3.2.2).

Equation satisfied by r_w. We finally establish that the function $r_w(t)$ satisfies the backward initial value problem (2.19).

Proposition 3.6.2. Assume (H.1) through (H.5). Then, the function r_w defined by (3.6.3) satisfies

$$(\dot{r}_w(t), x)_Y = -(A_P^* r_w(t), x)_Y - (Pw(t), x)_Y, \quad \forall x \in \mathcal{D}(A). \tag{3.6.12}$$

Proof. Let $x \in \mathcal{D}(A)$. We start from the defining formula (3.6.3) for r_w and differentiate in t, thus obtaining

$$(\dot{r}_w(t), x)_Y = (\dot{p}_w(t; y_0 = 0) - P \dot{y}_w^0(t; y_0 = 0), x)_Y, \quad x \in \mathcal{D}(A). \tag{3.6.13}$$

Optimal Control and Algebraic Riccati Equations

Next, we substitute Eqn. (2.18) for $\dot{p}_w$ [already proved above] and Eqn. (3.6.7) for $\dot{y}_w^0$. We obtain from (3.6.13):

$$(\dot{r}_w(t), x)_Y = -(A^* p_w(t; y_0 = 0), x)_Y - (R^* R y_w^0(t; y_0 = 0), x)_Y$$

$$- (P[A_P y_w^0(t; y_0 = 0) - BB^* r_w(t) + w], x)_Y \quad (3.6.14)$$

$$= -(p_w(t; y_0 = 0), Ax)_Y - (y_w^0(t; y_0 = 0), [R^* R x + A_P^* P x])_Y$$

$$+ (B^* r_w(t), B^* P x)_U - (Pw, x)_Y, \quad x \in \mathcal{D}(A). \quad (3.6.15)$$

Notice that each term in (3.6.15) is well-defined (by the regularity of p_w and $B^* r_w$ in (3.6.2) and (3.6.6), respectively; by $B^* P \in \mathcal{L}(Y; U)$ in (2.6) or (3.2.10)) and, moreover, that $R^* R x + A_P^* P x = -P A x$, $x \in \mathcal{D}(A)$ by (3.4.2a). Using this identity in (3.6.14), we rewrite it as

$$(\dot{r}_w(t), x)_Y = -(p_w(t; y_0 = 0), Ax)_Y + (y_w^0(t; y_0 = 0), PAx)_Y$$

$$+ (r_w(t), B(B^* P) x)_Y - (Pw, x)_Y \quad (3.6.16)$$

$$= -([p_w(t; y_0 = 0) - P y_w^0(t; y_0 = 0)], Ax)_Y$$

$$+ (r_w(t), B(B^* P) x)_Y - (Pw, x)_Y, \quad x \in \mathcal{D}(A), \quad (3.6.17)$$

where each term in (3.6.17) is well-defined. Recalling the definition of $r_w(t)$ given by (3.6.3) in the first term of the right side of (3.6.17), we rewrite it as

$$(\dot{r}_w(t), x)_Y = -(r_w(t), Ax)_Y + (r_w(t), B(B^* P) x)_Y - (Pw, x)_Y \quad (3.6.18)$$

$$= -(r_w(t), [A - BB^* P] x)_Y - (Pw, x)_Y, \quad x \in \mathcal{D}(A) \quad (3.6.19)$$

$$= -(r_w(t), A_P x)_Y - (Pw, x)_Y, \quad x \in \mathcal{D}(A). \quad (3.6.20)$$

Finally, from (3.6.20) we obtain

$$(\dot{r}_w(t), x)_Y = -(A_P^* r_w(t), x)_Y - (Pw, x), \quad x \in \mathcal{D}(A), \quad (3.6.21)$$

which is precisely (3.6.12). □

4 Illustrations. Structural acoustic problems satisfying (H.1), (H.2), (H.3) with $\gamma = \frac{3}{8} + \epsilon$ or $\gamma = \frac{3}{4} + \epsilon$

In this section we provide two structural acoustic models which, once written abstractly as in (1.1), satisfy all the required assumptions (H.1), (H.2), (H.3) = (1.3), the latter one with either $0 < \gamma < \frac{1}{2}$, typically $\gamma = \frac{3}{8} + \epsilon$ (Example 4.1), or else $\gamma = \frac{3}{4} + \epsilon > \frac{1}{2}$ (Example 4.2).

Example 4.1: **A class of structural acoustic problems with constant $0 < \gamma < \frac{1}{2}$**

The model. We consider the following class of structural acoustic problems, where Ω is an acoustic chamber with flexible (elastic) wall Γ_0, assumed *flat* and rigid wall Γ_1. Thus, let $\Omega \subset \mathbb{R}^n$, $n = 2, 3$, be an open bounded domain with boundary $\Gamma = \overline{\Gamma_0 \cup \Gamma_1}$, where Γ_0 and Γ_1 are open, connected, and disjoint parts, $\Gamma_0 \cap \Gamma_1 = \phi$, in $\mathbb{R}^{n-1}$, and Γ_0 is flat. We allow either Γ to be sufficiently smooth (say, C^2), or else Ω to be convex: this assumption will then guarantee that solutions to classical elliptic equations with $L_2(\Omega)$-non-homogeneous terms be in $H^2(\Omega)$ [Gr.2]. Let z denote the velocity potential of the acoustic medium within the chamber. For simplicity of notation, we take equal to 1 both the density of the fluid and the speed of sound in the fluid. Then z_t is the acoustic pressure. Let v denote the displacement of the flat flexible wall Γ_0, modeled by an elastic beam or plate equation ($n = 2$, or $n = 3$). The structural acoustic model here considered is as follows:

$$\begin{cases} \text{acoustic} \\ \text{chamber:} \end{cases} \begin{cases} z_{tt} = \Delta z - d_1 z_t + f & \text{in } (0, T] \times \Omega, \quad (4.1\text{a}) \\ \dfrac{\partial z}{\partial \nu} + d_2 z_t = 0 & \text{in } (0, T] \times \Gamma_1, \quad (4.1\text{b}) \\ \dfrac{\partial z}{\partial \nu} + d_3 z_t = \pm v_t & \text{in } (0, T] \times \Gamma_0, \quad (4.1\text{c}) \end{cases}$$

$$\text{elastic wall} \quad \mathcal{M}_k v_{tt} + \mathcal{A} v_t + \mathcal{A} v \pm z_t = \mathcal{B}u \text{ in } (0, T] \times [\mathcal{D}(\mathcal{A}^{\frac{1}{2}})]' \quad (4.1\text{d})$$

$$z(0, \cdot) = z_0, \; z_t(0, \cdot) = z_1, \; v(0, \cdot) = v_0, \; v_t(0, \cdot) = v_0 \text{ in } \Omega. \quad (4.1\text{e})$$

Assumptions. In (4.1a–c), $f \in L_2(0, T; L_2(\Omega))$ denotes the deterministic external noise within the chamber; and the non-negative constant d_i, when positive, introduce interior/boundary damping in the model. Equation (4.1d) is an abstract version encompassing several 'concrete' elastic models, as documented below. At the abstract level, we make the following assumptions:
(a1)

$$\begin{cases} \mathcal{M}_k : L_2(\Gamma_0) \supset \mathcal{D}(\mathcal{M}_k) \to L_2(\Gamma_0); \quad \mathcal{A} : L_2(\Gamma_0) \supset \mathcal{D}(\mathcal{A}) \to L_2(\Gamma_0) \\ \text{are two positive, self-adjoint operators} \end{cases} \quad (4.2)$$

(the stiffness operator, and the elastic operator, respectively, the first depending on a non-negative parameter $k \geq 0$; in concrete situations, if $k > 0$, the elastic model on Γ_0 accounts for rotational forces);
(a2)

$$\mathcal{D}(\mathcal{A}^{\frac{1}{2}}) \subset \mathcal{D}(\mathcal{M}_k^{\frac{1}{2}}); \quad (4.3)$$

Optimal Control and Algebraic Riccati Equations

(a3) there is a positive constant $\rho < \frac{1}{2}$ such that

$$\mathcal{A}^{-\rho}\mathcal{B} \in \mathcal{L}(\mathcal{U}; L_2(\Gamma_0)), \text{ equivalently } \mathcal{B}: \text{ continuous } \mathcal{U} \to [\mathcal{D}(\mathcal{A}^\rho)]', \tag{4.4}$$

where $[\ ]'$ denotes duality with respect to $L_2(\Gamma_0)$ as a pivot space, and $\mathcal{U}$ is the control Hilbert space.

(a4) Moreover, the parameter ρ in (4.4) satisfies either one of the following additional assumptions:

(a4i) either

$$\begin{cases} \rho \leq \dfrac{5}{12}, & \text{if } \Omega \text{ is a smooth domain,} \\ \rho \leq \dfrac{7}{16}, & \text{if } \Omega \text{ is a parallelopiped,} \end{cases} \tag{4.5a}$$

(a4ii) or else

$$\mathcal{D}(\mathcal{M}_k^{\frac{1}{2}}) \subset H^{\frac{1}{3}}(\Gamma_0). \tag{4.5b}$$

Remark 4.1.1 (on the control operator $\mathcal{B}$). In concrete PDE examples of the structural acoustic problems, such as they arise in smart material technology, all of the above assumptions are satisfied. First, in this case, the control operator $\mathcal{B}$ is given by

$$\mathcal{B}u = \sum_{j=1}^J a_i u_i \delta'_{\xi_j}, \quad u = [u_1, \ldots, u_J] \in \mathbb{R}^J = \mathcal{U}, \tag{4.6}$$

where: (i) if $\dim \Gamma_0 = 1$ ($\dim \Omega = 2$), then ξ_j are points on Γ_0, a_j are constants, and δ'_{ξ_j} are derivatives of the Dirac distribution supported at ξ_j; (ii) if $\dim \Gamma_0 = 2$ ($\dim \Omega = 3$), then ξ_j denote closed regular curves on Γ_0, a_j are smooth functions and δ'_{ξ_j} denotes the normal derivative supported at ξ_j:

$$(\delta'_{\xi_j}, \phi)_{L_2(\Gamma_0)} = \begin{cases} -\phi'(\xi_j), \ \dim \Gamma_0 = 1 & (4.7a) \\ -\displaystyle\int_{\xi_j} \nabla \phi \cdot \nu \, d\xi_j, \ \dim \Gamma_0 = 2 & (4.7b) \end{cases} ; \ \forall \phi \in H^{\frac{3}{2}+\epsilon}(\Gamma_0),$$

where ν is the unit outer normal vector to the closed curve ξ_j, and $\epsilon > 0$ is arbitrary. Thus, from (4.7) we have that, a-fortiori, $(\delta'_{\xi_j}, \mathcal{A}^{-(\frac{3}{8}+\frac{\epsilon}{4})}\psi)_{L_2(\Gamma_0)}$ is well defined $\forall \psi \in L_2(\Gamma_0)$, since $\mathcal{A}^{-(\frac{3}{8}+\frac{\epsilon}{4})}\psi \in \mathcal{D}(A^{\frac{3}{8}+\frac{\epsilon}{4}}) \subset H^{\frac{3}{2}+\epsilon}(\Gamma_0)$. Hence the operator $\mathcal{B}$ defined by (4.6) satisfies assumption (a3) = (4.4) with $\rho = \frac{3}{8} + \epsilon$, $\forall \epsilon > 0$ small [Las.3], [L-T.2, vol.2, p. 907]. Then, such $\rho = \frac{3}{8} + \epsilon$ satisfies both conditions (a4i) = (4.5a) as well.

Remark 4.1.2 (on the stiffness operator $\mathcal{M}_k$). In concrete PDE examples where the parameter $k > 0$ (and so the elastic beam/plate-model on Γ_0 accounts for rotational forces),

$\mathcal{M}_k$ is the translation by the identity of the realization of $(-\Delta)$ on Γ_0 subject to appropriate boundary conditions. Thus, $\mathcal{D}(\mathcal{M}_k^{\frac{1}{2}}) \subset H^1(\Gamma_0) \subset H^{\frac{1}{3}}(\Gamma_0)$ and assumption (a4ii) = (4.5b) is satisfied as well. □

The above structural acoustic model (4.1), subject to assumptions (a1) through (a4), satisfies assumption (H.3) = (1.3) with $\gamma = \rho < \frac{1}{2}$. The following claims are shown in [Las.3], [A-L.1]. The structural acoustic problem (4.1) can be rewritten in the abstract form (1.1), with operators A and B explicitly identified, and $w = [0, f, 0, 0]$. Moreover, the operator A is the generator of a s.c. contraction semigroup e^{At} on an appropriate finite energy space Y_k given by

$$Y_k \equiv H^1(\Omega) \times L_2(\Omega) \times \mathcal{D}(\mathcal{A}^{\frac{1}{2}}) \times \mathcal{D}(\mathcal{M}_k^{\frac{1}{2}}), \tag{4.8}$$

for the variables $[z(t), z_t(t), v(t), v_t(t)] = e^{At}[z_0, z_1, v_0, v_1]$.

Finally, such operators A and B do satisfy assumption (H.3) = (1.3) for $\gamma = \rho < \frac{1}{2}$, the constant in assumptions (a3) and (a4). In particular, if the control operator $\mathcal{B}$ is defined by (4.6), then we have $\gamma = \rho = \frac{3}{8} + \epsilon$, $\forall \epsilon > 0$.

Remark 4.1.3 (on the uniform stability of problem (4.1)). There are several configurations—that is choices of the damping constants d_i in (4.1a–c) and corresponding geometrical conditions—which ultimately yield uniform stability on Y_k of the associated s.c. semigroup e^{At} [L.2], as required by assumption (H.1) = (1.2). They include the following cases: (1) $d_2 = d_3 = 0$, $d_1 > 0$ (viscous damping); (2) $d_1 = d_3 = 0$, $d_2 > 0$ (boundary damping on rigid wall Γ_1), with no geometrical conditions; (3) $d_1 = d_2 = 0$, $d_3 > 0$ (boundary damping on flexible wall Γ_0) under the geometrical condition that Ω is convex and there exists a point $x_0 \in \mathbb{R}^n$ such that $(x - x_0) \cdot \nu(x) \leq 0$, $\forall x \in \Gamma_1$ [L-T-Z.1, Appendices]. Additional cases are also possible [L.3].

Remark 4.1.4. At the price introducing heavy notation, it would be possible to include into problem (4.1), also the case where the elastic wall Γ_0 is *curved* and, accordingly, modeled by a shell equation to be written abstractly as in (4.1d), see [Las.3]. □

Concrete illustrations of the abstract elastic equation (4.1d). As canonical illustrations of the abstract elastic equation (4.1d)—say with no coupling term z_t and with no control: $u \equiv 0$—we may take the classical Euler-Bernoulli equation on Γ_0 ($k = 0$) or the corresponding Kirchhoff equation on Γ_0 ($k > 0$):

$$v_{tt} - k\Delta v_{tt} + \Delta^2 v + \Delta^2 v_t = 0 \quad \text{on } (0, T] \times \Gamma_0, \tag{4.9}$$

under a variety of B.C. on $(0, T] \times \partial \Gamma_0$: hinged, clamped, free B.C., etc. [A-L.1], [Las.3], [L-T.3]. Then, $\mathcal{A}$ is the realization of Δ^2 on $L_2(\Gamma_0)$ subject to the appropriate B.C. Finally, $\mathcal{M}_k = I + k\mathcal{A}_1$ where $\mathcal{A}_1$ is the realization of $(-\Delta)$ on $L_2(\Gamma_0)$ under suitable B.C.

Conclusion. Under the above assumptions, including those of Remark 4.1.3, model (4.1) is covered by the abstract theory of Sections 1–3, with $\gamma < \frac{1}{2}$.

Optimal Control and Algebraic Riccati Equations

Example 4.2: A class of structural acoustic problems with constant $\frac{1}{2} < \gamma < 1$

The model. We use, when possible, the same notation as in Example 4.1. We consider again an acoustic chamber Ω endowed with a rigid wall Γ_1 and a flexible wall Γ_0, where, now, however, we introduce two main changes over Example 4.1: (i) the wave equation in z displays a 'strong' damping on the wall Γ_0 (much stronger than damping on z_t in (4.1c) of Example 4.1: see operator D in (4.10c) below); (ii) the flexible wall Γ_0 accounts now also for thermal effects, and is therefore modeled by a thermoelastic beam or plate, where w and θ denote displacement and temperature. Accordingly, the new model is now given by

$$\begin{cases} \text{acoustic chamber:} \begin{cases} z_{tt} = \Delta z - d_1 z + f & \text{in } (0,T] \times \Omega, \quad (4.10a) \\ \alpha \dfrac{\partial z}{\partial \nu} + z = 0 & \text{in } (0,T] \times \Gamma_1, \quad (4.10b) \\ \alpha = 0 : \text{Dirichlet B.C.}; \alpha > 0: \text{Robin B.C.} \\ \dfrac{\partial z}{\partial \nu} + D z_t = v_t & \text{in } (0,T] \times \Gamma_0, \quad (4.10c) \end{cases} \\ \text{thermo-elastic wall} \begin{cases} v_{tt} - k\Delta v_{tt} + \Delta^2 v + \Delta\theta + z_t = \mathcal{B}u & \text{in } (0,T] \times \Gamma_0, \quad (4.10d) \\ \theta_t - \Delta\theta - \Delta v_t = 0 & \text{in } (0,T] \times \Gamma_0, \quad (4.10e) \\ \text{plus Boundary Conditions} \end{cases} \\ z(0,\cdot) = z_0, \; z_t(0,\cdot) = z_1, \; v(0,\cdot) = v_0, \; v_t(0,\cdot) = v_1 \quad \text{in } \Omega. \quad (4.10f) \end{cases}$$

where

$$z_0 \in H^1_{\Gamma_1}(\Omega) \equiv \{\phi \in H^1(\Omega): \phi = 0 \text{ on } \Gamma_1\}; \; z_1 \in L_2(\Omega); \quad (4.11a)$$

$$v_0 \in H^2(\Gamma_0); \; v_1 \in H^1(\Gamma_0;k) \equiv \begin{cases} L_2(\Gamma_0) & \text{if } k = 0 \\ H^1(\Gamma_0) & \text{if } k > 0 \end{cases}; \; \theta_0 \in L_2(\Gamma_0) \quad (4.11b)$$

where, for $k > 0$, $H^1(\Gamma_0;k)$ is topologized by

$$\|\psi\|^2_{H^1(\Gamma_0;k)} \equiv \|\psi\|^2_{L_2(\Gamma_0)} + k\||\nabla\psi|\|^2_{L_2(\Gamma_0)}. \quad (4.11c)$$

In (4.10b-c), ν is the unit outward vector to the boundary $\Gamma = \partial\Omega$. By contrast, here below, when dim $\Omega = 3$ and so dim $\Gamma_0 = 2$, we shall let $\check{\nu}$ be the unit outward normal to $\partial\Gamma_0$ as the boundary of Γ_0; and, $\check{\tau}$ be the unit tangent vector to $\partial\Gamma_0$, oriented counterclockwise. System (4.10) is supplemented with Boundary Conditions (B.C.). We shall consider explicitly three sets of B.C. for the thermoelastic component:

Hinged B.C.:
$$v = \Delta v = \theta = 0 \quad \text{on } (0,T] \times \partial\Gamma_0; \tag{4.12}$$

Clamped B.C.:
$$v = \frac{\partial v}{\partial \tilde{\nu}} = \theta = 0 \quad \text{on } (0,T] \times \partial\Gamma_0. \tag{4.13}$$

Free B.C. when dim $\Omega = 3$, dim $\Gamma_0 = 2$:
$$\begin{cases} \Delta v + B_1 v + \theta = 0 & \text{(4.14a)} \\ \dfrac{\partial}{\partial \tilde{\nu}} \Delta v + B_2 v - \gamma \dfrac{\partial}{\partial \tilde{\nu}} v_{tt} + \dfrac{\partial \theta}{\partial \tilde{\nu}} = 0 & \text{on } (0,T] \times \partial\Gamma_0 \quad \text{(4.14b)} \\ \dfrac{\partial \theta}{\partial \tilde{\nu}} + \lambda \theta = 0, \ \lambda > 0, & \text{(4.14c)} \end{cases}$$

where, with constant $0 < \mu < 1$, we have

$$\begin{aligned} B_1 v &= (1-\mu)(2\nu_1\nu_2 v_{xy} - \nu_1^2 v_{yy} - \nu_2^2 v_{xx}), \quad \tilde{\nu} = [\nu_1, \nu_2]; \\ B_2 v &= (1-\mu)\left\{ \frac{\partial}{\partial \tilde{\tau}}[(\nu_1^2 - \nu_2^2)v_{xy} + \nu_1\nu_2(v_{yy} - v_{xx})] + \ell v \right\}, \ \ell > 0 \end{aligned} \tag{4.14d}$$
$$\tag{4.14e}$$

Remark 4.2.1. The parameter $k \geq 0$ in (4.10d) is critical in describing the character of the dynamics of the uncoupled free thermoelastic system (4.10d–e) [that is with no coupling term z_t and with $u \equiv 0$]: for $k = 0$, such thermoelastic problem generates a s.c. analytic semigroup ('parabolic' case) [L-T.2, Chapter 3, Appendices 3E-3I], [L-T.4-6], [L-L.1], while for $k > 0$ the corresponding s.c. semigroup is 'hyperbolic-dominated' in a technical sense [L-T.3]. □

Throughout this example, we let

$$\mathcal{A} = \text{realization in } L_2(\Gamma_0) \text{ of } \Delta^2 \text{ subject to hinged, or} \tag{4.15}$$
$$\text{clamped, or free homogeneous B.C.}$$

Regarding the control operator $\mathcal{B}$ in (4.10d), we shall assume the same hypothesis (A3) = (4.4), here restated as

(h1) there exists a positive constant $\rho < \frac{1}{2}$, such that

$$\mathcal{A}^{-\rho}\mathcal{B} \in \mathcal{L}(\mathcal{U}; L_2(\Gamma)), \text{ equivalently } \mathcal{B} : \text{continuous } \mathcal{U} \to [\mathcal{D}(\mathcal{A}^\rho)]' \tag{4.16}$$

[an assumption satisfied with $\rho = \frac{3}{8} + \epsilon$, if $\mathcal{B}$ is the operator defined in (4.6)]. In addition, we make the following assumption on the tangential positive self-adjoint operator D occurring in (4.10c):

(h2) With ρ as in (4.16), there exist positive constants δ_1, δ_2, such that $\mathcal{D}(D^{\frac{1}{2}}) = \mathcal{D}(\mathcal{A}^{\rho_0})$ and

$$\delta_1 \|z\|^2_{\mathcal{D}(\mathcal{A}^{\rho_0})} \leq (Dz, z)_{L_2(\Gamma_0)} \leq \delta_2 \|z\|^2_{\mathcal{D}(\mathcal{A}^{\rho_0})'}, \quad \forall z \in \mathcal{D}(D^{\frac{1}{2}}) \tag{4.17}$$

where

$$\begin{cases} \text{if } \frac{1}{4} \leq \rho < \frac{1}{2}, & \text{then: } \rho - \frac{1}{4} \leq \rho_0 \leq \frac{1}{4}; & (4.18a) \\ \text{if } \rho < \frac{1}{4}, & \text{then: } \rho_0 = 0. & (4.18b) \end{cases}$$

A typical example of such tangential operator D is a realization of the Laplace-Beltrami operator on $L_2(\Gamma_0)$.

Remark 4.2.2 (on assumption (h2)). (a) If the constant ρ in (4.16) satisfies $\rho < \frac{1}{4}$, then the damping operator D may be taken to be the identity operator on $L_2(\Gamma_0)$.

(b) If, however, $\frac{1}{4} \leq \rho < \frac{1}{2}$ [as in the case of the control operator $\mathcal{B}$ in (4.6)], then a stronger, unbounded damping operator D is needed. More precisely:

(b1) Let $\frac{1}{4} \leq \rho \leq \frac{3}{8}$ so that $\rho - \frac{1}{4} \leq \frac{1}{8}$ and we can take $\rho_0 = \frac{1}{8}$. Then $4\rho_0 - \frac{1}{2} = 0$ and then $\mathcal{D}(\mathcal{A}^{\rho_0}) = \mathcal{D}(\mathcal{A}^{\frac{1}{8}})$ is topologically equivalent to $H^{\frac{1}{2}}(\Gamma_0)$ subject to appropriate B.C. [e.g., in the case of hinged or clamped B.C., then $\mathcal{D}(\mathcal{A}^{\frac{1}{8}}) = H_{00}^{\frac{1}{2}}(\Gamma_0)$].

(b2) Let $\frac{3}{8} < \rho < \frac{1}{2}$ so that $\frac{1}{8} = \frac{3}{8} - \frac{1}{4} < \rho - \frac{1}{4} < \frac{1}{4}$, and we can take ρ_0 satisfying $\frac{1}{8} < \rho_0 \leq \frac{1}{4}$. Then, (*): $0 < 4\rho_0 - \frac{1}{2} \leq \frac{1}{2}$. Then, the following two subcases need to be considered.

(b2i) Assume either hinged or clamped B.C. for the operator $\mathcal{A}$ in (4.15), see (4.12) or (4.13). Then (*) above implies [Gr.1]

$$\mathcal{D}(\mathcal{A}^{\rho_0}) = H_0^{4\rho_0}(\Gamma_0) \subset H^1(\Gamma_0). \tag{4.19}$$

Thus, in this case, Dz_t well-defined requires, by assumptions (4.17) and (4.19), that $z|_{\partial \Gamma_0} \equiv 0$. To ensure this, we then take $\alpha = 0$ in (4.10b), so that the z-problem is endowed with Dirichlet, rather than Robin, B.C.

(b2ii) Assume now free B.C. for the operator $\mathcal{A}$ in (4.15), see (4.15). Then, (*) above implies

$$\mathcal{D}(\mathcal{A}^{\rho_0}) = H^{4\rho_0}(\Gamma_0) \subset H^1(\Gamma_0), \tag{4.20}$$

and then we can allow $\alpha \geq 0$ in (4.10b): that is either Robin or Dirichlet B.C.

The above structural acoustic model (4.10), subject to assumptions (h1), (h2), and $k = 0$ (no rotational forces accounted for) satisfies assumption (H.3) = (1.3) with $\gamma = 2\rho$ [Las.3]. Thus, if $\mathcal{B}$ is the control operator defined by (4.6), then $\frac{1}{2} < \gamma = 2(\frac{3}{8} + \epsilon) = \frac{3}{4} + 2\epsilon < 1$.

When $k = 0$ in (4.10d) [Euler-Bernoulli rather than Kirchhoff equation] and assumptions (h1), (h2) above are in force, then the structural acoustic model (4.10) can be rewritten in the abstract form (1.1), with operators A and B explicitly identified. Moreover, the operator A is the generator of a s.c. contraction semigroup e^{At} on an appropriate finite energy space Y given by

$$Y \equiv H^1_{\Gamma_1}(\Omega) \times L_2(\Omega) \times \mathcal{D}(\mathcal{A}^{\frac{1}{2}}) \times L_2(\Gamma_0) \times L_2(\Gamma_0) \tag{4.21}$$

for the variables $[z(t), z_t(t), v(t), v_t(t), \theta(t)]$. The abstract deterministic disturbance w in Eqn. (1.1) is now $w = [0, f, 0, 0, 0]$, with f the disturbance in (4.10a). Finally, the s.c. semigroup e^{At} is uniformly stable on Y [L-T.2, Chapter 3, Sections 3.11–3.13]. (In the case of free B.C., this is due to the term $(1 - \mu)\ell v$ in (4.14e) with coefficient $(1 - \mu)\ell < 0$.)

Conclusion. Under the above assumptions, model (4.10) is covered by the abstract theory of Sections 1–3.

References

[A-L.1]   G. Avalos and I. Lasiecka, Differential Riccati equations for the active control of a problem in structural acoustics, *JOTA* 91 (1996), 695–728.

[B-D-D-M] A. Bensoussan, G. Da Prato, M. Delfour, and S. Mitter, *Representation and Control of Infinite Dimensional Systems*, vols. 1 and 2, Birkhäuser, 1993.

[Gr.1]   P. Grisvard, Characterisation del quelques espaces d'interpolation, *Arch. Rational Mech. Anal.* 25 (1967), 40–63.

[Gr.2]   P. Grisvard,

[L.2]   I. Lasiecka, Optimization problems for structural acoustic models with thermoelasticity and smart materials, *Discussiones Mathematicae, Differential Inclusions, Control and Optimization*, vol. 20 (2000), 113–140.

[L.3]   I. Lasiecka, Mathematical control theory of coupled PDEs systems, *SIAM*.

[L-T.1]   I. Lasiecka and R. Triggiani, *Differential and Algebraic Riccati Equations with Applications to Boundary/Point Control Problems: Continuous and Approximation Theory*, LNICS, Springer Verlag, 1991, 160 pp.

[L-T.2]   I. Lasiecka and R. Triggiani, *Control Theory for Partial Differential Equations*, vol. I, Cambridge University Press, Encyclopedia of Mathematics and its Applications, January 2000, 668 pp.

[L-T.3]   I. Lasiecka and R. Triggiani, Structural decomposition of thermo-elastic semigroups with rotational forces, *Semigroup Forum* 60 (2000), 16–66.

[L-T.4]   I. Lasiecka and R. Triggiani, Two direct proofs on the analyticity of the s.c. semigroup arising in abstract thermo-elastic equations, *Advances Diff. Eqns.* 3(3) (May 1998), 387–416.

[L-T.5]   I. Lasiecka and R. Triggiani, Analyticity of thermo-elastic semigroups with coupled hinged/Neumann B.C., *Abstract Appl. Anal.* 3(1–2) (1998), 153–169.

[L-T.6]   I. Lasiecka and R. Triggiani, Analyticity of thermo-elastic semigroups with free B.C., *Annali Scuola Normale Superiore*, Pisa, Cl. Sci. (4), XXXVII (1998), 457–482.

[L-T-Z.1] I. Lasiecka, R. Triggiani, and X. Zhang, Nonconservative wave equations with unobserved Neumann B.C.: Global uniqueness and observability in one shot, *Contemporary Mathematics*, vol. 267, to appear.

[L-R.1] K. Liu and M. Renardy, A note on equations of a thermoelastic plates, *Appl. Math. Letters* 8 (1995), 1–6.

Solving Identification Problems for the Wave Equation by Optimal Control Methods

Suzanne Lenhart
Mathematics Department
University of Tennessee
Knoxville, TN 37996-1300

Vladimir Protopopescu
Oak Ridge National Laboratory
Computer Science and Mathematics Division
Oak Ridge, TN 37831-6355

Abstract: Inverse problems of identification type for the wave equation are approximated via optimal control methods. The sought "unknown" coefficients are treated as controls and the goal is to drive the model solution close to the observation data by adjusting these controls. Tikhonov regularization is coupled with optimal control techniques and illustrated for three examples.

Research supported in part by the U.S. Department of Energy, Office of Basic Energy Sciences, under contract No. AC05-00OR22725 with U. T.– Battelle, LLC.

" This submitted manuscript has been authored by a contractor of the U.S. Government under Contract No. DE-AC05-00OR22725. Accordingly, the U.S. Government retains a nonexclusive, royalty-free license to publish or reproduce the published form of this contribution, or allow others to do so, for U.S. Government purpose".

0. Introduction

We survey our recent results on the approximation of inverse problem of identification type for wave equations by optimal control methods and we illustrate this approach with three examples.

The wave propagation problem (state system) has the following form:

$$\begin{aligned} u_{tt} &= \Delta u + hu + f & \text{on} \quad & Q = \Omega \times (0,T) \\ u(x,0) &= u_0(x), \quad u_t(x,0) = u_1(x), & \text{on} \quad & \Omega \times \{t=0\} \\ \tfrac{\partial u}{\partial \eta} + \sigma u &= g & \text{on} \quad & \partial\Omega \times (0,T) \end{aligned} \qquad (0.1)$$

where $\Omega \subset \mathbb{R}^n$ is a bounded spatial domain with C^1 boundary. The boundary conditions are of nonhomogeneous Robin type which includes Dirichlet and Neumann as particular cases. To streamline the presentation, we consider an isotropic homogeneous medium and normalize the speed of propagation to one, but anisotropic, inhomogeneous media can be treated without significant difficulty [5, 6, 7].

We denote by ϕ the coefficient/function to be identified. For instance, ϕ can be the source f, the dispersion coefficient h, the surface reflection coefficient σ, or other function occurring in the problem. Starting from actual observations z of the solution of the wave problem, $u = u(\phi)$, one seeks to minimize the objective functional

$$J(\phi) = \int_{W_1} |u(\phi) - z|^2 \, dx \, dt$$

over a class of functions ϕ in a control set U, where W_1 is a subset of Q or $\partial\Omega \times (0,T)$,– depending on the type of observations. The inverse problem is solved by finding $\phi \in U$, for which $u(\phi)$ is as close as possible to the observations z in the L^2 sense.

To approximate this identification problem, we introduce the following optimal control problem for $\beta > 0$:

$$\min_{\phi \in U} J_\beta(\phi)$$

with

$$J_\beta(\phi) = \frac{1}{2}\left(\int_{W_1} |u(\phi) - z|^2\, dx\, dt + \beta \int_{W_2} \phi^2\, dx\, dt\right)$$

where W_2 is a subset of Q, Ω, or $\partial\Omega$, depending on the type of control. This type of approximation is called Tikhonov regularization [12]. For each $\beta > 0$, the coefficient to be identified, ϕ, is viewed as a control and is adjusted to get the corresponding solution, $u(\phi)$, close to the observations z. For a fixed $\beta > 0$, the optimal control, h_β, that minimizes $J_\beta(\phi)$, will be explicitly characterized in terms of solutions of the optimality system, which consists of the state problem coupled with an adjoint problem. Taking a sequence of β_n that converges to zero, the corresponding sequence of optimal controls, ϕ_{β_n}, is shown to converge to a "solution", ϕ^*, of our identification problem. The interpretation of this "solution" is as follows: If the identification problem is known to have a solution, then our method will find a solution to that identification problem. If the observations are imprecise and thus the identification problem may not have a solution, then our method finds a projection onto an appropriate range space.

This interpretation holds for the case of one observation (one input yielding one output), but in the third example, we discuss its extension to multiple observations coming form multiple inputs.

The majority of traditional approaches to inverse identification problems [1, 2, 3] couple Tikhonov's regularization with an optimization algorithm. Puel and Yamamoto [10, 11, 13] have obtained uniqueness and stability results for reconstruction algorithms using exact controllability for various wave equation problems. Our approach has the advantage of an explicit characterization of the approximate coefficients; moreover, this characterization leads to a natural numerical algorithm in solving the resulting optimality system.

1. First Example

This example identifies the dispersive coefficient, $h(x,t)$, from observations of the solution of a wave equation on a set $Q' \subset Q = \Omega \times (0,T)$. Consider the wave equation:

$$\begin{aligned} u_{tt} &= \Delta u + hu + f && \text{in} \quad Q \\ u &= u_0, \quad u = u_1 && \text{on} \quad \Omega \times \{t=0\} \\ u &= 0, && \text{on} \quad \partial\Omega \times (0,T), \end{aligned} \quad (1.1)$$

where $f \in L^2(Q), u_0 \in H_0^1(\Omega), u_1 \in L^2(\Omega)$ and $\Omega \subset \mathbb{R}^n$ has a C^2 boundary. The identification problem is to find bounded h such that the corresponding solution $u = u(h)$ of system (1.1) is close to the observations z on Q'. We consider the control set,

$$U = \{h \in L^\infty(Q) | -M \leq h(x,t) \leq M\},$$

and the approximate functional is:

$$J_\beta(h) = \frac{1}{2}\left(\int_{Q'}(u(h)-z)^2 \, dx\, dt + \beta \int_Q h^2 \, dx\, dt\right).$$

We seek $h_\beta \in U$ such that

$$J_\beta(h_\beta) = \inf_{h \in U} J_\beta(h).$$

See Liang [8] for some further results on this type of control problem. The joint work with Liang [9] gives more details on this particular example.

The solution space is

$$u \in L^2(0,T; H_0^1(\Omega)), \quad u_t \in L^2(Q), \quad u_{tt} \in L^2(0,T; H^{-1}(\Omega)).$$

From differentiating the maps

$$h \to u(h)$$

and

$$h \to J_\beta(h),$$

with respect to h and *a priori* estimates, we obtain

Theorem 1.1. *There exists a control $h_\beta \in U$ and corresponding state $u_\beta = u(h_\beta)$, that minimizes the functional $J_\beta(h)$ over U. Furthermore, there exists a weak solution p in $L^2(0,T; H_0^1(\Omega))$ to the adjoint problem:*

$$\begin{aligned} p_{tt} &= \Delta p + hp + (u_\beta - z)\chi_{Q'} &&\text{in} && Q \\ p &= p_t = 0, &&\text{on} && \Omega \times \{t = T\} \\ p &= 0 &&\text{on} && \partial\Omega \times (0,T), \end{aligned} \quad (1.2)$$

where $\chi_{Q'}$ is the characteristic function of the set Q', and, h_β satisfies

$$h_\beta = \max\left(-M, \min\left(\frac{-u_\beta p}{\beta}, M\right)\right). \quad (1.3)$$

This sequence h_β of optimal controls approximates h, the desired coefficient as $\beta \to 0$.

If the measurements are inaccurate or affected by noise, then the observations z may either come from a solution of (1.1) that does not represent the actual scenario or from a function that is not a solution of (1.1) for any $h \in U$. Thus we do not assume that z is the range of the map

$$h \in U \to u(h)|_{Q'}. \quad (1.4)$$

In this case we can prove:

Theorem 1.2. *There exists:*

(i) *a sequence $\beta_n \to 0$*
(ii) *corresponding optimal controls, h_{β_n}, for the functionals $J_{\beta_n}(h)$,*
(iii) *$h^* \in U$ and*
(iv) *$u^* = u(h^*)$ such that*

$$h_{\beta_n} \rightharpoonup h^* \quad \text{weakly in} \quad L^2(Q),$$

$$u(h_{\beta_n}) \rightharpoonup u^* \quad \text{weakly in} \quad L^2(0,T; H_0^1(\Omega))$$

and

$$\int_{Q'} (u^* - z)^2 \, dx \, dt = \inf_{h \in U} \int_{Q'} (u(h) - z)^2 \, dx \, dt.$$

Note the limit u^* can be interpreted as a (not necessarily unique) projection of z onto the range of the map (1.4).

2. Second Example

In this example, we seek to identify the reflection coefficient σ of part of the spatial boundary, $\partial\Omega$. For bounded domain $D \subset \mathbb{R}^2$ with C^1 boundary, define the spatial domain

$$\Omega = \{(x,y,z) | (x,y) \in D, w(x,y) < z < 0\}$$

where $w: D \to (-\infty, 0)$ is a C^2 function. Assume the region Ω contains a certain medium (like water in a section of the ocean) and denote, as before, $Q = \Omega \times (0,T)$. We seek to identify the reflection coefficient from that set

$$U = \{\sigma \in L^\infty(D) | 0 \leq \sigma(x,y) \leq K\}.$$

Consider the solution, $u = u(\sigma)$, of the acoustic wave equation:

$$\begin{aligned} u_{tt} &= \Delta u + f & \text{in} \quad & Q \\ u &= 0, & \text{on} \quad & \Sigma \times (0,T), \text{ sides of spatial domain} \\ \tfrac{\partial u}{\partial \eta} &= 0 & \text{on} \quad & D \times \{z=0\} \times (0,T), \text{ top of spatial domain} \\ \tfrac{\partial u}{\partial \nu} + \sigma u &= 0 & \text{on} \quad & \Gamma \times (0,T), \text{ bottom of spatial domain} \\ u &= u_0, \quad u_t = u_1, & \text{on} \quad & \Omega \times \{0\}. \end{aligned} \qquad (2.1)$$

Where

$$\Gamma = \{(x,y,w(x,y)) | (x,y) \in D\}$$
$$\Sigma = \{(x,y,z) | (x,y) \in \partial D, w(x,y) < z < 0\},$$

and $\dfrac{\partial u}{\partial \nu} = \nabla w \cdot \eta$ denotes the outward co-normal derivative. To approximate the identification problem, we consider the objective functional

$$J_\beta(\sigma) = \frac{1}{2}\left(\int_{G \times (0,T)} (u-z)^2 \, dx \, dy \, dz \, dt + \beta \int_\Gamma (\sigma(x,h))^2 \, ds\right),$$

where $G \subset \Omega$ is a set with positive measure. We seek to identify σ from observations z on the set $G \times (0, T)$, resulting from a single source f. To define our solution space, let

$$V = \{v \in H^1(\Omega) | v < 0 \text{ on } \Sigma\}$$

with norm

$$\left(\int_\Omega |\nabla v|^2 \, dx \, dy \, dz\right)^{1/2}.$$

The solution space for the state system (2.1) is defined by

$$u \in L^2(0, T; V)$$
$$u_t \in L^2(Q)$$
$$u_{tt} \in L^2(0, T; V'),$$

where V' denotes the dual space of V.

Assume:

$$f, f_t \in L^2(Q), \quad u_0 \in H^2(\Omega) \cap V, \quad u_1 \in H^1(\Omega).$$
$$G \subset \Omega \quad \text{with positive Lebesgue measure}$$
$$z \in L^2(G \times (0, T)), \quad w \in C^2(D), \quad w(x, y) < 0.$$

From [6, 7], we state the combined existence and characterization result for the approximate functional $J_\beta(\sigma)$.

Theorem 2.1. *There exists a unique control σ_β in U and corresponding state $u_\beta = u(\sigma_\beta)$, that minimizes the functional $J_\beta(\sigma)$ over U. Furthermore, there exists an adjoint solution* in $L^2(0,T;V)$ such that

$$\begin{aligned}
p_{tt} &= \Delta p + (u_\beta - h)\chi_G & \text{in} \quad & Q \\
p &= 0 & \text{on} \quad & \Sigma \times (0,T) \\
\tfrac{\partial p}{\partial \nu} &= 0 & \text{on} \quad & D \times \{z=0\} \times (0,T) \\
\tfrac{\partial p}{\partial \nu} + \sigma_\beta p &= 0 & \text{on} \quad & \Gamma \times (0,T) \\
p &= p_t = 0, & \text{on} \quad & \Omega \times \{T\},
\end{aligned} \quad (2.2)$$

and

$$\sigma_\beta(x,y) = \min\left\{ \left(\frac{1}{\beta}\int_0^T u_\beta p(x,y,u(x,y),t)\,dt\right)^+, K \right\}. \qquad (2.3)$$

We now let $\beta \to 0$ and the sequence of optimal controls σ_β converge to the desired coefficient [7].

Theorem 2.2. *Suppose the inverse problem has a solution, i.e., there exists $\sigma^* \in U$ such that $u^* = u(\sigma^*)$ satisfies $u^* = z$ a.e. on $G \times (0,T)$. Then there exists $\sigma_0 \in U$ such that on a subsequence $\beta_n \to 0$, we have*

$$\begin{aligned}
\sigma_{\beta_n} &\rightharpoonup \sigma_0 & \text{in} \quad & L^2(\Gamma) \\
w_{\beta_n} = w(\beta_n) &\rightharpoonup w_0 & \text{in} \quad & L^2(0,T;V)
\end{aligned}$$

and

$$w_0 = z \quad a.\ e.\ on \quad G \times (0,T).$$

We also solved a more complicated problem along this line, namely we identified at the same time the shape of part of the boundary, $w(x,y)$, and the reflection coefficient, $\sigma(x,y)$ [5].

3. THIRD EXAMPLE

We apply the optimal control techniques to reconstruct the dispersive coefficient in a wave equation from a single input of Neumann data and possibly noisy observation (output) of

Solving Identification Problems for the Wave Equation

Dirichlet data. For $\Omega \subset \mathbb{R}^n$ and $Q = \Omega \times (0,T)$, consider the wave equation:

$$\begin{aligned} u_{tt} &= \Delta u + hu + f &&\text{in} \quad Q \\ u &= u_0, \quad u_t = u_1, &&\text{on} \quad \Omega \times \{t = 0\} \\ \tfrac{\partial u}{\partial \eta} &= g &&\text{on} \quad \partial\Omega \times (0,T). \end{aligned} \quad (3.1)$$

Given Neumann data, $g \in H^{1/2}(\partial\Omega \times (0,T))$, we seek to identify the dispersive coefficient h, from observations z on $\partial\Omega \times (t_1, t_2)$. This reconstruction is done from a *single* Neumann-to-Dirichlet type measurement.

The approximate control problem treats h as a control in

$$U = \{h \in L^\infty(\Omega) | -M \leq H(x) \leq M\}$$

and seeks to minimize for $\beta > 0$,

$$J_\beta(h) = \frac{1}{2}\left(\int_{\partial\Omega \times (t_1, t_2)} (u(h) - z)^2 \, ds \, dt + \beta \int_\Omega h^2 \, dx\right).$$

We make the following assumptions:

$$\partial\Omega \in C^2$$

$$z \in L^2(\partial\Omega \times (t_1, t_2)) \quad \text{with } 0 \leq t_1 < t_2 \leq T$$

$$g \in H^{1/2}(\partial\Omega \times (0,T)), \quad f \in L^2(Q)$$

$$u_0 \in H^1(\Omega), \quad u_1 \in L^2(\Omega).$$

The solution space for the state problem is

$$u \in H^1(Q) \quad \text{with} \quad u_{tt} \in L^2\left(0, T; (H^1)^*(\Omega)\right).$$

We gave a complete characterization of the optimal control h_β in [4].

Theorem 3.1. *There exists an optimal control h_β and corresponding state, $u_\beta = u(h_\beta)$, minimizing $J_\beta(h)$ over U. Furthermore, there exists a solution p in $H^{1/2}(Q)$ solving the* adjoint problem:

$$p_{tt} = \Delta p + h_\beta p \quad \text{in} \quad Q$$
$$p = 0, \quad p_t = 0, \quad \text{on} \quad \Omega \times \{T\} \qquad (3.2)$$
$$\frac{\partial p}{\partial \eta} = (u_\beta - z)\chi_{(t_1,t_2)} \quad \text{on} \quad \partial\Omega \times (0,T)$$

where $\chi_{(t_1,t_2)}$ is the characteristic function of the time interval (t_1, t_2), and

$$h_\beta = \min\left(\max\left(-\frac{1}{\beta}\int_0^T u_\beta p(x,t)\,dt, -M\right), M\right). \qquad (3.3)$$

Note that the solution space for the adjoint problem is weaker than the state solution space due to the less regular Neumann data.

Under additional regularity assumptions of f, z, and g, we can prove the uniqueness of the optimal control h_β.

Now as $\beta \to 0$, we do not assume the identification problem has a solution, which allows for noisy or inaccurate observations.

Theorem 3.2. *There exists*

(i) *a sequence $\beta_n \to 0$*

(ii) *a sequence of corresponding optimal controls, h_{β_n}, for the functionals $J_{\beta_n}(h)$,*

(iii) *$h^* \in U$ and $u^* = u(h^*)$, such that*

$$\begin{aligned} h_n &\rightharpoonup h^* & \text{weakly in} & \quad L^2(\Omega) \\ u(h_{\beta_n}) &\rightharpoonup u^* & \text{weakly in} & \quad H^1(Q) \\ u(\beta_n)_{tt} &\rightharpoonup u^*_{tt} & \text{weak}^* \text{ in} & \quad L^2(0,T; H^1(Q)^*) \end{aligned}$$

and

$$\int_{\partial\Omega \times (t_1,t_2)} (u^* - z)^2 \, ds\, dt = \inf_{h \in U} \int_{\partial\Omega \times (t_1,t_2)} (u(h) - z)^2 \, ds\, dt.$$

We do not assume z is in the range of the maps ND_h:

$$ND_h : H^{1/2}(\partial\Omega \times (0,T)) \to L^2(\partial\Omega \times (t_1,t_2))$$
$$ND_h(g) = u(h)|_{\partial\Omega \times (t_1,t_2)}.$$

Solving Identification Problems for the Wave Equation

Thus the limit u^* can be interpreted as a projection of z onto the range of these maps. Numerical illustrations related to Example 3 can be found in Ref. [4].

To consider multiple inputs g and corresponding observations $z(g)$, define the input set

$$G = \{g \in H^{1/2}(\partial\Omega \times (0,T))|\ \|g\|_{H^{1/2}(\partial\Omega\times(0,T))} \leq K_2\},$$

and the set of observations

$$\{z \in H^{1/2}(\partial\Omega \times (0,T))|\ \|z\|_{H^{1/2}(\partial\Omega\times(0,T))} \leq K_2\},$$

where K_2 depends on K_1. We assume the map

$$g \in G \to z(g)$$

is compact, i.e.

$$g_n \rightharpoonup g^* \quad \text{in } H^{1/2}(\partial\Omega \times (0,T))$$

implies $z(g_n) \to z(g^*)$ in $L^2(\partial\Omega \times (0,T))$. We note that the solution u of the wave equation depends on h, g; i.e.,

$$u = u(h, g).$$

Then we can prove:

Theorem 3.3. *There exists $u^* \in L^2(0,T; H^1(\Omega))$ with $u_{tt} \in L^2(0,T; H^1(\Omega)^*), h^* \in U$ and $g^* \in H^{1/2}(\partial\Omega \times (0,T))$, such that $u^* = u(h^*, g^*)$ and*

$$\int_{\partial\Omega\times(t_1,t_2)} (u^* - z(g^*))^2\, ds\, dt = \min_{g\in G}\min_{h\in U} \int_{\partial\Omega\times(t_1,t_2)} |u(h,g) - z(g)|^2\, ds\, dt.$$

References

[1] Banks, H.T. and K.K. Kunish, *Estimation Techniques for Distributed Parameter Systems*, Birkhäuser, Boston, 1989.

[2] Borggard, J. and J. Burns, *PDE Sensitivity Equation Methods for Optimal Aerodynamic Design*, ICASE Report 96-44, NASA Langley Research Center.

[3] Isakov, V., *Inverse Problems for Partial Differential Equations*, Springer Verlag, Berlin, 1998.

[4] Feng, X., S. Lenhart, V. Protopopescu, L. Rachele, and B. Sutton, *Identification Problem for the Wave Equation with Neumann Data Input and Dirichlet Data Observations*, submitted to IMA Journal on Applied Math.

[5] Lenhart, S., V. Protopopescu, and J. Yong, *Identification of Boundary Shape and Reflexivity in a Wave Equation by Optimal Control Techniques*, Diff. and Int. Eqs., 13(2000), 941-972.

[6] Lenhart, S., V. Protopopescu, and J. Yong, *Identification of of a Reflection Boundary Coefficient in an Acoustic Wave Equation*, ISAACS Conference Proceedings, Direct and Inverse Problems of Mathematical Physics, Kluwer Publishers, 2000, 251-266.

[7] Lenhart, S., V. Protopopescu, and J. Yong, *Optimal Control of a Reflection Boundary Coefficient in an Acoustic Wave Equation*, Applicable Analysis 69(1998), 179-194.

[8] Liang, M., *Bilinear Optimal Control for a Wave Equation*, Math. Models and Methods in Applied Sciences, 9(1999), 45-68.

[9] Liang, M., S. Lenhart, and V. Protopopescu, *Identification Problem for a Wave Equation via Optimal Control, Control of Distributed Parameter and Stochastic Systems*, Kluwer Academic Press, Boston, 1999, 79-84.

[10] Puel, J. P., and M. Yamamoto, *Applications of Exact Controllability to Some Inverse Hyperbolic Problems* C. R. Acad. Sci. Paris, 320(1995), Series 1, 1171-1176.

[11] Puel, J. P., and M. Yamamoto, *On a Global Estimate in a Linear Inverse Hyperbolic Problems*, Inverse Problems 12(1996), 995-1002.

[12] Tikhonov, A. N., and V.Y. Arsenin, *Solutions of Ill-posed Problems*, John Wiley, New York, 1977.

[13] Yamamoto, M., *Stability, Reconstruction Formula, and Regularization for an Inverse Source Hyperbolic Problem by a Control Method*, Inverse Problems 11(1995), 481-496.

Singular Perturbations and Approximations for Integrodifferential Equations

J. Liu, J. Sochacki, P. Dostert
Department of Mathematics, James Madison University, Harrisonburg, VA 22807.

Abstract

Let $\varepsilon > 0$ and consider

$$\begin{cases} \varepsilon^2 u''(t;\varepsilon) + u'(t;\varepsilon) = Au(t;\varepsilon) + \int_0^t b(t-s)Au(s;\varepsilon)ds + f(t;\varepsilon), & t \geq 0, \\ u(0;\varepsilon) = u_0(\varepsilon), \quad u'(0;\varepsilon) = u_1(\varepsilon), \end{cases}$$

and

$$w'(t) = Aw(t) + \int_0^t b(t-s)Aw(s)ds + f(t), \quad t \geq 0, \quad w(0) = w_0,$$

in a Banach space X. Here the unbounded operator A is the generator of a strongly continuous cosine family and a strongly continuous semigroup, and $b(\cdot)$ is a continuous scalar function. We will look at the singular perturbations when $\varepsilon \to 0$, and approximate the above two integrodifferential equations with two corresponding systems of ordinary differential equations for which the numerical solutions can be carried out easily. An application to partial differential equations with numerical solutions is given.

1 INTRODUCTION.

We study the integrodifferential equations

$$\begin{cases} \varepsilon^2 u''(t;\varepsilon) + u'(t;\varepsilon) = Au(t;\varepsilon) + \int_0^t b(t-s)Au(s;\varepsilon)ds + f(t;\varepsilon), & t \geq 0, \\ u(0;\varepsilon) = u_0(\varepsilon), \quad u'(0;\varepsilon) = u_1(\varepsilon), \end{cases} \quad (1.1)$$

and

$$\begin{cases} w'(t) = Aw(t) + \int_0^t b(t-s)Aw(s)ds + f(t), \quad t \geq 0, \\ w(0) = w_0, \end{cases} \quad (1.2)$$

in a Banach space X, where the unbounded operator A is the generator of a strongly continuous cosine family and a strongly continuous semigroup, and $b(\cdot)$ is a continuous scalar function. We regard Eq.(1.2) as the limiting equation of Eq.(1.1) as $\varepsilon \to 0$. Now, Eq.(1.2) is of lower order in derivative of t, in this sense we say that we are dealing with the singular perturbation problems.

In an early study in Liu [9], it was shown that under some convergence conditions on the initial data and $f(t;\varepsilon)$, one has $u(\cdot;\varepsilon) \to w(\cdot)$ as $\varepsilon \to 0$. In this paper, we will apply the techniques we developed in [10] to approximate the two integrodiffernetial equations (1.1) and (1.2) with two corresponding systems of ordinary differential equations for which the numerical solutions can be derived easily. This way, we can provide a very useful procedure to numerically approximate the integrodifferential equation (1.1) arising from engineering with some simpler system of ordinary differential equations. Also, we will be able to see how the singular perturbations are carried out as $\varepsilon \to 0$. An application with numerical solutions will be given to a partial differential equation.

2 APPROXIMATION METHODS.

In this paper we make the following hypotheses ([6]) :

(H1). The operator A generates a strongly continuous cosine family and a strongly continuous semigroup.

(H2). $b(\cdot) \in C^2(R^+, R)$, $R^+ = [0, \infty)$.

(H3). $f(\cdot; \varepsilon)$ and $f \in C^1(R^+, X)$, $\varepsilon > 0$.

(H4). $u_0(\varepsilon), w_0 \in D(A), u_0(\varepsilon) \to w_0$, $\varepsilon^2 u_1(\varepsilon) \to 0$, as $\varepsilon \to 0$.

(H5). For any $T > 0, f(\cdot; \varepsilon) \to f(\cdot)$ in $L^1([0, T], X)$ as $\varepsilon \to 0$.

We say that $u : R^+ \to X$ is a solution of Eq.(1.1) if $u \in C^2(R^+, X), u(t) \in D(A)$ (domain of A) for $t \geq 0$ and Eq.(1.1) is satisfied on R^+. Solutions of Eq.(1.2) are defined in a similar way. Note that the existence and uniqueness of solutions of Eqs.(1.1) and (1.2) are obtained in [3, 5, 7, 14, 15], and we are only interested in singular perturbations and approximations in this paper, thus we assume that Eqs.(1.1) and (1.2) have unique solutions $u(t; \varepsilon)$ and $w(t)$ respectively for every $\varepsilon > 0$, when the initial data satisfy certain conditions.

From the singular perturbations results in [9], we have

Theorem 2.1. [9] Assume that hypotheses (H1) – (H5) are satisfied and let $T > 0$ be fixed. Then as $\varepsilon \to 0$, one has $u(t; \varepsilon) \to w(t)$ in X uniformly for $t \in [0, T]$.

Next, we modify the techniques we developed in [10] so as to approximate Eq.(1.1) with a simpler system of ordinary differential equations.

First, as in [10], we invert $Au(\cdot)$ so that Eq.(1.1) can be transformed into a form with continuous kernel, that is, the unbounded operator A will not appear in the integral.

Theorem 2.2. Equation (1.1) is equivalent to

$$\begin{cases} \widetilde{\varepsilon}^2 u''(t; \widetilde{\varepsilon}) + u'(t; \widetilde{\varepsilon}) = \widetilde{A} u(t; \widetilde{\varepsilon}) + \int_0^t \widetilde{b}(t-s) u(s; \widetilde{\varepsilon}) ds + \widetilde{f}(t; \widetilde{\varepsilon}), & t \geq 0, \\ u(0; \widetilde{\varepsilon}) = u_0(\widetilde{\varepsilon}), \ u'(0; \widetilde{\varepsilon}) = u_1(\widetilde{\varepsilon}), \end{cases} \qquad (2.1)$$

where $\widetilde{b}(\cdot)$ is a continuous scalar function determined by $b(\cdot)$.

Proof. Define

$$R * H(t) = \int_0^t R(t-s) H(s) ds \text{ and } \delta * H = H.$$

Then we can find a C^2 solution F of $F + b + F * b = 0$ (see, e.g., [2, 4, 8, 11, 12]) such that

$$(\delta + F) * (\delta + b) = \delta. \qquad (2.2)$$

Now, write Eq.(1.1) as

$$\varepsilon^2 u''(\varepsilon) + u'(\varepsilon) = (\delta + b) * Au(\varepsilon) + f(\varepsilon).$$

Then we have

$$(\delta + F) * \left[\varepsilon^2 u''(\varepsilon) + u'(\varepsilon) \right] = Au(\varepsilon) + (\delta + F) * f(\varepsilon).$$

Hence
$$\varepsilon^2 u''(\varepsilon) + u'(\varepsilon) = Au(\varepsilon) + (\delta + F) * f(\varepsilon) - F * \left[\varepsilon^2 u''(\varepsilon) + u'(\varepsilon)\right].$$

Integration by parts yields
$$F * u'(t;\varepsilon) = \int_0^t F'(t-s)u(s;\varepsilon)ds + F(0)u(t;\varepsilon) - F(t)u_0(\varepsilon),$$

$$F * u''(t;\varepsilon) = \int_0^t F''(t-s)u(s;\varepsilon)ds + F(0)u'(t;\varepsilon) - F(t)u_1(\varepsilon) + F'(0)u(t;\varepsilon) - F'(t)u_0(\varepsilon).$$

Therefore Eq.(1.1) can be replaced by
$$\begin{aligned}\varepsilon^2 u''(t;\varepsilon) + [1 + \varepsilon^2 F(0)]u'(t;\varepsilon) &= [A - F(0) - \varepsilon^2 F'(0)]u(t;\varepsilon) \\ &\quad + \int_0^t [-F'(t-s) - \varepsilon^2 F''(t-s)]u(s;\varepsilon)ds \\ &\quad + [(\delta + F) * f(t;\varepsilon) + F(t)u_0(\varepsilon) + \varepsilon^2 F'(t)u_0(\varepsilon) \\ &\quad + \varepsilon^2 F(t)u_1(\varepsilon)], \end{aligned} \quad (2.3)$$

hence we complete the proof by dividing $[1 + \varepsilon^2 F(0)]$. $\square$

Next, for Eq.(2.1), we approximate the continuous scalar function $\tilde{b}(\cdot)$ by a polynomial $P_n(\cdot)$ of degree n, and then use
$$\begin{cases} \tilde{\varepsilon}^2 v_n''(t;\tilde{\varepsilon}) + v_n'(t;\tilde{\varepsilon}) = \tilde{A}v_n(t;\tilde{\varepsilon}) + \int_0^t P_n(t-s)v_n(s;\tilde{\varepsilon})ds + \tilde{f}(t;\tilde{\varepsilon}), & t \geq 0, \\ v_n(0;\tilde{\varepsilon}) = u_0(\tilde{\varepsilon}), \quad v_n'(0;\tilde{\varepsilon}) = u_1(\tilde{\varepsilon}), \end{cases} \quad (2.4)$$

to approximate Eq.(2.1). To do that, we need the following inequality.

Lemma 2.1.[1] Let $u(t)$ be a nonnegative continuous function for $t \geq \alpha$, and suppose that
$$u(t) \leq C_1 + \int_\alpha^t k(t,s)u(s)ds + \int_\alpha^t \left(\int_\alpha^s h(t,s,r)u(r)dr\right)ds, \quad t \geq \alpha,$$

where $C_1 \geq 0$ is a constant and $k(t,s)$ and $h(t,s,r)$ are nonnegative continuous functions for $\alpha \leq r \leq s \leq t$. If the functions $k(t,s)$ and $h(t,s,r)$ are nondecreasing in t for fixed s, r, then
$$u(t) \leq C_1 \exp\left\{\int_\alpha^t k(t,s)ds + \int_\alpha^t \left(\int_\alpha^s h(t,s,r)dr\right)ds\right\}, \quad t \geq \alpha.$$

Now, we can prove the following

Theorem 2.3. Let $T_0 > 0$ be fixed. For any $\delta > 0$, there is a polynomial $P_n(\cdot)$ of degree n such that for the solution $v_n(\cdot;\tilde{\varepsilon})$ of Eq.(2.4) and the solution $u(\cdot;\tilde{\varepsilon})$ of Eq.(2.1) (or Eq.(1.1)), one has, uniformly for $0 < \tilde{\varepsilon} \leq K$ where K is any given constant,
$$\max_{t \in [0,T_0]} \|u(t;\tilde{\varepsilon}) - v_n(t;\tilde{\varepsilon})\| \leq \delta. \quad (2.5)$$

Proof. Note that $\tilde{A}$ also generates a strongly continuous cosine family, which we denote by

$C(\cdot)$. Then we can use the results in [6] to get, for $t \geq 0$,

$$\begin{aligned} u(t;\widetilde{\varepsilon}) &= e^{-t/2\widetilde{\varepsilon}^2}C(t/\widetilde{\varepsilon})u_0(\widetilde{\varepsilon}) + \frac{1}{2}R(t;\widetilde{\varepsilon})u_0(\widetilde{\varepsilon}) \\ &\quad + G(t;\widetilde{\varepsilon})\left[\frac{1}{2}u_0(\widetilde{\varepsilon}) + \widetilde{\varepsilon}^2 u_1(\widetilde{\varepsilon})\right] \\ &\quad + \int_0^t G(t-s;\widetilde{\varepsilon})\left[\int_0^s \widetilde{b}(s-r)u(r;\widetilde{\varepsilon})dr + \widetilde{f}(s;\widetilde{\varepsilon})\right]ds, \end{aligned} \quad (2.6)$$

$$\begin{aligned} v_n(t;\widetilde{\varepsilon}) &= e^{-t/2\widetilde{\varepsilon}^2}C(t/\widetilde{\varepsilon})u_0(\widetilde{\varepsilon}) + \frac{1}{2}R(t;\widetilde{\varepsilon})u_0(\widetilde{\varepsilon}) \\ &\quad + G(t;\widetilde{\varepsilon})\left[\frac{1}{2}u_0(\widetilde{\varepsilon}) + \widetilde{\varepsilon}^2 u_1(\widetilde{\varepsilon})\right] \\ &\quad + \int_0^t G(t-s;\widetilde{\varepsilon})\left[\int_0^s P_n(s-r)v_n(r;\widetilde{\varepsilon})dr + \widetilde{f}(s;\widetilde{\varepsilon})\right]ds, \end{aligned} \quad (2.7)$$

where $R(\cdot;\widetilde{\varepsilon}), G(\cdot;\widetilde{\varepsilon})$ are linear operators defined in [6] using the Bessel functions; and they have the following properties: For some independent constants $M > 1$ and $a > 0$,

(P1). $\|G(t;\widetilde{\varepsilon})\| \leq Me^{at}$, $t \geq 0$, $\widetilde{\varepsilon} > 0$.

(P2). $\|e^{-t/2\widetilde{\varepsilon}^2}C(t/\widetilde{\varepsilon})u_0(\widetilde{\varepsilon}) + \frac{1}{2}R(t;\widetilde{\varepsilon})u_0(\widetilde{\varepsilon}) + G(t;\widetilde{\varepsilon})\left[\frac{1}{2}u_0(\widetilde{\varepsilon}) + \widetilde{\varepsilon}^2 u_1(\widetilde{\varepsilon})\right]\|$

$\leq Me^{at}\left[\|u_0(\widetilde{\varepsilon})\| + \widetilde{\varepsilon}^2\|u_1(\widetilde{\varepsilon})\|\right]$, $t \geq 0$, $\widetilde{\varepsilon} > 0$.

Using (P2) and Lemma 2.1, we can verify that there is a constant $\overline{C}$ independent of n and $\widetilde{\varepsilon}$ such that $\|v_n(t;\widetilde{\varepsilon})\| < \overline{C}$ for $0 < \widetilde{\varepsilon} \leq K$, $n = 1, 2, ...$, and $t \in [0, T_0]$. (See the treatment of $u(t;\widetilde{\varepsilon}) - v_n(t;\widetilde{\varepsilon})$ below for details.) Therefore,

$$\begin{aligned} \|u(t;\widetilde{\varepsilon}) - v_n(t;\widetilde{\varepsilon})\| &\leq \left\|\int_0^t G(t-s;\widetilde{\varepsilon})\left[\int_0^s \left(\widetilde{b}(s-r)u(r;\widetilde{\varepsilon}) - P_n(s-r)v_n(r;\widetilde{\varepsilon})\right)dr\right]ds\right\| \\ &\leq \left\|\int_0^t G(t-s;\widetilde{\varepsilon})\left[\int_0^s \left\{\widetilde{b}(s-r)[u(r;\widetilde{\varepsilon}) - v_n(r;\widetilde{\varepsilon})]\right.\right.\right. \\ &\quad \left.\left.\left. + [\widetilde{b}(s-r) - P_n(s-r)]v_n(r;\widetilde{\varepsilon})\right\}dr\right]ds\right\| \end{aligned} \quad (2.8)$$

$$\begin{aligned} &\leq \int_0^t \int_0^s Me^{a(t-s)}|\widetilde{b}(s-r) - P_n(s-r)|\|v_n(r;\widetilde{\varepsilon})\|drds \\ &\quad + \int_0^t \int_0^s Me^{a(t-s)}|\widetilde{b}(s-r)|\|u(r;\widetilde{\varepsilon}) - v_n(r;\widetilde{\varepsilon})\|drds \end{aligned} \quad (2.9)$$

$$\begin{aligned} &\leq Me^{aT_0}T_0^2\overline{C}\max_{l\in[0,T_0]}|\widetilde{b}(l) - P_n(l)| \\ &\quad + \int_0^t \int_0^s Me^{a(t-s)}|\widetilde{b}(s-r)|\|u(r;\widetilde{\varepsilon}) - v_n(r;\widetilde{\varepsilon})\|drds. \end{aligned} \quad (2.10)$$

Now, we can apply Lemma 2.1 with $k(t,s) = 0$ and $h(t,s,r) = Me^{a(t-s)}|\widetilde{b}(s-r)|$ to obtain

$\|u(t;\widetilde{\varepsilon}) - v_n(t;\widetilde{\varepsilon})\| \leq$

$\left(\max_{l\in[0,T_0]}|\widetilde{b}(l) - P_n(l)|\right)Me^{aT_0}T_0^2\overline{C}\exp\left\{\int_0^t \int_0^s Me^{a(t-s)}|\widetilde{b}(s-r)|drds\right\}.$ (2.11)

Singular Perturbations for Integrodifferential Equations

Therefore we can find P_n to approximate $\tilde{b}$ to complete the proof. □

Now, for this polynomial $P_n(\cdot)$ of degree n, the $(n+1)$th derivative is zero. So we are able to rewrite (2.4) as a system of ordinary differential equations.

Theorem 2.4. Eq.(2.4) is equivalent to an ordinary differential equation

$$\begin{cases} Y'(t) = GY(t) + H(t), & t \geq 0, \\ Y(0) = Y_0, \end{cases} \quad (2.12)$$

on X^{n+3}, where

$$G = \begin{bmatrix} 0 & I & 0 & 0 & \cdots & 0 & 0 \\ \frac{1}{\tilde{\varepsilon}^2}\tilde{A} & -\frac{1}{\tilde{\varepsilon}^2}I & \frac{1}{\tilde{\varepsilon}^2}I & 0 & \cdots & 0 & 0 \\ P_n(0)I & 0 & 0 & I & \cdots & 0 & 0 \\ P_n'(0)I & 0 & 0 & 0 & \cdots & 0 & 0 \\ \cdot & & & & \cdots & 0 & 0 \\ \cdot & & & & \cdots & 0 & 0 \\ \cdot & & & & \cdots & I & 0 \\ P_n^{(n-1)}(0)I & 0 & 0 & 0 & \cdots & 0 & I \\ P_n^{(n)}(0)I & 0 & 0 & 0 & \cdots & 0 & 0 \end{bmatrix}, \quad H(t) = \begin{bmatrix} 0 \\ \frac{1}{\tilde{\varepsilon}^2}\tilde{f}(t;\tilde{\varepsilon}) \\ 0 \\ 0 \\ \cdot \\ \cdot \\ \cdot \\ 0 \\ 0 \end{bmatrix}, \quad Y_0 = \begin{bmatrix} u_0(\tilde{\varepsilon}) \\ u_1(\tilde{\varepsilon}) \\ 0 \\ 0 \\ \cdot \\ \cdot \\ \cdot \\ 0 \\ 0 \end{bmatrix},$$

with I the identity operator. And G generates a strongly continuous semigroup on X^{n+3}. That is, Eq.(2.12) is wellposed on X^{n+3}.

Proof. In equation (2.4), define

$$y_1(t;\tilde{\varepsilon}) = v_n(t;\tilde{\varepsilon}), \quad (2.13)$$
$$y_2(t;\tilde{\varepsilon}) = v_n'(t;\tilde{\varepsilon}). \quad (2.14)$$

Then

$$y_2'(t;\tilde{\varepsilon}) = v_n''(t;\tilde{\varepsilon}) \quad (2.15)$$
$$= -\frac{1}{\tilde{\varepsilon}^2}y_2(t;\tilde{\varepsilon}) + \frac{1}{\tilde{\varepsilon}^2}\tilde{A}y_1(t;\tilde{\varepsilon}) + \frac{1}{\tilde{\varepsilon}^2}\int_0^t P_n(t-s)y_1(s;\tilde{\varepsilon})ds + \frac{1}{\tilde{\varepsilon}^2}\tilde{f}(t;\tilde{\varepsilon}). \quad (2.16)$$

Now define

$$y_3(t;\tilde{\varepsilon}) = \int_0^t P_n(t-s)y_1(s;\tilde{\varepsilon})ds. \quad (2.17)$$

Then

$$y_3'(t;\tilde{\varepsilon}) = P_n(0)y_1(t;\tilde{\varepsilon}) + \int_0^t P_n'(t-s)y_1(s;\tilde{\varepsilon})ds. \quad (2.18)$$

Next, define

$$y_4(t;\tilde{\varepsilon}) = \int_0^t P_n'(t-s)y_1(s;\tilde{\varepsilon})ds, \quad (2.19)$$

to get

$$y_4'(t;\tilde{\varepsilon}) = P_n'(0)y_1(t;\tilde{\varepsilon}) + \int_0^t P_n''(t-s)y_1(s;\tilde{\varepsilon})ds. \quad (2.20)$$

Continuing in this way we obtain

$$y_k(t;\widetilde{\varepsilon}) = \int_0^t P_n^{(k-3)}(t-s)y_1(s;\widetilde{\varepsilon})ds, \tag{2.21}$$

$$y_k'(t;\widetilde{\varepsilon}) = P_n^{(k-3)}(0)y_1(t;\widetilde{\varepsilon}) + \int_0^t P_n^{(k-2)}(t-s)y_1(s;\widetilde{\varepsilon})ds, \tag{2.22}$$

and finally, we have

$$y_{n+3}(t;\widetilde{\varepsilon}) = \int_0^t P_n^{(n)}(t-s)y_1(s;\widetilde{\varepsilon})ds, \tag{2.23}$$

$$y_{n+3}'(t;\widetilde{\varepsilon}) = P_n^{(n)}(0)y_1(t;\widetilde{\varepsilon}). \tag{2.24}$$

Therefore, Eq.(2.4) is equivalent to Eq.(2.12). Next, G generates a strongly continuous semigroup on X^{n+3} by using the perturbation results from the semigroup theory. This completes the proof. □

Now we see that by applying Theorems 2.2 – 2.4, the solution of the integrodifferential equation (1.1) can be approximated by $y_1(\cdot;\widetilde{\varepsilon}) = v_n(\cdot;\widetilde{\varepsilon})$, the first component of the solution of the system of ordinary differential equations (2.12) when $0 < \widetilde{\varepsilon} \leq K$, where K is any given constant.

Next, the results concerning the singular perturbations in [9] implies that the solution u of Eq.(1.1) and w of Eq.(1.2) are almost the same when $\varepsilon \approx 0$. Now, from [10], Eq.(1.2) can be replaces by

$$w'(t) = \Big(A + b(0)I\Big)w(t) - \int_0^t F'(t-s)w(s)ds + \widetilde{f}(t), \quad t \geq 0, \quad w(0) = w_0, \tag{2.25}$$

where F is from (2.2) and $\widetilde{f}(t)$ is determined by $f(t)$. Also from [10], $w(\cdot)$ of (2.25) can be approximated by $z_1(\cdot)$, the first component of the solution of

$$\begin{aligned} Z'(t) &= G_1 Z(t) + F_1(t), \quad t \geq 0, \\ Z(0) &= Z_0, \end{aligned} \tag{2.26}$$

on X^{n+2}, with

$$G_1 = \begin{bmatrix} A+b(0)I & I & 0 & \cdots & 0 & 0 \\ \hat{P}_n(0)I & 0 & I & \cdots & 0 & 0 \\ \hat{P}_n'(0)I & 0 & 0 & \cdots & 0 & 0 \\ \vdots & & & & 0 & 0 \\ \vdots & & & & 0 & 0 \\ \vdots & & & & I & 0 \\ \hat{P}_n^{(n-1)}(0)I & 0 & 0 & \cdots & 0 & I \\ \hat{P}_n^{(n)}(0)I & 0 & 0 & \cdots & 0 & 0 \end{bmatrix}, \quad F_1(t) = \begin{bmatrix} \widetilde{f}(t) \\ 0 \\ 0 \\ \vdots \\ \vdots \\ \vdots \\ 0 \\ 0 \end{bmatrix}, \quad Z_0 = \begin{bmatrix} w_0 \\ 0 \\ 0 \\ \vdots \\ \vdots \\ \vdots \\ 0 \\ 0 \end{bmatrix}, \tag{2.27}$$

where $\hat{P}_n$ is a polynomial of degree n approximating $-F'$. Therefore, these results imply that for ε small, z_1 from Eq.(2.26) can be used to approximate u of Eq.(1.1).

Summarizing the above results, we have the following methods to approximate the solution u of Eq.(1.1): If ε is small, that is, if the singular perturbation is considered, then the first

Singular Perturbations for Integrodifferential Equations 239

component of the solution Z of Eq.(2.26) can be used. Otherwise, if ε is not small, then the first component of the solution Y of Eq.(2.12) can be used. In either case, we are able to approximate the integrodifferential equation (1.1) (which is hard to solve) by using a simpler system of ordinary differential equations.

3 AN APPLICATION.

Let us consider the following partial differential equation from engineering,

$$\begin{cases} \rho u_{tt}(t,x;\rho) + \alpha u_t(t,x;\rho) &= u_{xx}(t,x;\rho) + \int_0^t b(t-s)u_{xx}(s,x;\rho)ds + f(t,x;\rho), \\ u(t,0;\rho) &= u(t,1;\rho) = 0, \\ u(0,x;\rho) &= u_0(x;\rho), \; u_t(0,x;\rho) = u_1(x;\rho), \quad t \geq 0, \; x \in [0,1], \end{cases} \quad (3.1)$$

in $L^2[0,1]$, where u is the displacement of an object, ρ is the density per unit area, and α is the coefficient of viscosity of the medium. Divide α and change variables if necessary, we may assume that $\alpha = 1$. Therefore Eq.(3.1) is given by Eq.(1.1) with $\varepsilon^2 = \rho$ and $A = \frac{\partial^2}{\partial x^2}$ with (domain) $D(A) = W_0^{1,2}[0,1] \cap W^{2,2}[0,1]$. Thus, the results in [9] implies that, with some convergence conditions on the initial data and $f(t,x;\varepsilon)$, when the density $\rho \to 0$, solutions of (3.1) will converge to solutions of the "limiting" heat equation

$$\begin{cases} \alpha w_t(t,x) &= w_{xx}(t,x) + \int_0^t b(t-s)w_{xx}(s,x)ds + f(t,x), \\ w(t,0) &= w(t,1) = 0, \; w(0,x) = w_0(x), \quad t \geq 0, \; x \in [0,1]. \end{cases} \quad (3.2)$$

Next, let's look at how to approximate equations (3.1) and (3.2) with systems of ordinary differential equations. First, from Theorem 2.2 and a corresponding result from [10], we see that the integrals in Eq.(3.1) and Eq.(3.2) can be replaced by integrals with continuous kernels. Thus, without loss of generality, and also for computational reasons, we may consider the following equations with a continuous scalar kernel $E(\cdot)$,

$$\begin{cases} \varepsilon^2 u_{tt}(t,x;\varepsilon) + u_t(t,x;\varepsilon) &= u_{xx}(t,x;\varepsilon) + \int_0^t E(t-s)u(s,x;\varepsilon)ds + f(t,x;\varepsilon), \\ u(t,0;\varepsilon) &= u(t,1;\varepsilon) = 0, \\ u(0,x;\varepsilon) &= u_0(x;\varepsilon), \; u_t(0,x;\varepsilon) = u_1(x;\varepsilon), \quad t \geq 0, \; x \in [0,1], \end{cases} \quad (3.3)$$

and

$$\begin{cases} w_t(t,x) &= w_{xx}(t,x) + \int_0^t E(t-s)w(s,x)ds + f(t,x), \\ w(t,0) &= w(t,1) = 0, \; w(0,x) = w_0(x), \quad t \geq 0, \; x \in [0,1]. \end{cases} \quad (3.4)$$

Now, applying Theorems 2.3 – 2.4 to Eq.(3.3), that is, let $P_n(\cdot)$ be the approximation of $E(\cdot)$, we see that, for $Y(\varepsilon) = (y_1(\cdot,\cdot;\varepsilon), y_2(\cdot,\cdot;\varepsilon), ..., y_{n+3}(\cdot,\cdot;\varepsilon))$ in X^{n+3}, Eq.(2.12) for Eq.(3.3) becomes

$$\begin{cases} \frac{\partial}{\partial t}y_1(t,x;\varepsilon) &= y_2(t,x;\varepsilon), \\ \frac{\partial}{\partial t}y_2(t,x;\varepsilon) &= \frac{1}{\varepsilon^2}\frac{\partial^2}{\partial x^2}y_1(t,x;\varepsilon) - \frac{1}{\varepsilon^2}y_2(t,x;\varepsilon) + \frac{1}{\varepsilon^2}y_3(t,x;\varepsilon) + \frac{1}{\varepsilon^2}f(t,x;\varepsilon), \\ \frac{\partial}{\partial t}y_3(t,x;\varepsilon) &= P_n(0)y_1(t,x;\varepsilon) + y_4(t,x;\varepsilon), \\ \frac{\partial}{\partial t}y_4(t,x;\varepsilon) &= P_n'(0)y_1(t,x;\varepsilon) + y_5(t,x;\varepsilon), \\ \quad \vdots \\ \frac{\partial}{\partial t}y_{n+2}(t,x;\varepsilon) &= P_n^{(n-1)}(0)y_1(t,x;\varepsilon) + y_{n+3}(t,x;\varepsilon), \\ \frac{\partial}{\partial t}y_{n+3}(t,x;\varepsilon) &= P_n^{(n)}(0)y_1(t,x;\varepsilon), \end{cases} \quad (3.5)$$

with the corresponding initial and boundary conditions. Now the first component $y_1(t,x;\varepsilon)$ of the solution can be used to approximate the solution u of Eq.(3.3).

For Eq.(3.4), the kernel $E(\cdot)$ is the same as in Eq.(3.3), thus the P_n for Eq.(3.3) can be used as $\hat{P}_n$ for Eq.(3.4). Therefore, for $Z = (z_1(\cdot,\cdot), z_2(\cdot,\cdot), ..., z_{n+2}(\cdot,\cdot))$ in X^{n+2}, Eq.(2.26) for Eq.(3.4) becomes

$$\begin{cases} \frac{\partial}{\partial t} z_1(t,x) &= \frac{\partial^2}{\partial x^2} z_1(t,x) + z_2(t,x) + f(t,x), \\ \frac{\partial}{\partial t} z_2(t,x) &= P_n(0) z_1(t,x) + z_3(t,x), \\ \frac{\partial}{\partial t} z_3(t,x) &= P'_n(0) z_1(t,x) + z_4(t,x), \\ \quad \vdots \\ \frac{\partial}{\partial t} z_{n+1}(t,x) &= P_n^{(n-1)}(0) z_1(t,x) + z_{n+2}(t,x), \\ \frac{\partial}{\partial t} z_{n+2}(t,x) &= P_n^{(n)}(0) z_1(t,x), \end{cases} \quad (3.6)$$

with the corresponding initial and boundary conditions. From [10], the first component $z_1(t,x)$ of the solution can be used to approximate the solution w of Eq.(3.4).

The discussion at the end of the previous section indicates that when ε is small, z_1 from Eq.(3.6) can be used to approximate u of Eq.(3.3); otherwise, $y_1(\varepsilon)$ from Eq.(3.5) can be used. To check these results out, we look at the following example. But note that the solution u of Eq.(3.3) is difficult to obtain, thus we will demonstrate that for ε small, z_1 from Eq.(3.6) and $y_1(\varepsilon)$ from Eq.(3.5) are close, which, according to Theorem 2.3, implies that z_1 from Eq.(3.6) and u of Eq.(3.3) are close.

Example. Let $E(t) = \sin t$ in Eq.(3.3) and in Eq.(3.4) and let $f(t,x;\varepsilon) = 0$ in Eq.(3.3) and let $f(t,x) = 0$ in Eq.(3.4). Next, let $u_0(x;\varepsilon) = w_0(x) = x(1-x)$, and let $u_1(x;\varepsilon) = 0$. Then (H4) and (H5) are satisfied. We use $P_n(t) = \sum_{i=0}^{(n-1)/2} (-1)^i \frac{t^{2i+1}}{(2i+1)!}$ (when n is odd) as the approximation of $\sin t$. Now Eq.(3.5) becomes

$$\begin{cases} \frac{\partial}{\partial t} y_1(t,x;\varepsilon) &= y_2(t,x;\varepsilon), \\ \frac{\partial}{\partial t} y_2(t,x;\varepsilon) &= \frac{1}{\varepsilon^2} \frac{\partial^2}{\partial x^2} y_1(t,x;\varepsilon) - \frac{1}{\varepsilon^2} y_2(t,x;\varepsilon) + \frac{1}{\varepsilon^2} y_3(t,x;\varepsilon), \\ \frac{\partial}{\partial t} y_3(t,x;\varepsilon) &= y_4(t,x;\varepsilon), \\ \frac{\partial}{\partial t} y_4(t,x;\varepsilon) &= y_1(t,x;\varepsilon) + y_5(t,x;\varepsilon), \\ \quad \vdots \\ \frac{\partial}{\partial t} y_{n+2}(t,x;\varepsilon) &= y_{n+3}(t,x;\varepsilon), \\ \frac{\partial}{\partial t} y_{n+3}(t,x;\varepsilon) &= (-1)^{(n-1)/2} y_1(t,x;\varepsilon), \end{cases} \quad (3.7)$$

with the initial conditions

$$y_1(0,x;\varepsilon) = x(1-x), \quad y_j(0,x;\varepsilon) = 0, \quad j = 2,3,...,n+3, \quad x \in [0,1], \quad (3.8)$$

and the boundary conditions $y_j(t,0;\varepsilon) = y_j(t,1;\varepsilon) = 0$, $j = 1,...,n+3$, $t \geq 0$. Similarly,

Singular Perturbations for Integrodifferential Equations

Eq.(3.6) becomes

$$\begin{cases} \frac{\partial}{\partial t} z_1(t,x) &= \frac{\partial^2}{\partial x^2} z_1(t,x) + z_2(t,x), \\ \frac{\partial}{\partial t} z_2(t,x) &= z_3(t,x), \\ \frac{\partial}{\partial t} z_3(t,x) &= z_1(t,x) + z_4(t,x), \\ \quad \vdots \\ \frac{\partial}{\partial t} z_{n+1}(t,x) &= z_{n+2}(t,x), \\ \frac{\partial}{\partial t} z_{n+2}(t,x) &= (-1)^{(n-1)/2} z_1(t,x), \end{cases} \quad (3.9)$$

with the initial conditions

$$z_1(0,x) = x(1-x), \quad z_j(0,x) = 0, \quad j = 2, 3, ..., n+2, \quad x \in [0,1], \quad (3.10)$$

and the boundary conditions $z_j(t,0) = z_j(t,1) = 0$, $j = 1, ..., n+2$, $t \geq 0$.

In the following, we solve Eq.(3.7) and Eq.(3.9) numerically. We use Forward Euler in t and Centered Differences in x, and apply the modified Picard method presented in [13]. Therefore Eq.(3.7) becomes

$$\begin{cases} y_{1,k}(t_{j+1}, x_i; \varepsilon) &= y_{1,k}(t_j, x_i; \varepsilon) + \int_{t_j}^{t_{j+1}} y_{2,k-1}(s, x_i; \varepsilon) ds, \\ y_{2,k}(t_{j+1}, x_i; \varepsilon) &= y_{2,k}(t_j, x_i; \varepsilon) + \frac{1}{\varepsilon^2} \int_{t_j}^{t_{j+1}} \left[\frac{y_{1,k-1}(s, x_{i+1}; \varepsilon) - 2y_{1,k-1}(s, x_i; \varepsilon) + y_{1,k-1}(s, x_{i-1}; \varepsilon)}{(\Delta x)^2} \right. \\ &\qquad \left. - y_{2,k-1}(s, x_i; \varepsilon) + y_{3,k-1}(s, x_i; \varepsilon) \right] ds, \\ y_{3,k}(t_{j+1}, x_i; \varepsilon) &= y_{3,k}(t_{j+1}, x_i; \varepsilon) + \int_{t_j}^{t_{j+1}} y_{4,k-1}(s, x_i; \varepsilon) ds, \\ \quad \vdots \\ y_{n+2,k}(t_{j+1}, x_i; \varepsilon) &= y_{n+2,k}(t_j, x_i; \varepsilon) + \int_{t_j}^{t_{j+1}} y_{n+3,k-1}(s, x_i; \varepsilon) ds, \\ y_{n+3,k}(t_{j+1}, x_i; \varepsilon) &= y_{n+3,k}(t_j, x_i; \varepsilon) + \int_{t_j}^{t_{j+1}} (-1)^{(n-1)/2} y_{1,k-1}(s, x_i; \varepsilon) ds, \end{cases}$$

and Eq.(3.9) becomes

$$\begin{cases} z_{1,k}(t_{j+1}, x_i) &= z_{1,k}(t_j, x_i) + \int_{t_j}^{t_{j+1}} \left[\frac{z_{1,k-1}(s, x_{i+1}) - 2z_{1,k-1}(s, x_i) + z_{1,k-1}(s, x_{i-1})}{(\Delta x)^2} \right. \\ &\qquad \left. + z_{2,k-1}(s, x_i) \right] ds, \\ z_{2,k}(t_{j+1}, x_i) &= z_{2,k}(t_j, x_i) + \int_{t_j}^{t_{j+1}} z_{3,k-1}(s, x_i) ds, \\ \quad \vdots \\ z_{n+1,k}(t_{j+1}, x_i) &= z_{n+1,k}(t_j, x_i) + \int_{t_j}^{t_{j+1}} z_{n+2,k-1}(s, x_i) ds, \\ z_{n+2,k}(t_{j+1}, x_i) &= z_{n+2,k}(t_j, x_i) + \int_{t_j}^{t_{j+1}} (-1)^{(n-1)/2} z_{1,k-1}(s, x_i) ds. \end{cases}$$

In the figures below we present solutions to Eq.(3.7) and Eq.(3.9) using the above Picard iterations with $n = 5$, $x_i = i\Delta x$, $i = 1, ..., 20$, $\Delta x = 0.05$, $t_{j+1} - t_j = \Delta t = 0.00125$, and let $t \to 5$.

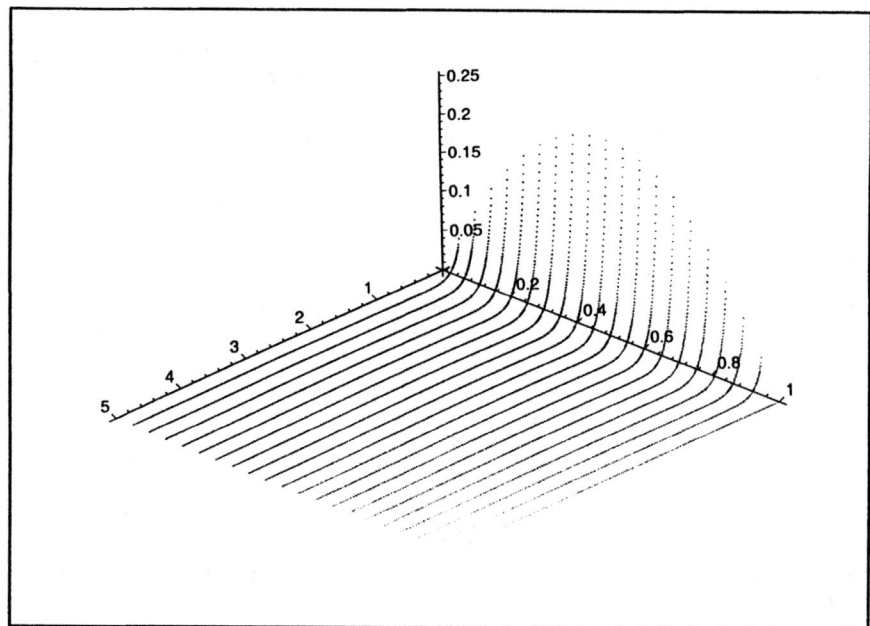

Fig. 1 $\epsilon = 0$

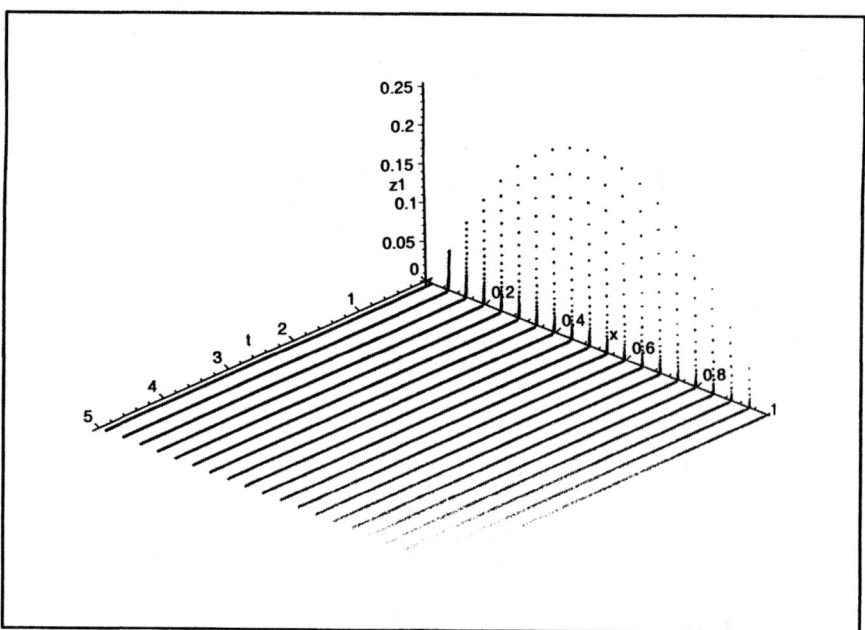

Fig. 2 $\epsilon = 0.015625$

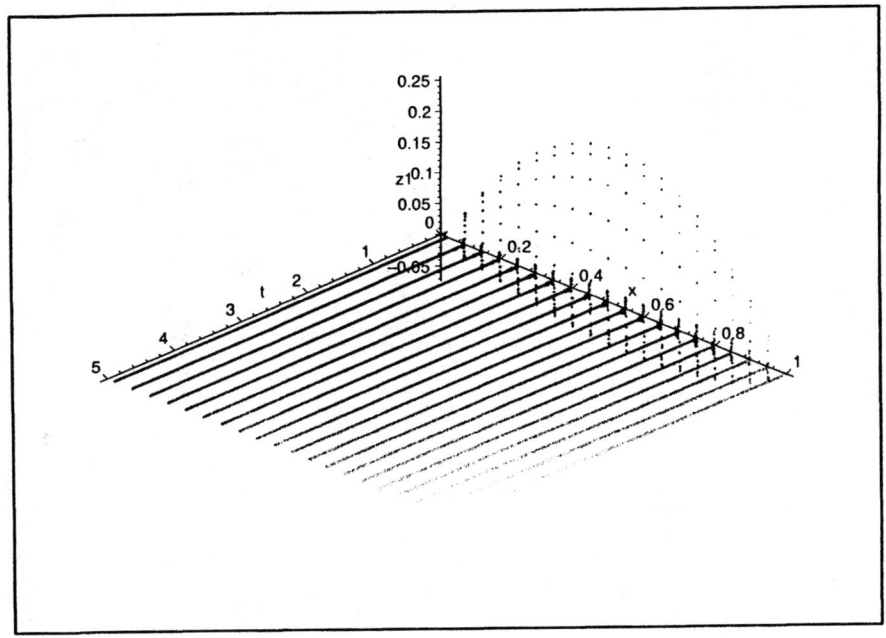

Fig. 3 $\epsilon = 0.0625$

References

[1] D. Bainov and P. Simeonov, *Integral inequalities and applications*, Kluwer Academic Publishers, Boston, 1992.

[2] W. Desch, R. Grimmer, *Propagation of singularities for integrodifferential equations*, J. Diff. Eq., **65**(1986), 411-426.

[3] W. Desch, R. Grimmer and W. Schappacher, *Some considerations for linear integrodifferential equations*, J. Math. Anal. & Appl., **104**(1984), 219-234.

[4] W. Desch, R. Grimmer and W. Schappacher, *Propagation of singularities by solutions of second order integrodifferential equations*, Volterra Integrodifferential Equations in Banach Spaces and Applications, G. Da Prato and M. Iannelli (eds.), Pitman Research Notes in Mathematics, Series 190, 101- 110.

[5] W. Desch and W. Schappacher, *A semigroup approach to integrodifferential equations in Banach space*, J. Integ. Eq., **10**(1985), 99-110.

[6] H. Fattorini, *Second order linear differential equations in Banach spaces*, North - Holland, 1985, 165-237.

[7] R. Grimmer and J. Liu, *Integrodifferential equations with nondensely defined operators*, Differential Equations with Applications in Biology, Physics, and Engineering, J. Goldstein, F. Kapple, and W. Schappacher (eds.), Marcel Dekker, Inc., New York, 1991, 185-199.

[8] G. Gripenberg, S-O. Londen and O. Staffans, *Volterra integral and functional equations*, Cambridge University Press, Cambridge, 1990.

[9] J. Liu, *Singular perturbations of integrodifferential equations in Banach space*, Proceedings of the American Mathematical Society, **122**(1994), 791-799.

[10] J. Liu, E. Parker, J. Sochacki, A. Knutsen, *Approximation methods for integrodifferential equations*, Proceedings of Dynamic Systems and Applications, Vol. III, to appear.

[11] R. MacCamy, *An integro - differential equation with application in heat flow*, Q. Appl. Math., **35**(1977), 1-19.

[12] R. MacCamy, *A model for one - dimensional nonlinear viscoelasticity*, Q. Appl. Math., **35**(1977), 21-33.

[13] E. Parker and J. Sochacki, *Implementing the Picard iteration*, Neural, Parallel & Scientific Computations, **4**(1996), 97-112.

[14] K. Tsuruta, *Bounded linear operators satisfying second order integrodifferential equations in Banach space*, J. Integ. Eq., **6**(1984), 231-268.

[15] C. Travis and G. Webb, *An abstract second order semi - linear Volterra integrodifferential equation*, SIAM J. Math. Anal., **10**(1979), 412-424.

Remarks on Impulse Control Problems for the Stochastic Navier-Stokes Equations

J.L. MENALDI
Wayne State University
Department of Mathematics
Detroit, Michigan 48202, USA
(e-mail jlm@math.wayne.edu)

S.S. SRITHARAN
US Navy
SPAWAR SSD – Code D73H
San Diego, CA 92152-5001, USA
(e-mail srith@spawar.navy.mil)

Abstract

In this paper we will review certain recent developments in impulse control problems for the stochastic Navier-Stokes equation. The dynamic programming equations for the optimal impulse control problem arises as a quasi-variational inequality in infinite dimensions which is resolved in a weak sense using the semigroup approach.

1 Introduction

During the past decade several fundamental advances have been made in optimal control of fluid mechanics by a number of researchers [6]. In this paper we study impulse control theory for turbulence. In optimal weather prediction the task of updating the initial data optimally at strategic times can be reformulated precisely as an impulse control problem for the primitive cloud equations.

Variational technique to treat impulse control problems has been adapted to Gauss-Sobolev spaces (e.g., Chow and Menaldi [2]) with partial results. However, because of the technical difficulties associated with the domain of the generator we prefer to follow the semigroup approach.

The dynamic programming approach is used to discuss a simple optimal stopping time problem for the Navier-Stokes equation. We are forced to use sufficiently weak conditions on the data because our final objective is the optimal impulse control problems.

In order to facilitate the use of the semigroup technique we first consider the 2-D Navier-Stokes equation with random (Gaussian) forcing field. Several approaches have been proposed in the literature (see Sritharan [7] for a complete reference list). We then proceed to treat the infinite dimensional quasi-variational inequality to deal with the optimal impulse control problem in a weak sense.

Complete proofs of the results stated in this paper can be found elsewhere (cf. [4] and [5]), here only the main ideas are given.

2 Stochastic 2-D Navier-Stokes Equation

Let $\mathcal{O} \subset \mathbb{R}^2$ be a bounded domain with smooth boundary and $\mathbf{u}$ the velocity field. The Navier-Stokes problem can be written in the abstract form as follows

$$\partial_t \mathbf{u} + A\mathbf{u} + B(\mathbf{u}) = \mathbf{f} \quad \text{in } \mathbb{L}^2(0,T;\mathbb{V}'), \tag{2.1}$$

with the initial condition

$$\mathbf{u}(0) = \mathbf{u}_0 \quad \text{in } \mathbb{H}, \tag{2.2}$$

where $\mathbf{u}_0$ belong to $\mathbb{H}$ and the force field $\mathbf{f}$ is in $\mathbb{L}^2(0,T;\mathbb{H})$. Let us begin by defining some standard function spaces,

$$\mathbb{V} = \{\mathbf{v} \in \mathbb{H}_0^1(\mathcal{O}, \mathbb{R}^2); \nabla \cdot \mathbf{v} = 0 \text{ a.e. in } \mathcal{O}\}, \tag{2.3}$$

with the norm

$$\|\mathbf{v}\|_{\mathbb{V}} = \left(\int_{\mathcal{O}} |\nabla \mathbf{v}|^2 dx\right)^{1/2} = \|\mathbf{v}\|, \tag{2.4}$$

and $\mathbb{H}$ is the closure of $\mathbb{V}$ in the $\mathbb{L}^2$-norm

$$\|\mathbf{v}\|_{\mathbb{H}} = \left(\int_{\mathcal{O}} |\mathbf{v}|^2 dx\right)^{1/2} = |\mathbf{v}|. \tag{2.5}$$

We will also define the following linear operators

$$\begin{aligned} &P_{\mathbb{H}} : \mathbb{L}^2(\mathcal{O}, \mathbb{R}^2) \longrightarrow \mathbb{H} \quad \text{is the Helmhotz-Hodge orthogonal projection and} \\ &A : \mathbb{H}^2(\mathcal{O}, \mathbb{R}^2) \cap \mathbb{V} \longrightarrow \mathbb{H}, \quad A\mathbf{u} = -\nu P_{\mathbb{H}} \Delta \mathbf{u}, \ \nu > 0, \text{ is the Stokes operator,} \end{aligned} \tag{2.6}$$

where ν is the coefficient of kinematics viscosity. The inertia term is represented by the nonlinear operator

$$B : \mathcal{D}_B \subset \mathbb{H} \times \mathbb{V} \longrightarrow \mathbb{H}, \quad B(\mathbf{u}, \mathbf{v}) = P_{\mathbb{H}}(\mathbf{u} \cdot \nabla \mathbf{v}), \tag{2.7}$$

with the notation $B(\mathbf{u}) = B(\mathbf{u}, \mathbf{u})$. The domain of B requires that $(\mathbf{u} \cdot \nabla \mathbf{v})$ belongs to the Lebesgue space $\mathbb{L}^2(\mathcal{O}, \mathbb{R}^2)$.

Let us consider the Navier-Stokes equation subject to a random (Gaussian) term i.e., the forcing field $\mathbf{f}$ has a mean value still denoted by $\mathbf{f}$ and a noise denoted by $\dot{\mathbf{G}}$. We can write (to simplify notation we use time-invariant forces) $\mathbf{f}(t) = \mathbf{f}(x,t)$ and the noise process $\dot{\mathbf{G}}(t) = \dot{\mathbf{G}}(x,t)$ as a series $d\mathbf{G}_k = \sum_k \mathbf{g}_k(x,t) dw_k(t)$, where $\mathbf{g} = (\mathbf{g}_1, \mathbf{g}_2, \cdots)$ and $w = (w_1, w_2, \ldots)$ are regarded as ℓ^2-valued functions. The stochastic noise process represented by $\mathbf{g}(t) dw(t) = \sum_k \mathbf{g}_k(x,t) dw_k(t,\omega)$ is normal distributed in $\mathbb{H}$ with a trace-class co-variance operator denoted by $\mathbf{g}^2 = \mathbf{g}^2(t)$ and given by

$$\begin{cases} (\mathbf{g}^2(t)\mathbf{u}, \mathbf{v}) = \sum_k (\mathbf{g}_k(t), \mathbf{u})(\mathbf{g}_k(t), \mathbf{v}) \\ \text{Tr}(\mathbf{g}^2(t)) = \sum_k |\mathbf{g}_k(t)|^2 < \infty. \end{cases} \tag{2.8}$$

Impulse Control for Stochastic Navier-Stokes Equations

We interpret the stochastic Navier-Stokes equation as an Itô stochastic equation in variational form

$$d(\mathbf{u}(t),\mathbf{v}) + \langle A\mathbf{u}(t) + B(\mathbf{u}(t)),\mathbf{v}\rangle\, dt = (\mathbf{f},\mathbf{v})\, dt + \sum_k (\mathbf{g}_k,\mathbf{v})\, dw_k(t), \tag{2.9}$$

in $(0,T)$, with the initial condition

$$(\mathbf{u}(0),\mathbf{v}) = (\mathbf{u}_0,\mathbf{v}), \tag{2.10}$$

for any $\mathbf{v}$ in the space $\mathbb{V}$.

A finite-dimensional (Galerkin) approximation of the stochastic Navier-Stokes equation can be defined as follows. Let $\{\mathbf{e}_1, \mathbf{e}_2, \ldots\}$ be a complete orthonormal system (i.e., a basis) in the Hilbert space $\mathbb{H}$ belonging to the space $\mathbb{V}$ (and $\mathbb{L}^4$). Denote by $\mathbb{H}_n$ the n-dimensional subspace of $\mathbb{H}$ and $\mathbb{V}$ of all linear combinations of the first n elements $\{\mathbf{e}_1, \mathbf{e}_2, \ldots, \mathbf{e}_n\}$. Consider the following stochastic ODE in $\mathbb{R}^n$

$$\begin{cases} d(\mathbf{u}^n(t),\mathbf{v}) + \langle A\mathbf{u}^n(t) + B(\mathbf{u}^n(t)),\mathbf{v}\rangle\, dt = \\ \qquad = (\mathbf{f}(t),\mathbf{v})\, dt + \sum_k (\mathbf{g}_k(t),\mathbf{v})\, dw_k(t), \end{cases} \tag{2.11}$$

in $(0,T)$, with the initial condition

$$(\mathbf{u}(0),\mathbf{v}) = (\mathbf{u}_0,\mathbf{v}), \tag{2.12}$$

for any $\mathbf{v}$ in the space $\mathbb{H}_n$. The coefficients involved are locally Lipschitz and we need some *a priori* estimate to show global existence of a solution $\mathbf{u}^n(t)$ as an adapted process in the space $C^0(0,T,\mathbb{H}_n)$.

Proposition 2.1 (energy estimate). *Under the above mathematical setting let*

$$\mathbf{f} \in L^2(0,T;\mathbb{H}), \mathbf{g} \in L^2(0,T;\ell_2(\mathbb{H})) \quad \text{and} \quad \mathbf{u}_0 \in \mathbb{H}. \tag{2.13}$$

Let $\mathbf{u}^n(t)$ be an adapted process in $C^0(0,T,\mathbb{H}_n)$ which solves the stochastic ODE (2.11). Then we have the energy equality

$$\begin{cases} d|\mathbf{u}^n(t)|^2 + 2\nu\, |\nabla \mathbf{u}^n(t)|^2 dt = \left[2\,(\mathbf{f}(t),\mathbf{u}^n(t)) + \mathrm{Tr}(\mathbf{g}^2(t))\right] dt + \\ \qquad\qquad + 2 \sum_k (\mathbf{g}_k(t),\mathbf{u}^n(t))\, dw_k(t), \end{cases} \tag{2.14}$$

which yields the following estimate for any $\varepsilon > 0$

$$\begin{cases} E\{|\mathbf{u}^n(t)|^2\} e^{-\varepsilon t} + 2\nu \int_0^T E\{|\nabla \mathbf{u}^n(t)|^2\} e^{-\varepsilon t} dt \leq \\ \qquad\qquad \leq |\mathbf{u}(0)|^2 + \int_0^T \left[\frac{1}{\varepsilon} |\mathbf{f}(t)|^2 + \mathrm{Tr}(\mathbf{g}^2(t))\right] e^{-\varepsilon t} dt, \end{cases} \tag{2.15}$$

for any $0 \leq t \leq T$. Moreover, if we suppose

$$\mathbf{f} \in L^p(0,T;\mathbb{H}), \qquad \mathbf{g} \in L^p(0,T;\ell_2(\mathbb{H})) \tag{2.16}$$

then we also have

$$\begin{cases} E\{\sup_{0\leq t\leq T} |\mathbf{u}^n(t)|^p e^{-\varepsilon t} + p\nu \int_0^T |\nabla \mathbf{u}^n(t)|^2 |\mathbf{u}^n(t)|^{p-2} e^{-\varepsilon t} dt\} \leq \\ \qquad \leq |\mathbf{u}(0)|^p + C_{\varepsilon,p,T} \int_0^T \left[|\mathbf{f}(t)|^p + \text{Tr}(\mathbf{g}^2(t))^{p/2}\right] e^{-\varepsilon t} dt, \end{cases} \tag{2.17}$$

for some constant $C_{\varepsilon,p,T}$ depending only on $\varepsilon > 0$, $1 \leq p < \infty$ and $T > 0$.

Proposition 2.2 (uniqueness). *Let $\mathbf{u}$ be a solution of the stochastic Navier-Stokes equation (SPDE) with the regularity*

$$\mathbf{u} \in L^2(\Omega; C^0(0,T;\mathbb{H}) \cap L^2(0,T;\mathbb{V})), \quad \mathbf{u} \in \mathbb{L}^4(\mathcal{O} \times (0,T)) \tag{2.18}$$

and let the data $\mathbf{f}$, $\mathbf{g}$ and $\mathbf{u}_0$ satisfy the condition

$$\mathbf{f} \in L^2(0,T;\mathbb{V}'), \quad \mathbf{g} \in L^2(0,T;\ell_2(\mathbb{H})), \quad \mathbf{u}_0 \in \mathbb{H}. \tag{2.19}$$

If $\mathbf{v}$ in $L^2(\Omega; C^0(0,T,\mathbb{H}) \cap L^2(0,T,\mathbb{V}))$ is another solution then

$$|\mathbf{u}(t) - \mathbf{v}(t)|^2 \exp\left[-\frac{32}{\nu^3} \int_0^t \|\mathbf{u}(s)\|_{\mathbb{L}^4(\mathcal{O})}^4 ds\right] \leq |\mathbf{u}(0) - \mathbf{v}(0)|^2, \tag{2.20}$$

with probability 1 for any $0 \leq t \leq T$.

Proof. Indeed if $\mathbf{u}$ and $\mathbf{v}$ are two solutions then $\mathbf{w} = \mathbf{v} - \mathbf{u}$ solves the deterministic equation

$$\partial_t \mathbf{w} + A\mathbf{w} = B(\mathbf{u}) - B(\mathbf{v}) \quad \text{in } \mathbb{L}^2(0,T;\mathbb{V}'),$$

and setting $r(t) = \frac{32}{\nu^3} \int_0^t \|\mathbf{u}(s)\|_{\mathbb{L}^4(\mathcal{O})}^4 ds$ we have

$$d(e^{-r(t)}|\mathbf{w}(t)|^2) + \nu e^{-r(t)} \|\mathbf{w}(t)\|^2 dt = -\dot{r}(t) e^{-r(t)} |\mathbf{w}(t)|^2 dt$$
$$\qquad - \nu e^{-r(t)} \|\mathbf{w}(t)\|^2 dt - 2e^{-r(t)} \langle B(\mathbf{v}(t)) - B(\mathbf{u}(t)), \mathbf{w}(t)\rangle dt \leq 0,$$

and integrating in t, we conclude. □

Each solution $\mathbf{u}$ in the space $L^2(\Omega; L^\infty(0,T;\mathbb{H}) \cap L^2(0,T;\mathbb{V}))$ of the stochastic Navier-Stokes equation actually belongs to $L^2(\Omega; C^0(0,T;\mathbb{H}) \cap \mathbb{L}^4(\mathcal{O} \times (0,T)))$ in 2-D, $\mathcal{O} \subset \mathbb{R}^2$. Thus in 2-D, the uniqueness holds in the space $L^2(\Omega; L^2(0,T;\mathbb{V}))$.

If a given adapted process $\mathbf{u}$ in $L^2(\Omega; L^\infty(0,T;\mathbb{H}) \cap L^2(0,T;\mathbb{V}))$ satisfies

$$d(\mathbf{u}(t), \mathbf{v}) = \langle \mathbf{f}(t), \mathbf{v}\rangle dt + (\mathbf{g}(t), \mathbf{v}) dw(t), \tag{2.21}$$

for any function $\mathbf{v}$ in $\mathbb{V}$ and some $\mathbf{f}$ in $L^2(0,T;\mathbb{V}')$ and $\mathbf{g}$ in $L^2(0,T;\ell_2(\mathbb{H}))$, then we can find a version of $\mathbf{u}$ (still denoted by $\mathbf{u}$) in $L^2(\Omega; C^0(0,T;\mathbb{H}))$ satisfying the energy equality

$$d|\mathbf{u}(t)|^2 = \left[2\langle \mathbf{f}(t), \mathbf{u}(t)\rangle + \text{Tr}(\mathbf{g}^2(t))\right] dt + 2(\mathbf{g}(t), \mathbf{u}(t)) dw(t) \tag{2.22}$$

see e.g. Gyongy and Krylov [3].

Proposition 2.3 (2-D existence). *Let* **f**, **g** *and* u_0 *be such that*

$$\mathbf{f} \in L^p(0,T;\mathbb{V}'), \quad \mathbf{g} \in L^p(0,T;\ell_2(\mathbb{H})), \quad u_0 \in \mathbb{H}, \tag{2.23}$$

for some $p \geq 4$. Then there is an adapted process $\mathbf{u}(t,x,\omega)$ with the regularity

$$\mathbf{u} \in L^p(\Omega; C^0(0,T;\mathbb{H})) \cap L^2(\Omega; L^2(0,T;\mathbb{V})) \tag{2.24}$$

which solves the stochastic Navier-Stokes equation and the following a priori bound holds

$$\begin{cases} E\{\sup_{0 \leq t \leq T} |\mathbf{u}(t)|^p + \int_0^T |\nabla \mathbf{u}(t)|^2 |\mathbf{u}(t)|^{p-2} dt\} \leq \\ \leq C_p E\{|\mathbf{u}(0)|^p + \int_0^T [\|\mathbf{f}(t)\|_\mathbb{V}^p + \mathrm{Tr}(\mathbf{g}^2(t))^{p/2}] dt\}, \end{cases} \tag{2.25}$$

for some constant $C_p = C(T,\nu,p)$ depending only on the numbers $T > 0$, $\nu > 0$ and $p \geq 2$. □

The proof of this result can be found in our previous work [4, 7]. Our proof is based on the $\mathbb{L}^4$–monotonicity of the nonlinear Navier-Stokes operator (and it generalizes to other cases, including multiplicative noise). If we denote by $\mathbb{B}_r$ the (closed) $\mathbb{L}^4$-ball in $\mathbb{V}$

$$\mathbb{B}_r = \{\mathbf{v} \in \mathbb{V}\, ; \|\mathbf{v}\|_{\mathbb{L}^4(O,\mathbb{R}^2)} \leq r\}, \tag{2.26}$$

then the nonlinear operator $u \mapsto Au + B(u)$ is monotone in the convex ball $\mathbb{B}_r$ i.e.,

$$\langle A\mathbf{w}, \mathbf{w}\rangle + \langle B(\mathbf{u}) - B(\mathbf{v}), \mathbf{w}\rangle + \frac{32r^4}{\nu^3}|\mathbf{w}|^2 \geq \frac{\nu}{2}\|\mathbf{w}\|^2, \tag{2.27}$$

$\forall \mathbf{u} \in \mathbb{V}, \mathbf{v} \in \mathbb{B}_r$ and $\mathbf{w} = \mathbf{u} - \mathbf{v}$.

3 Markov-Feller Process

In what follows for the sake of simplicity we assume that the processes $\mathbf{f}(x,t,\omega)$ and $\mathbf{g}(x,t,\omega)$ are independent of t, i.e.,

$$\mathbf{f} \in \mathbb{V}' \quad \text{and} \quad \mathbf{g} \in \ell_2(\mathbb{H}) \tag{3.1}$$

and we denote by $\mathbf{u}(t;u_0)$ the semiflow, i.e., the solution of Navier-Stokes equation. Also usually we substitute u_0 with $\mathbf{v}$.

Proposition 3.1 (continuity). *Under the previous conditions the stochastic semiflow $\mathbf{u}(t;\mathbf{v})$ is locally uniformly continuous in $\mathbf{v}$, locally uniformly for t in $[0,\infty)$. Moreover,* *for any $p > 0$ and $\alpha > 0$ there is a positive constant λ sufficiently large such that the following estimate*

$$E\{e^{-\alpha t}(\lambda + |\mathbf{u}(t;\mathbf{v})|^2)^{p/2}\} \leq (\lambda + |\mathbf{v}|^2)^{p/2}, \tag{3.2}$$

$\forall t \geq 0$, $\mathbf{v} \in \mathbb{H}$ *holds, also for any stopping time $t = \tau$. Furthermore, if $\mathbf{f}$ and $\mathbf{g}$ belong to $\mathbb{H}$ and $\ell_2(\mathbb{V})$ respectively, then the semiflow is also locally uniformly continuous in t, locally uniformly for $\mathbf{v}$ in $\mathbb{V}$.* □

The Navier-Stokes semigroup $(\Phi(t), t \geq 0)$ defined by $\Phi(t)h(\mathbf{v}) = E\{h(\mathbf{u}(t;\mathbf{v}))\}$, is indeed a Markov-Feller semigroup on the space $C_b(\mathbb{H})$ (of continuous and bounded real function on $\mathbb{H}$ endowed with the sup-norm). Since the base space $\mathbb{H}$ is not locally compact, the Navier-Stokes semigroup is not *strongly continuous*.

In our approach, it is convenient to work with unbounded functions. Let $C_p(\mathbb{H})$ be the space of real uniformly continuous functions on any ball and with a growth bounded by the norm to the $p \geq 0$ power, in another words, the space of real functions h on $\mathbb{H}$ such that $\mathbf{v} \mapsto h(\mathbf{v})(1 + |\mathbf{v}|^2)^{-p/2}$ is bounded and locally uniformly continuous, with the weighted sup-norm

$$\|h\| = \|h\|_{C_p} = \sup_{x \in \mathbb{H}}\{|h(x)|(\lambda + |x|^2)^{-p/2}\}, \tag{3.3}$$

where λ is a positive constant sufficiently large to so that

$$\alpha \geq \alpha_0(p), \quad p \geq 0. \tag{3.4}$$

It is clear that $C_b(\mathbb{H}) = C_0(\mathbb{H})$ and $C_q(\mathbb{H}) \subset C_p(\mathbb{H})$ for any $0 \leq q < p$.

Then for any $\alpha \geq 0$, (linear) Navier-Stokes semigroup $(\Phi_\alpha(t), t \geq 0)$ with an α-exponential factor is defined as follows

$$\Phi_\alpha(t) : C_p(\mathbb{H}) \longrightarrow C_p(\mathbb{H}), \quad \Phi_\alpha(t)h(\mathbf{v}) = E\{e^{-\alpha t}h[\mathbf{u}(t;\mathbf{v})]\}. \tag{3.5}$$

Proposition 3.2 (semigroup). *Under the above assumptions the Navier-Stokes semigroup $(\Phi_\alpha(t), t \geq 0)$ is a weakly continuous Markov-Feller semigroup in the space $C_p(\mathbb{H})$.* □

Since the Navier-Stokes semigroup is not strongly continuous, we cannot consider the *strong* infinitesimal generator as acting on a dense domain in $C_p(\mathbb{H})$. However, this Markov-Feller semigroup $(\Phi_\alpha(t), t \geq 0)$ may be considered as acting on real Borel functions with p-polynomial growth, which is Banach space with the sup-weighted norm and denoted by $B_p(\mathbb{H})$. It is convenient to define the family of semi-norms on $B_p(\mathbb{H})$

$$p_0(h,\mathbf{v}) = E\{\sup_{s \geq 0}|h(\mathbf{u}(s;\mathbf{v}))|e^{-\alpha_0 s}\}, \quad \forall \mathbf{v} \in \mathbb{H}, \tag{3.6}$$

where λ is sufficiently large. Now, if a sequence $\{h_n\}$ of equi-bounded functions in $B_p(\mathbb{H})$ satisfies $p_0(h_n - h, \mathbf{v}) \to 0$ for any $\mathbf{v}$ in $\mathbb{H}$, we say that $h_n \to h$ boundedly pointwise convergence relative to the above family of semi-norms. It is clear that $p_0(\Phi_\alpha(t)h - h, \mathbf{v}) \to 0$ as $t \to 0$, for any function h in $C_p(\mathbb{H})$ and any $\mathbf{v}$ in $\mathbb{H}$.

Definition 3.3. Let $\bar{C}_p(\mathbb{H})$ be the subspace of functions $\bar{h}$ in $B_p(\mathbb{H})$ such that the mapping given by $t \mapsto \bar{h}[\mathbf{u}(t;\mathbf{v})]$ is almost surely continuous on $[0,+\infty)$ for any $\mathbf{v}$ in $\mathbb{H}$ and satisfies

$$\lim_{t \to 0} p_0(\Phi_\alpha(t)\bar{h} - \bar{h}, \mathbf{v}) = 0, \quad \forall \mathbf{v} \in \mathbb{H} \tag{3.7}$$

where $p_0(\cdot,\cdot)$ is given above. □

This is the space of function (uniformly) continuous over the flow $\mathbf{u}(\cdot,\mathbf{v})$, relative to the family of semi-norms and it is independent of α. Hence, we may consider the Navier-Stokes semigroup on the Banach space $\bar{C}_p(\mathbb{H})$, endowed with the sup-weighted norm. The *weak* infinitesimal generator $-\bar{A}_\alpha$ with domain $\mathcal{D}_p(\bar{A}_\alpha)$ (as a subspace of $\bar{C}_p(\mathbb{H})$) is defined by boundedly pointwise limit $[h - \Phi_\alpha(t)h]/t \to \bar{A}_\alpha h$ as $t \to 0$, relative to the family of semi-norms. Notice that $p_0(\Phi_\alpha(t)\bar{h}, \mathbf{v}) \leq p_0(\bar{h}, \mathbf{v})$ for any $t \geq 0$, $\bar{h}$ in $\bar{C}_p(\mathbb{H})$ and $\mathbf{v}$ in $\mathbb{H}$.

Proposition 3.4 (density). *If the above assumptions hold, then $C_p(\mathbb{H}) \subset \bar{C}_p(\mathbb{H})$, the Navier-Stokes semigroup leaves invariant the space $\bar{C}_p(\mathbb{H})$ and for any function $\bar{h}$ in $\bar{C}_p(\mathbb{H})$, there is a equi-bounded sequence $\{\bar{h}_n\}$ of functions in the domain $\mathcal{D}_p(\bar{A}_\alpha)$ satisfying $p_0(\bar{h}_n - \bar{h}, \mathbf{v}) \to 0$ for any $\mathbf{v}$ in $\mathbb{H}$.* □

¿From above results it is clear that given $\alpha > 0$, $p \geq 0$, λ sufficiently large and a function $\bar{h}$ in $\bar{C}_p(\mathbb{H})$ there is another function $\bar{u}$ in $\mathcal{D}_p(\bar{A}_\alpha)$ such that $\bar{A}_\alpha \bar{u} = \bar{h}$, where

$$\bar{u} = \int_0^\infty \Phi_\alpha(t)\bar{h}\, dt. \tag{3.8}$$

The right-hand side is called the *weak* resolvent operator and denoted by either $\mathcal{R}_\alpha = \bar{A}_\alpha^{-1}$ or $\mathcal{R}_\alpha = (\bar{A}_0 + \alpha I)^{-1}$. Moreover, if $\alpha_0 = \alpha_0(\lambda)$ then for any $p > 0$ we have $\alpha_0(\lambda) \to 0$ as $\lambda \to \infty$, and for any stopping time τ,

$$\begin{cases} \dfrac{p\nu}{2} E\Big\{ \int_0^\tau |\nabla \mathbf{u}(t;\mathbf{v})|^2 (\lambda + |\mathbf{u}(t;\mathbf{v})|^2)^{p/2-1} e^{-\alpha_0 t} dt \Big\} + \\ \qquad + E\big\{ e^{-\alpha_0 \tau}(\lambda + |\mathbf{u}(\tau;\mathbf{v})|^2)^{p/2} \big\} \leq (\lambda + |\mathbf{v}|^2)^{p/2}, \end{cases} \tag{3.9}$$

$\forall \mathbf{v} \in \mathbb{H}$, and then for any $\alpha > \alpha_0$ we obtain

$$\|\Phi_\alpha(t)\bar{h}\| \leq e^{-(\alpha-\alpha_0)t}\|\bar{h}\|, \qquad p_0(\Phi_\alpha(t)\bar{h}, \mathbf{v}) \leq e^{-(\alpha-\alpha_0)t} p_0(\bar{h}, \mathbf{v}), \tag{3.10}$$

for any $t \geq 0$, and

$$\|\mathcal{R}_\alpha \bar{h}\| \leq \frac{1}{\alpha - \alpha_0} \|\bar{h}\|, \qquad p_0(\mathcal{R}_\alpha \bar{h}, \mathbf{v}) \leq \frac{1}{\alpha - \alpha_0} p_0(\bar{h}, \mathbf{v}), \tag{3.11}$$

for any $\mathbf{v}$ in $\mathbb{H}$ and where the norm $\|\cdot\|$ and the semi-norms $p_0(\cdot, \mathbf{v})$.

4 Impulse Control Problem

Let us now consider the problem of sequentially controlling the evolution of the stochastic process $\mathbf{u}(t;\mathbf{v})$ by changing the initial condition $\mathbf{v}$. For this purpose we consider a controlled Markov chain $\mathbf{q}_k(i)$ in $\mathbb{H}$ with transition operator $Q(k)$ and a control parameter k which belongs to a compact metric space K. For a sequence $(\zeta_i, i = 1, 2, \ldots)$ of independent identically distributed $\mathbb{H}$–valued random variables we have

$$\begin{cases} \mathbf{q}_k(i+1) = \mathbf{q}(\mathbf{q}_k(i), \zeta_i \,|\, k), & \forall i = 1, 2, \ldots, \\ E\{h(\mathbf{q}(\mathbf{v}, \zeta_1 \,|\, k))\} = Q(k)h(\mathbf{v}), & \forall \mathbf{v} \in \mathbb{H}, \end{cases} \quad (4.1)$$

for any initial value $\mathbf{q}(1)$, any bounded and measurable real-valued function h on $\mathbb{H}$ and any k in K. For the sake of simplicity, this Markov chain (i.e., each random variable ζ_i) is assumed to be independent of the Wiener process $w = (w_1, w_2, \ldots)$ used to model the disturbances in dynamic equation.

A sequence $\{\tau_i, k_i; i = 1, 2, \ldots\}$ of stopping times τ_i and decisions k_i such that τ_i approaches infinity is called an impulse control. At time $t = \tau_i$ the system has an impulse described by the (controlled) Markov chain $\mathbf{q}_k(i)$ with $k = k_i$. Between two consecutive times $\tau_i \leq t < \tau_{i+1}$, the evolution follows the Navier-Stokes equation:

$$\begin{cases} \mathbf{u}(t) = \mathbf{u}(t, \tau_i; \mathbf{u}(\tau_i)), & \text{if } \tau_i \leq t < \tau_{i+1}, \\ \text{and} \quad \mathbf{u}(\tau_i) = \mathbf{q}(\mathbf{u}(\tau_i-), \zeta_i \,|\, k_i), \end{cases} \quad (4.2)$$

where $\mathbf{u}(t, s; \mathbf{v})$ is the solution of Navier-Stokes eqaution with initial value v at time s. Since $\tau_i \to \infty$, we can construct the process $\mathbf{u}(t)$ by iteration, for any impulse control $\{\tau_i, k_i; i = 1, 2, \ldots\}$ and initial condition $\mathbf{v}$ in $\mathbb{H}$. It is clear that τ_i is an stopping time with respect to the Wiener process enlarged by the σ–algebras generated by the random variables $\zeta_1, \zeta_2, \ldots, \zeta_{i-1}$. Also, the decision random variables k_i are measurable with respect to the σ–algebra generated by τ_i.

To each impulse we associate a strictly positive cost known as *cost-per-impulse* and given by the functional $L(\mathbf{v}, k)$. The total cost for an impulse control $\{\tau_i, k_i; i = 1, 2, \ldots\}$ and initial condition $\mathbf{v}$ is given by

$$J(\mathbf{v}, \{\tau_i, k_i\}) = E\left\{ \int_0^\infty F(\mathbf{u}(t))e^{-\alpha t} dt + \sum_i L(\mathbf{u}(\tau_i-), k_i) e^{-\alpha \tau_i} \right\} \quad (4.3)$$

and the optimal cost

$$\hat{U}(\mathbf{v}) = \inf_{\{\tau_i, k_i\}} J(\mathbf{v}, \{\tau_i, k_i\}), \quad (4.4)$$

where the infimum is taken over all impulse controls, and $\mathbf{u}(t)$ is the evolution constructed as above, with initial condition $\mathbf{v}$.

Let us follow a *hybrid control* setting as in Bensoussan and Menaldi [1]. The dynamic programming principle yields to the following problem. Find U in $\bar{C}_p(\mathbb{H})$ such that

$$U \leq MU, \quad \bar{A}_\alpha U \leq F, \quad \text{and} \quad \bar{A}_\alpha U = F \quad \text{in } [U < MU], \quad (4.5)$$

where $\bar{\mathcal{A}}_\alpha$ is interpreted in the *martingale (semigroup or weak) sense* and M is the following nonlinear operator on $\bar{C}(\mathbb{H})$ given by

$$Mh(\mathbf{v}) = \inf_k \{L(\mathbf{v}, k) + Q(k)h(\mathbf{v})\}, \quad \forall \mathbf{v} \in \mathbb{H}, \tag{4.6}$$

where $Q(k)h(\mathbf{v}) = E\{h(\mathbf{q}(\mathbf{v}, \zeta_1 \,|\, k))\}$ is the transition operator. This problem is called a *quasi-variational inequality* (QVI).

To solve the QVI we define by induction the sequence of variational inequalities (VI)

$$\begin{cases} \hat{U}^{n+1} \in \bar{C}_p(\mathbb{H}) \quad \text{such that} \quad \hat{U}^{n+1} \leq M\hat{U}^n, \quad \bar{\mathcal{A}}_\alpha \hat{U}^{n+1} \leq F \quad \text{and} \\ \bar{\mathcal{A}}_\alpha \hat{U}^{n+1} = F \quad \text{in} \quad [\hat{U}^{n+1} < M\hat{U}^n], \end{cases} \tag{4.7}$$

where $\hat{U}^0 = U^0$ solves the equation $\bar{\mathcal{A}}_\alpha U^0 = F$. This VI can be formulated as a maximum sub-solution problem

$$U^{n+1} \in \bar{C}_p(\mathbb{H}) \quad \text{such that} \quad U^{n+1} \leq MU^n, \quad \bar{\mathcal{A}}_\alpha U^{n+1} \leq F, \tag{4.8}$$

for any $n \geq 0$. In view of the Theorem for the VI in [5], we need only to assume that M operates on the space $\bar{C}_p(\mathbb{H})$ to define the above sequence $\hat{U}^n$ of functions. This means that first, we impose the condition

$$\begin{cases} \|L(\cdot, k)\| \leq C, \quad \forall k \in K, \\ \lim_{t \to 0} \sup_k \{p_0(\Phi_\alpha(t)L(\cdot, k) - L(\cdot, k), \mathbf{v})\} = 0, \end{cases} \tag{4.9}$$

$\forall \mathbf{v} \in \mathbb{H}$, and next

$$\begin{cases} E\{|\mathbf{q}(\mathbf{v}, \zeta_1 \,|\, k)|^m\} \leq C_m(1 + |\mathbf{v}|^m), \\ \lim_{t \to 0} \sup_k \{p_0(\Phi_\alpha(t)Q(k)h - Q(k)h, \mathbf{v})\} = 0, \end{cases} \tag{4.10}$$

$\forall k \in K$, $\mathbf{v} \in \mathbb{H}$, $\forall h \in \bar{C}_p(\mathbb{H})$, for any $m \geq 0$, some positive constant C_m and where the norm $\|\cdot\|$ and the semi-norms $p_0(\cdot, \mathbf{v})$. One of the main differences between impulse and continuous type control is the positive cost-per-impulse, i.e., the requirement

$$L(\mathbf{v}, k) \geq \ell_0 > 0, \quad \forall \mathbf{v} \in \mathbb{H}, \, k \in K, \tag{4.11}$$

which forbids the accumulation of impulses. We also need

$$F \in \bar{C}_p(\mathbb{H}), \quad F(\mathbf{v}) \geq 0, \quad \forall \mathbf{v} \in \mathbb{H}, \tag{4.12}$$

to set up the above sequence.

An important role is played by the function $\hat{U}^0 = U^0$ which solves $\bar{\mathcal{A}}_\alpha U^0 = F$, and by the function $\hat{U}_0 = U_0$, which is defined as the solution of the following variational inequality

$$\begin{cases} \hat{U}_0 \in \bar{C}_p(\mathbb{H}) \quad \text{such that} \quad \hat{U}_0 \leq \inf_k L(\cdot, k), \quad \bar{\mathcal{A}}_\alpha \hat{U}_0 \leq F \quad \text{and} \\ \bar{\mathcal{A}}_\alpha \hat{U}_0 = F \quad \text{in} \quad [\hat{U}_0 < \inf_k L(\cdot, k)], \end{cases} \tag{4.13}$$

or as the maximum sub-solution of the problem

$$U_0 \in \bar{C}_p(\mathbb{H}) \quad \text{such that} \quad U_0 \leq \inf_k L(\cdot, k), \quad \bar{\mathcal{A}}_\alpha U_0 \leq F. \tag{4.14}$$

Consider the quasi-variational inequality (QVI)

$$\begin{cases} \hat{U} \in \bar{C}_p(\mathbb{H}) \quad \text{such that} \quad \hat{U} \leq M\hat{U}, \quad \bar{\mathcal{A}}_\alpha \hat{U} \leq F \quad \text{and} \\ \bar{\mathcal{A}}_\alpha \hat{U} = F \quad \text{in} \quad [\hat{U} < M\hat{U}], \end{cases} \tag{4.15}$$

or the maximum sub-solution of the problem

$$U \in \bar{C}_p(\mathbb{H}) \quad \text{such that} \quad U \leq MU, \quad \bar{\mathcal{A}}_\alpha U \leq F. \tag{4.16}$$

Theorem 4.1 (QVI). *Let the above assumptions hold. Then the VI defines a (pointwise) decreasing sequence of functions $\hat{U}^n(\mathbf{v})$ which converges to the optimal cost $\hat{U}(\mathbf{v})$, for any $\mathbf{v}$ in $\mathbb{H}$. Moreover, if the condition*

$$\text{there exists} \quad r \in (0,1] \quad \text{such that} \quad r\, U^0(\mathbf{v}) \leq U_0(\mathbf{v}), \quad \forall \mathbf{v} \in \mathbb{H} \tag{4.17}$$

is satisfied then we have the estimate

$$0 \leq \hat{U}^n - \hat{U}^{n+1} \leq (1-r)^n \hat{U}^0, \quad \forall n = 0, 1, \ldots, \tag{4.18}$$

the automaton impulse control $\{\hat{\tau}_i, \hat{k}_i\}$, generated by the the continuation region $[\hat{U} < M\hat{U}]$ and defined by $\hat{\tau}_0 = 0$,

$$\begin{cases} \hat{\tau}_i = \inf\{t \geq \hat{\tau}_{i-1} \ \hat{U}[\mathbf{u}(t; \mathbf{u}(\tau_{i-1}))] = M\hat{U}[\mathbf{u}(t; \mathbf{u}(\tau_{i-1}))]\}, \\ \hat{k}_i = \arg\min_{k \in K}\{L(\mathbf{u}(\tau_i), k) + Q(k)\hat{U}(\mathbf{u}(\tau_i))\} \end{cases} \tag{4.19}$$

is optimal, i.e., $\hat{U}(\mathbf{v}) = J(\mathbf{v}, \{\hat{\tau}_i, \hat{k}_i\})$, and the optimal cost $\hat{U}$ is the unique solution of the QVI or the maximum sub-solution of the problem. □

If we impose

$$L(\mathbf{v}, k) \geq \ell_0 (1 + |\mathbf{v}|^2)^{p/2} > 0, \quad \forall \mathbf{v} \in \mathbb{H}, \ k \in K, \tag{4.20}$$

then assumption (4.17) holds for any $0 < r < 1$ such that $r\|F\| \leq \ell_0(\alpha - \alpha_0)$.

References

[1] A. Bensoussan and J.L. Menaldi, Hybrid Control and Dynamic Programming, *Dynamics of Continuous Discrete and Impulsive Systems*, **3** (1997), 395–442.

[2] P.L. Chow and J.L. Menaldi, Variational Inequalities for the Control of Stochastic Partial Differential Equations, in Proceedings of the Stochastic Partial Differential Equations and Application II, Trento, Italy, 1988, *Lecture Notes in Math.*, **1390** (1989), 42–51.

[3] I. Gyongy and N. V. Krylov, On stochastic equations with respect to semimartingales Itô formula in Banach spaces, *Stochastics*, **6** (1982), 153–173.

[4] J.L. Menaldi and S.S. Sritharan, Stochastic 2-D Navier-Stokes Equation, *Appl. Math. Optim.*, submitted.

[5] J.L. Menaldi and S.S. Sritharan, Impulse Control of Stochastic Navier-Stokes Equations, *Nonlinear Analysis, Methods, Theory and Application*, submitted.

[6] S.S. Sritharan, (Editor) *Optimal Control of Viscous Flow*, SIAM, Philadelphia, 1998.

[7] S.S. Sritharan, Deterministic and Stochastic Control of Navier-Stokes Equation with Linear, Monotone and Hyper Viscosities, *Appl. Math. Optim.*, **41** (2000), 255–308.

Recent Progress on the Lavrentiev Phenomenon with Applications

Victor J. Mizel   Carnegie-Mellon University, Philadelphia, Pennsylvania

The Lavrentiev phenomenon is associated with the sensitivity of the infimum of a variational problem to the smoothness of the class of admissible functions. Since the determination by Tonelli that the class of absolutely continuous functions is an appropriate class of admissible functions with which to obtain existence of minimizers to variational problems on a real interval by direct methods, it was shown that many of the classical problems yield to this approach. Moreover, in many of the classical problems the minimizers were actually Lipschitz or better, so that the problems were insensitive to the particular subclass of absolutely continuous admissible functions chosen for applying the direct method for existence. Therefore it was quite surprising when in 1926 M. Lavrentiev published, in response to a challenge issued by Tonelli during a lecture in Moscow, an example [L] of a functional of the form

$$J[y] = \int_a^b f(x, y(x), y'(x))dx \quad \text{with } y : [a, b]$$

subject to certain constraints and smoothness conditions, in which the infimum of J over the class of all absolutely continuous functions subject to certain boundary conditions at a and b is strictly smaller than its infimum over the class of all C^1 functions meeting the same boundary conditions. Thereafter in 1934 B. Mania published an example [Ma] involving a much simpler (polynomial) integrand. The presentation of an example involving a functional J possessing a strictly elliptic integrand occurred only in 1985 [B&M] during the course of an investigation stimulated by the possible relevance of such questions to the theory of (multidimensional) hyperelastic materials.

The present article reports on recent progress in the study of this phenomenon. We adopt the following notations for clarity. For each $p \in [1, \infty]$ $W^{1,p}[a,b] = \{y : [a,b] \to \mathbb{R} | y \in AC[a,b], y' \in L^p(a,b)\}$, so that for example with $[a,b] = [0,1]$ and $y(x) = x^\beta$, $\beta \in (0,1), y \in W^{1,p}[0,1] \Leftrightarrow p \in [1, 1/(1-\beta))$. For J as above we put $i(p) = \inf\{J[y]| y \in W^{1,p}[a,b] + BC's\}$, $p \in [1, \infty]$, so that $p_1 \leq p_2 \Rightarrow i(p_1) \leq i(p_2)$. Then if f is such that $i(p_1) < i(p_2)$ for some $p_1 < p_2$ we have the Lavrentiev Phenomenon Λ. (Cf [Bu&M],[Bu&B] for a relaxation view.)

I. The first topic we study is a description of the possible boundary conditions which can lead to Λ, as well as the issue of whether this can occur when the integrand f in J is strictly elliptic and coercive. To facilitate this discussion we introduce for $a, b, A, B \in \mathbb{R}$, $p \in [1, \infty]$ the following notation:

  (i) Both ends pinned [Lagrange problem] $\mathcal{A}_2(p) = \{y \in W^{1,p}[a,b] | y(a) = A, y(b) = B\}$
  (ii) One end pinned $\mathcal{A}_1(p) = \{y \in W^{1,p}[a,b] | y(a) = A\}$
  (iii) No ends pinned $\mathcal{A}_0(p) = W^{1,p}[a,b]$,

and we denote the respective infima by $i_2(p), i_1(p)$, and $i_0(p)$. The matter will be clarified by consideration of the following:

---

[1] Research partially supported by the US National Science Foundation.

A. We take as our integrand $f(x,y,z) = (y^5 - x^3)^2(1+z^{20})$ so that $J[y] = \int_0^1 (y^5 - x^3)^2(1+(y')^{20})dx$, and we take $y \in \mathcal{A}_1(p) = \{y \in W^{1,p}[0,1] | y(0) = 0\}$. Clearly, $J[y] \geq 0$, while for $y_0(x) = x^{3/5}$ $J[y_0] = 0$. Hence by the computation on the preceding page, $i_1(p) = 0$, if $p \in [1, 5/2)$. We now show that $i_1(5/2) > 0$ whence Λ occurs. The argument consists in treating two cases (cf[M1],[DHM]):
Case 1: For some $x^* \in (0,1], y(x) < [\frac{1}{2}(x)^{3/2}]^{1/5}, x \in (0,x^*), y(x^*) = [\frac{1}{2}(x^*)^{3/2}]^{1/5}$. Then by use of Holder's and Jensen's inequalities and the chain rule $J[y] \geq k > 0$, for $k = (2/3)^{20}(1/2)^6$;
Case 2: $y(x) < [\frac{1}{2}(x)^{3/2}]^{1/5}$ for all $x \in (0,1]$. In this case a direct computation yields, with k as above, $J[y] \geq k > 0$.
Although the integrand f is convex in z ["elliptic"] it is neither strictly elliptic nor coercive in z, but use of a device from [B&M] leads to the construction of a perturbed integrand $f^*(x,y,z) = f(x,y,z) + \epsilon z^2$ which satisfies both requirements and retains the Lavrentiev phenomenon Λ $i_1^*(p) < i_1^*(5/2)$, for all $p \in [1, 5/2)$.

B. With f as above we take $J^\sim[y] = \int_{-1}^1 (y^5 - x^3)^2(1+(y')^{20})dx$, $y \in \mathcal{A}_0(p)$. By considering separately **Case 1** $y(0) \geq 0$, **Case 2** $y(0) < 0$, one deduces that Λ $i_0^\sim(p) = 0$, if $p \in [1, 5/2)$, $i_0^\sim(5/2) \geq k > 0$ (with k as in A). Once again one can construct a modified strictly elliptic coercive integrand f^* with the Lavrentiev phenomenon Λ $i_0^*(p) < i_1^*(5/2)$, $p \in [1, 5/2)$ (cf. [DHM]).

II. The second topic concerns the role of Λ for autonomous integrands f, for integrands f which have no y-dependence, and for integrands which are jointly convex in (y,z). For first order problems it has been shown under various regularity assumptions that Λ cannot occur in any of these cases (cf. [C&V],[A&C], [Da], [S&M]).

Therefore we begin by considering an autonomous f which involves second order as well as lower order derivatives. We take as our integrand $f(x,y,z,w) = f(y,z,w) = [(\frac{5}{3}y)^2 - (z+4)^2(\frac{3}{5}(z-6))^3]^2 w^{16}$, so that with obvious notation for the derivative of the functions $y' \in AC[0,1]$ $J[y] = \int_0^1 [(\frac{5}{3}y)^2 - (y'+4)^2(\frac{3}{5}(y'-6))^3]^2 (y'')^{16} dx$, $y \in \mathcal{A}_{2,2}(p) = \{y \in W^{2,p}[0,1] | y(0) = 0, y'(0) = 6, y(1) = 7, y'(1) = \frac{23}{3}\}$. Clearly $J[y] \geq 0$, while $J[y_0] = 0$, for $y_0(x) = 6x + x^{5/3}$. A direct computation shows that $y_0 \in W^{2,p}[0,1]$, $p \in [1,3)$. However it can be shown by the use of Holder's and Jensen's inequalities and the chain rule (cf.[M1],[H]) that $i_2(3) > 0$. Thus we do have Λ $i_2(p) = 0, p \in [1,3), i_2(3) > 0$ in this autonomous second order context. Here, too, one can construct a modified strictly elliptic coercive autonomous integrand f^* which retains the Lavrentiev phenomenon Λ (cf [H]). On the other hand, one can raise the question of whether in higher dimensions and **first** order variational problems any one of the three restrictions on the integrand mentioned above retains its role of excluding Λ. To the contrary it can be shown that there is an integrand $f = f(F)$ for which **all three** restrictions mentioned above hold, such that the corresponding variational functional exhibits Λ with an elementary set of boundary conditions. This example is very closely related to that discussed in **IV** below (cf. [FHM2] for a detailed description).

III. The next topic concerns the possible structure of infimum functions $i(\cdot): [1, \infty] \to [0, \infty)$ exhibiting the Lavrentiev phenomenon. Although the simplest examples discussed above possessed right continuous jump discontinuities, it is not

Lavrentiev Phenomenon

evident that this is the only possibility. We consider first the following
$f(x, y, z) = (y^2 - 16x)^2 \{cz^6 + [\exp(-x^{-1/2})y^2 - 16\exp(-x^{-1/2})x/(\log x)^2]^2 \exp(13z^8)\}$
for $x \in [0, 1/e], c = 1/[6^6(16)^2]$.
Thus we have
$J[y] = \int_0^{1/e}(y^2-16x)^2\{c(y')^6+[\exp(-x^{-1/2})y^2-16\exp(-x^{-1/2})x/(\log x)^2]^2\exp(13y'^8)\}dx$
for $y \in \mathcal{A}_2(p) = \{y \in W^{1,p}[0, 1/e] | y(0) = 0, y(1) = 4e^{-1/2}\}$. It can be proved that this satisfies Λ $i_2(p) = 0, p \in [1, 2), i_2(2) \in (0, 1), i_2(p) = 1, p \in (2, \infty)$, so that i_2 is neither right nor left continuous at the jump point $p = 2$. Again it is possible to produce a modified strictly elliptic coercive integrand f^* for which the lack of one sided continuity of i_2^* at $p = 2$ is retained (cf.[S] for details and related results).

Next we consider a two-dimensional example. Given $p_0, p_1 \in (1, \infty)$ with $p_0 < p_1$ and a function $\alpha \in W^{1,1}[1, \infty)$ satisfying $\alpha(p) = 0, p \in [1, p_0], \alpha'(p) \geq 0, p \in [p_0, p_1], \alpha(p) = \alpha(p_1) > 0, p \in [p_1, \infty)$, consider the integrand
$f(x, y, u, z) = c(y)\alpha'(y)|u|^{(m-3y)/(y-1)}(|u|^{y/(y-1)} - x)^2|z|^m$
with $m \in [3p_1, \infty), c(y) =$ (m-3)(m-2)/2 $[((m-1)/m)(y/(y-1))]^m, x \in [0, 1], y \in [p_0, p_1]$
and take
$J[u] = \int_{\Sigma_{p_0,p_1}} c(y)\alpha'(y)|u|^{(m-3y)/(y-1)}(|u|^{y/(y-1)} - x)^2|u_{,x}|^m dxdy$, with
$\Sigma_{p_0,p_1} = [0, 1] \times [p_0, p_1]$ and $u \in \mathcal{A}_2(p) = \{u \in W^{1,p}(\Sigma_{p_0,p_1}) | u(0, \cdot) = 0, u(1, \cdot) = 1\}$.
The infimum for this problem satisfies Λ $i(p) = \alpha(p), p \in [1, \infty)$, so that the infimum function can be absolutely continuous. The argument makes use of a 1988 result (cf. [H&M]) involving Noether's theorem for invariant one-dimensional variational problems (cf. [F] for details and related results). An effort to present a **one-dimensional** variational problem possessing properties of both types encountered above is currently under way ([M2]).

IV. Our final discussion has to do with the topic which originally motivated the work by Ball and Mizel on the Lavrentiev phenomenon (cf. [B&M]), namely the relevance of this issue to the behavior of multidimensional nonlinearly elastic materials. Here we consider a homogeneous material in two dimensions with stored energy integrand $W = W_\epsilon : \text{Lin}^+(\mathbb{R}^2) \to [0, \infty)$ of the form
$W(F) = (||F||^2 - 2\det F)^4 + \epsilon[(\det F)^{-1} + (1 + ||F||^2)^{q/2}]$, for $q > 2, \epsilon > 0$, where $\text{Lin}^+(\mathbb{R}^2) \subset \text{Lin}(\mathbb{R}^2)$ denotes the set of linear operators on 2-dimensional space with positive determinant.

It is easy to verify that W has the following properties:
(a) W is smooth,
(b) W is **objective** and **isotropic**, i.e. $W(QF) = W(F) = W(FQ)$ for each orthogonal operator $Q \in \text{Lin}^+(\mathbb{R}^2)$,
(c) W is polyconvex, i.e. there is a convex mapping $g : \text{Lin}(\mathbb{R}^2) \times (0, \infty) \to \mathbb{R}$ such that $W(F) = g(F, \det F)$, for $F \in \text{Lin}^+(\mathbb{R}^2)$,
(d) $W(F) \to +\infty$ as $\det F \to 0^+$,
(e) $W(F) \geq c_1 ||F||^q - c_2, c_i > 0$,
(f) $W(F) = \psi(v_1, v_2) = |v_1 - v_2|^8 + \epsilon[1/(v_1 v_2) + (1 + (v_1)^2 + (v_2)^2)^{q/2}]$, where the v's are the singular values of F [principal stretches of a deformation]. Thus the function A defined by $A(\delta) = \psi(\sqrt{\delta}, \sqrt{\delta})$ is convex throughout $(0, \infty)$.

The variational problem to be considered involves the total stored energy
$E[u] = \int_\Omega W(\nabla u) dx$ for $\Omega = \{x \in \mathbb{R}^2 | x_1^2 + x_2^2 < 1, x_2 > 0\}$, with $u \in \mathcal{A}(p) = \{u \in W^{1,p}(\Omega; \mathbb{R}^2), \nabla u \in \text{Lin}^+ \text{ a.e.} | u(x_1, 0) \in [0,1] \times \{0\}, x_1 \in [0,1]; u(x_1, 0) \in \{0\} \times [0,1], x_1 \in [-1, 0]; u(x) = (\cos(\theta/2), \sin(\theta/2)), x = (\cos\theta, \sin\theta), \theta \in [0, \pi]\}$, $p > 2$, so that u maps the unit upper half disk Ω into the unit quarter disk Ω'. We have the following result

Theorem There is an $\epsilon_0(q) > 0$ such that if $\epsilon < \epsilon_0(q)$ and $2 < p_1 < 4 < p_2$ then Λ $i(p_1) = \inf\{E[u] | u \in \mathcal{A}(p_1)\} < i(p_2) = \inf\{E[u] | u \in \mathcal{A}(p_2)\}$

Notice that by the Sobolev imbedding theorem the displacements in both $\mathcal{A}(p_1)$ and $\mathcal{A}(p_2)$ are continuous on Ω, so that no cracks are created. Indeed, the convexity of A in (f) ensures that the stored energy of continuous deformations from Ω into Ω' cannot be lowered even by enlarging class $\mathcal{A}$ to include discontinuous deformations of bounded variation with the same average density (cf. [M3]). That is, this material has the property that opening "cracks" in a deformation cannot result in a lowering of the energy.

The proof depends critically on the following properties of the integrand W_0 [i.e., W_ϵ with $\epsilon = 0$]:

(i) $W_0 \geq 0$ is convex on $\text{Lin}(\mathbb{R}^2)$

(ii) solutions of the Euler-Lagrange system associated with W_0 are known explicitly via complex analysis. In fact, $u^*(x) = [(x_1)^2 + (x_2)^2]^{1/4}(\cos(\theta/2), \sin(\theta/2))$, where θ is the polar angle to x, yields $W_0(\nabla u^*) = 0$ and $u^* \in \mathcal{A}(p)$, for $p \in (2, 4)$, while $u^{**}(x) = [(x_1)^2 + (x_2)^2]^{11/28}(\cos(\theta/2), \sin(\theta/2))$ satisfies $u^{**} \in \mathcal{A}(p)$, for $p \in (2, \frac{28}{3})$ and $E_0[u^{**}] > 0$, with u^* and u^{**} both satisfying the Euler-Lagrange system for $E = E_0$ (cf. [FHM1] for further details and related results).

Bibliography

[A&C] G. Alberti and F. Serra Cassano, Nonoccurence of gap for one-dimensional autonomous functionals, Istituto di Matematiche Applicated, Universita di Pisa (1994)

[A&M] G. Alberti and P. Majer, Gap phenomenon for some autonomous functionals, J. Conv. Anal. **1** (1994), 31-45.

[B&M] J. M. Ball and V. J. Mizel, One-dimensional variational problems whose minimizers do not satisfy the Euler-Lagrange equation, Arch. Rational Mech. Anal. **90** (1985) 325-388.

[Bu&M] G. Buttazzo and V. J. Mizel, Interpretation of the Lavrentiev phenomenon by relaxation, J. Funct. Anal. **110** (1992) 434-460.

[Bu&B] G. Buttazzo and M. Belloni, A survey on old and recent results about the gap phenomenon in the calculus of variations, in Recent Developments in Well-Posed Variational Problems, R. Lucchetti and J. Revalski eds., Kluwer 1995.

[Ce] L. Cesari, Optimization-theory and Applications, Springer-Verlag, New York 1983.

[C&V] F. H. Clarke and R. B. Vinter, Regularity properties of solutions to the basic problem in the calculus of variations, Trans. Amer. Math. Soc. **289** (1985) 73-98.

[Da] A. M. Davie, Singular minimizers in the calculus of variations in one dimension, Arch. Rational Mech. Anal. **101** (1988) 161-177.

[DHM] K. Dani, W.J. Hrusa and V.J. Mizel, On the Lavrentiev phenomenon for totally unconstrained variational problems in one dimension, Nonlinear Diff. Equations and Appls. (in press).

[F] M. Foss, Examples of the Lavrentiev phenomenon with continuous Sobolev exponent dependence, Center for Nonlin. Anal. Dept. Math. Sciences CMU Res. Report 00-CNA-013 (10/2000).

[FHM1] M. Foss, W.J. Hrusa and V.J. Mizel, Occurrence of the Lavrentiev phenomenon in two dimensional nonlinear elasticity (in preparation).

[FHM2] _____, On types of integrands exhibiting the Lavrentiev phenomenon in dimensions greater than one, (in preparation).

[H] W.J. Hrusa, Lavrentiev's phenomenon for second-order autonomous variational problems in one dimension (preprint).

[H&M] A.C. Heinricher and V.J. Mizel, The Lavrentiev phenomenon for invariant variational problems, Arch. Rational Mech. Anal. **102** (1988) 57-93.

[L] M. Lavrentiev, Sur quelques problemes du calcul des variations, Ann. Mat. Pura Appl. **41** (1926) 107-124.

[Ma] M. Mania, Sopra un esempio di Lavrentieff, Boll. Un. Mat. Ital **13** (1934) 147-153.

[M1] V.J. Mizel, New developments concerning the Lavrentiev phenomenon, Technion 1998, in Calculus of Variations and Diff. Equations Chapman and Hall/CRC Research Notes in Math. #410,2000, A. Ioffe, S. Reich, I. Shafrir, eds.

[M2] _____, The Lavrentiev phenomenon in one dimension with general monotone exponent dependence (in preparation).

[M3] _____, On the ubiquity of fracture in nonlinear elasticity, J. of Elasticity **52** (1999) 257-266.

[S] A. Siegel, Two examples of Lavrentiev's phenomenon, Master's Thesis, Dept. of Math. Sciences, Carnegie Mellon U. 1999.

[S&M] M. Sychev and V.J. Mizel, A condition on the value function both necessary and sufficient for full regularity of minimizers of one-dimensional variational problems, Trans. Amer. Math. Soc. **350** (1998) 119-133.

Abstract Eigenvalue Problem for Monotone Operators and Applications to Differential Operators

Silviu Sburlan

Department of Mathematics, Ovidius University,
Bd. Mamaia 124, 8700-Constantza, Romania
E-mail:<ssburlan@univ-ovidius.ro>

Abstract

In this work we extend the multiple orthogonal sequence method, developed in [3], to the energetic space of an abstract linear monotone operator. This method leads to an abstract eingenvalue problem that it produces orthonormal bases in some nested Hilbert spaces, that they are suitable to develop abstract Fourier or projection methods. Some examples to diœerential operators are also given.

One also considers the abstract semilinear eingenvalue problem and some results, known for compact operators, are extended for monotone type operators with application to diœerential operators.

Let X be a real Hilbert space with inner product $(\cdot,\cdot)$ and the induced norm $\|\cdot\|$.

Consider a linear operator $B: D(B) \subset X \to X$, with $D(B)$ inønite dimensional, which is symmetric, i.e.,

$$(Bu, v) = (u, Bv), \ \forall u, v \in D(B) \tag{1}$$

and strongly monotone, that is, there exists $c > 0$ such that

$$(Bu, u) \geq c\|u\|^2, \ \forall u \in D(B) \tag{2}$$

We induce on $D(B)$ the energetic inner product

$$(u, v)_E := (Bu, v), \ \forall u, v \in D(B)$$

and the energetic norm

$$\|u\|_E := (u, u)_E^{1/2}, \ \forall u \in D(B).$$

Denote by E the completion in X of the linear subspace $D(B)$ with respect to the energetic norm an call it the energetic space of the operator B. It contains all $u \in X$ that are the limit points of Cauchy sequences $\{u_n\} \subset D(B)$ with respect to the energetic norm $\|\cdot\|_E$. Extending by continuity the energetic inner product, i.e.,

$$(u, v)_E := \lim(u_n, v_n)_E, \ \forall u, v \in E,$$

the energetic space E becomes a real Hilbert space containing $D(B)$ as a dense subset and the embedding $E \hookrightarrow X$ is continuous, namely

$$||u|| \leq c^{-\frac{1}{2}} ||u||_E, \ \forall u \in E.$$

The duality map $J : E \to E^*$, deøned through

$$< Ju, v > := (u,v)_E, \ \forall u, v \in E,$$

is a linear homeomorphism with

$$||Ju||_{E^*} = ||u||_E, \ \forall u \in E,$$

(see D. Pascali and S. Sburlan [7,p. 112]) and it is an extension of B, i.e.,

$$Ju = Bu, \ \forall u \in D(B).$$

The Friederichs extension $A : D(A) \subseteq X \to X$ of the operator B is deøned through

$$Au := Ju, \ \forall u \in D(A), \tag{3}$$

where $D(A) := \{u \in E | Ju \in X\}$. Observe that $u \in D(A)$ if and only if there exists an $f \in X$ such that

$$< Ju, v > = (f, v)_E, \ \forall v \in E$$

and $D(B) \subseteq E \subseteq X \subseteq E^*$, (see E. Zeidler [11,p.280]).

Remark that the Friederichs extension is in fact the maximal monotone extension of B in X since $D(A)$ is dense in X and A is closed, self-adjoint, bijective and strongly monotone, i.e.,

$$(Au, u) \geq c ||u||^2, \ \forall u \in D(A),$$

(see A. Haraux [2,p. 48]). Also, the inverse operator $A^{-1} : X \to X$ is linear continuous self-adjoint and compact, whenever the embedding $E \hookrightarrow X$ is compact. Therefore applying the Fredholm theory we can state following variant of the multiple orthogonal sequence theorem (G. Morosanu and S. Sburlan [3]):

Theorem 1: If the embedding $E \hookrightarrow X$ is compact, then there exists the sequences $\{e_n\} \subset E$ and $\{\lambda_n\} \subset (0, +\infty)$ that are eingensolutions of A, i.e.,

$$(Ae_n, v) = \lambda_n (e_n, v), \ \forall v \in X, \ n \in \mathbb{N} \tag{4}$$

and such that:

(i) $\{e_n\}$ is an orthonormal basis in E;
(ii) $\{\sqrt{\lambda_n}e_n\}$ is an orthonormal basis in X;
(iii) $\{\lambda_n e_n\}$ is an orthonormal basis in E^*;
(i) $\{\lambda_n\}$ is increasingly divergent to $+\infty$.

Proof. (i) Suppose that the embedding $E \hookrightarrow X$ is compact and let $\{u_n\} \in E$ be a bounded sequence, $\|u_n\|_E \leq c$. Then, passing eventually to a subsequence, we can suppose that $u_n \to u$ in X, and $u \in E$, because $\{u_n\}$ is strongly and also weakly convergent in E. Then by the compacity of the embedding

$$\|A^{-1}u_n - A^{-1}u\|_E = \sup\{< A^{-1}(u_n - u), v>; \|v\|_E \leq 1\} =$$
$$= \sup\{(u_n - u, v)\|v\|_E \leq 1\} = c\|u_n - u\| \to 0$$

and, thus, A^{-1} maps bounded sets into relative compact sets. By Hilbert - Schmidt theorem A^{-1} has a countable set of eingenvalues $\{k_n\} \subset (0, +\infty), k_n \geq k_{n+1} \to 0$, and the corresponding eingenvectors, say $\{e_n\}$,

$$A^{-1}e_n = k_n e_n, \quad \forall n \in \mathbb{N},$$

form a Hilbertean basis in E (see e.g. S. Sburlan & co [9, p. 186]). Hence

$$Ae_n = \lambda_n e_n, \quad \forall n \in \mathbb{N}$$

where the characteristic values, $\lambda_n = \frac{1}{k_n}$, are such that $\lambda_n \leq \lambda_{n+1} \to +\infty$, as stated in (iv), and therefore (4) holds.

iii) To show that $\{\sqrt{\lambda_n}e_n\}$ is a Hilbertean basis in X it suŒces to prove that it is complete in X because

$$\delta_{mn} = (e_m, e_n)_E = \lambda_n(e_m, e_n) \qquad (5)$$

as it results from (4). For this let $h \in X$ be such that

$$(h, \sqrt{\lambda_n}e_n) = 0, \quad \forall n \in \mathbb{N} \qquad (6)$$

by the Lax-Milgram theorem, with

$$a(u,v) := (u,v)_E, \quad \forall u, v \in E,$$

it exists only one $\tilde{u} \in E$ such that

$$(\tilde{u}, v)_E = (h, v), \quad \forall v \in E \qquad (7)$$

Therefore, from (6) and (7), it results

$$(\tilde{u}, e_n)_E = 0, \quad \forall n \in \mathbb{N},$$

and, since $\{e_n\}$ is complete in E, we have the implication $\tilde{u} = 0 \Rightarrow h = 0$, that is, the completeness in X of the system $\{\sqrt{\lambda_n}e_n\}$.

(iii) To show that $\{\lambda_n e_n\}$ is a Hilbertean basis in E^* we must prove that it is an orthogonal system complete in E^*.

Note that E^* is a Hilbert space with the norm

$$||v^*||_{E^*} := \sup\{v^*(v)|v \in E, ||v||_E \leq 1\}$$

and the inner product defined by

$$(v^*, u^*)_{E^*} := \frac{1}{4}\left(||v^* + u^*||_{E^*}^2 - ||v^* - u^*||_{E^*}^2\right).$$

For orthogonality we use the equation (4), namely

$$(\lambda_n e_n)(e_n) := <\lambda_n e_n, e_n> = (\lambda_n e_n, e_n) = (Ae_n, e_n) =$$
$$= (e_n, e_n)_E = ||e_n||_E = 1$$

and

$$(\lambda_n e_n, \lambda_m e_m)_{E^*} = (J^{-1}(\lambda_n e_n), J^{-1}(\lambda_m e_m))_E =$$
$$= (e_n, e_m)_E = 0, \forall m \neq n.$$

For completeness let $v^* \in E^*$ be such that $(v^*, \lambda_n e_n)_{E^*} = 0$ for all $n \in \mathbb{N}$. It then results

$$(v^*, h)_{E^*} = 0, \forall h \in X$$

because $\{\sqrt{\lambda_n}e_n\}$ is a complete system in X. Since X is dense in E^*, for all $\varepsilon > 0$ there exists an $h_\varepsilon \in X$ such that

$$\varepsilon > ||v^* - h_\varepsilon||_{E^*}^2 = ||v^*||_{E^*}^2 - 2(v^*, h_\varepsilon)_{E^*} + ||h_\varepsilon||_{E^*}^2 = ||v^*||_{E^*}^2 + ||h_\varepsilon||_{E^*}^2.$$

Hence, we have the implication

$$||v^*||_{E^*}^2 < \varepsilon, \forall \varepsilon > 0 \Rightarrow v^* = 0$$

as required. ∎

Direct consequence: Denote by $E_n := Sp\{e_1, e_2, ..., e_n\} \subset E$, $X_n = Sp\{\sqrt{\lambda_1}e_1, \sqrt{\lambda_2}e_2, ..., \sqrt{\lambda_n}e_n\} \subset X$ and $E_n^* := Sp\{\lambda_1 e_1, \lambda_2 e_2, ..., \lambda_n e_n\} \subset E^*$, the finite dimensional subspaces generated by the finite sequence $\{e_1, e_2, ..., e_n\}$. Then E_n, X_n and E_n^* are projectionally complete in E, X and E^*, respectively, that is $\pi_n u \to u$ in each space, where

$$\pi_n u := \sum_{k=1}^{n} \alpha_k \varphi_k, \forall n \geq 1,$$

with $\{\varphi_1, \varphi_2, ..., \varphi_n\}$ one of the above mentioned basis and $a_k, 1 \leq k \leq n$, the corresponding Fourier coefficients.

Eigenvalue Problem for Monotone Operators

These coordinate systems can be used either for abstract Galerkin projection method or for abstract Fourier series method (see S. Sburlan and G. Moroanu [10]).

Remark: Since $D(B^k) \subseteq D(B^{k-1}) \subseteq ... \subseteq D(B)$ and B^k is also symmetric we can apply the above method to produce orthogonal basis in the energetic spaces associated to $B^k, k \in \mathbb{N}$, and the corresponding dual spaces.

Define the spaces $E_k := D(A^{k/2}), k \in \mathbb{N}$, where A is the Friederichs extension of B. It is easily seen that E_k are Hilbert spaces with the inner products:

$$(u,v)_k := (A^{k/2}u, A^{k/2}v)_E, \quad \forall u, v \in E_k.$$

Of course, $E_1 = E$ and $(u,v)_1 = (u,v)_E$, $\forall u,v \in E$ (see E. Zeidler [11, p. 296]). As above we can easily show that $\{\lambda_n^{(-k+1)/2} e_n\}$ is an orthonormal basis in E_k and

$$\|y\|_k^2 := \sum_{n \geq 1}(y, \lambda^{(-k+1)/2}e_n)_E^2 = \sum_{n \geq 1}(y, A^k(\lambda^{(-k+1)/2}e_n)) = \sum_{n \geq 1} \lambda_n^{k+1}(y, e_n)^2 \tag{8}$$

Since $\{\lambda_n\}$ is an increasing sequence of positive numbers it results, by (8), that the embedding $E_{k+1} \hookrightarrow E_k$ is continuous. Also, identifying X by its dual space we get the inclusions

$$... \subseteq E_{k+1} \subseteq E_k \subseteq ... \subseteq E_1 \equiv E \subseteq X \subseteq E^* \equiv E_1^* \subseteq ... \subseteq E_k^* \subseteq E_{k+1}^* \subseteq ...$$

As the embedding $E \hookrightarrow X$ is compact, it then follows that the embeddings $E_{k+1} \hookrightarrow E_k$, $X \hookrightarrow E^*$ and $E_k^* \hookrightarrow E_{k+1}^*$ are also compact. The duality mapping $J_k : E_k \to E_k^*$ is defined by

$$< J_k u, v > = (u,v)_k, \quad \forall u, v \in E_k$$

and, obviously, $J_1 \equiv J$. It is easily seen that

$$J_k(\lambda_n^{(1-k)/2} e_n) = \lambda_n^{(k+1)/2} e_n, \quad \forall n, k \in \mathbb{N}. \tag{9}$$

and then $\{\lambda_n^{(k+1)/2} e_n\}$ is an orthonormal basis in E_k^*.

Example 1. Let Ω be a bounded domain in $\mathbb{R}^n$ with its boundary $\partial \Omega$ enough smooth to apply the Green's formula. Take $X := L^2(\Omega)$, that is a Hilbert space with the inner product

$$(u,v) := \int_\Omega u(x)v(x)dx, \quad \forall u, v \in L^2(\Omega),$$

and let $B := -\Delta$ be a defined on

$$D(B) := \{u \in C^2(\Omega) \cap C^1(\bar{\Omega}) | u = 0 \text{ on } \Gamma \subseteq \partial\Omega, \mu(\Gamma) > 0\}.$$

Then B is symmetric and strongly monotone having

$$E := \{v \in H^1(\Omega) | v = 0 \text{ on } \Gamma\}$$

as energetic space. Note that E is a Hilbert space with the inner product

$$(u,v)_E := \int_\Omega \nabla u(x) \cdot \nabla v(x) dx, \ \forall u,v \in E$$

and the embedding $E \hookrightarrow X$ is compact when $N \geq 2$ (see J. Neas [6,p.117]). Therefore we can apply the Theorem 1 and write (4) in the form

$$\int_\Omega \nabla e_n(x) \cdot \nabla v(x) dx = \lambda_n \int_\Omega e_n(x) v(x) dx, \ \forall v \in E \tag{9'}$$

Then by Green's formula we deduce that e_n are the weak solutions of the following eingenvalue problem

$$\begin{cases} -\Delta e_n(x) = \lambda_n e_n(x), x \in \Omega, \\ e_n(x) = 0 \qquad\qquad , x \in \Gamma, \\ \frac{\partial e_n}{\partial \nu}(x) = 0 \qquad\quad , x \in \partial\Omega \setminus \Gamma. \end{cases} \tag{10}$$

Remark the limit case $\Gamma = \phi$ when $E = H^1(\Omega)$ is a Hilbert space with another inner product

$$(u,v)_E := \int_\Omega [u(x)v(x) + \nabla u(x) \cdot \nabla v(x)] dx, \ \forall u,v \in E$$

that it modiøes (9') as follows

$$\int_\Omega \nabla e_n(x) \cdot \nabla v(x) dx = (\lambda_n - 1) \int_\Omega e_n(x) v(x) dx \tag{11}$$

and, thus, e_n are the weak solutions of the problem

$$\begin{cases} -\Delta e_n(x) = (\lambda_{n-1}) e_n(x), \ x \in \Omega, \\ \frac{\partial e_n}{\partial \nu}(x) = 0 \qquad\quad , \ x \in \partial\Omega. \end{cases} \tag{12}$$

Application: The deformation of a body $\mathcal{B}$, that occupies a bounded region $\Omega \subseteq \mathbb{R}^N$ ($N = 2$ or 3), is characterized by the displacement vector $\mathbf{u} : \Omega \to \mathbb{R}^N$ and the corresponding strain tensor $\varepsilon = \varepsilon(\mathbf{u})$. In the case of the small deformations, $\varepsilon(\mathbf{u})$ reduces to the symmetric part of the displacement gradient, i.e.,

$$\varepsilon(\mathbf{u}) := \{\varepsilon_{ij}(\mathbf{u}) | \varepsilon_{ij}(\mathbf{u}) = \frac{1}{2}\left(\frac{\partial u_i}{\partial x_j} + \frac{\partial u_j}{\partial x_i}\right), 1 \leq i,j \leq N\}. \tag{13}$$

The constitutive relation that characterizes the elasticity is a generally nonlinear dependence of the stress tensor

$$\sigma := \{\sigma_{ij} | \sigma_{ij} = \sigma_{ji}, \ 1 \leq i,j \leq N\} \tag{14}$$

on the strain, namely

$$\sigma = \sigma(\varepsilon) = \mathcal{A}\varepsilon + o(\varepsilon^2), \tag{15}$$

where $\mathcal{A} := \{a_{ijkl} \in \mathbb{R} | a_{ijkl} = a_{jikl} = a_{klij}, 1 \leq i,j,k,l \leq N\}$ are the elastic coefficients. In the linear case, adopting the Einstein's summing convention (i.e., the repeated index means summing over that index), we have the following relations

$$\sigma_{ij} = a_{ijkl}\varepsilon_{kl} \qquad (16)$$

known as the Hook's law. The coefficients $\mathcal{A}$ must depend continuously on the point in Ω or they are constants (the hyperelastic case) and they satisfy the ellipticity condition

$$a_{ijkl}\varepsilon_{ij}\varepsilon_{kl} \geq c|\varepsilon|^2, \ \forall \varepsilon \in \mathbb{R}^{N \times N}, \varepsilon = \varepsilon^T, \qquad (17)$$

where $|\cdot|$ means the Euclidean norm.

Let $X := [L^2(\Omega)]^N$ be the Hilbert space of square integrable vectorial functions endowed with the inner product

$$(\mathbf{u}, \mathbf{v}) = \int_\Omega u_i v_i dx$$

and define on

$$D(B) := \{\mathbf{v} \in [\mathcal{C}^2(\Omega) \cap \mathcal{C}^1(\bar{\Omega})]^N | \mathbf{v} = 0 \text{ on } \Gamma \subseteq \partial\Omega, \mu(\Gamma) > 0\}$$

the operator B of the linear elasticity

$$B\mathbf{v} = -div\sigma(\varepsilon(\mathbf{v})),$$

where the components of σ are given by (16).

By Green's formula we have

$$(B\mathbf{u}, \mathbf{v}) := \int_\Omega div\sigma(\varepsilon(\mathbf{u})) \cdot \mathbf{v} dx = \frac{1}{2}\int_\Omega a_{ijkl}\varepsilon_{kl}(\mathbf{u})\varepsilon_{ij}(\mathbf{v})dx =$$

$$= \frac{1}{2}\int_\Omega a_{klij}\varepsilon_{ij}(\mathbf{v})\varepsilon_{kl}(\mathbf{u})dx = -\int_\Omega div\sigma(\varepsilon(\mathbf{v}))\mathbf{u}dx =: (u, Bv), \ \forall \mathbf{u}, \mathbf{v} \in D(B).$$

Moreover, we have

$$(B\mathbf{u}, \mathbf{u}) = \frac{1}{2}\int_\Omega a_{ijkl}\varepsilon_{ij}(\mathbf{u})\varepsilon_{kl}(\mathbf{u})dx \geq \frac{c}{2}\int_\Omega \varepsilon_{ij}(\mathbf{u})\varepsilon_{ij}(\mathbf{u})dx = c\int_\Omega |\nabla \mathbf{u}|^2 dx.$$

The energetic space E is the completion on $D(B)$ with respect to the norm

$$||\mathbf{u}||_E^2 := \frac{1}{2}\int_\Omega \sigma_{ij}(\varepsilon(\mathbf{u}))\varepsilon_{ij}(\mathbf{u})dx = \frac{1}{2}\int_\Omega a_{ijkl}\varepsilon_{ij}(\mathbf{u})\varepsilon_{kl}(\mathbf{u})dx \qquad (18)$$

and it contains all vector functions from $[H^1(\Omega)]^N$ that vanishes on Γ.

Moreover, the norm (18) is equivalent with the following one

$$|||\mathbf{u}||| := \int_\Omega |\nabla \mathbf{u}|^2 dx,$$

and thus, by Sobolev-Kondrashov theorem, the embedding $E \hookrightarrow X$ is compact. Hence the duality mapping $J : E \to E^*$ is defined by

$$< J\mathbf{u}, \mathbf{v} > = \frac{1}{2} \int_\Omega a_{ijkl}\varepsilon_{ij}(\mathbf{u})\varepsilon_{kl}(\mathbf{v})dx \qquad (19)$$

and, by the Theorem 1:

$$\frac{1}{2}\int_\Omega a_{ijkl}\varepsilon_{ij}(\mathbf{e}_n)\varepsilon_{kl}(\mathbf{v})dx = \lambda_n \int_\Omega (\mathbf{e}_n)_i v_i dx.$$

By Green's formula we deduce that $\mathbf{e}_n$ are the weak solutions of the eingenvalue problem:

$$\begin{cases} -div\sigma(\varepsilon(\mathbf{e}_n)) = \lambda_n \mathbf{e}_n \text{ in } \Omega, \\ \mathbf{e}_n = \mathbf{0} \text{ on } \Gamma, \\ \sigma_{ij}(\varepsilon(\mathbf{e}_n))\nu_j = 0 \text{ on } \partial\Omega\backslash\Gamma. \end{cases} \qquad (20)$$
∎

Consider now the semilinear eingenvalue problem in X

$$Lu + \mu N(u) = f, \qquad (21)$$

where $L \in L(X)$ is compact and positive, N is a nonlinear perturbation of Leray-Schauder type (i.e., $N := I - T$ with T a compact operator), and $f \in X$ a given element. Since L is, in particular, hemicontinuous and monotone it is maximal monotone and, thus, $I + \lambda L$ is invertible for each $\lambda > 0$ and $\|(I + \lambda L)^{-1}\| \leq 1$ (see e.g. D. Pascali and S. Sburlan [7, p.106]).

Consequently we can write (21) equivalently as

$$(I - M(\lambda))u = g \qquad (22)$$

where $M(\lambda) := (I + \lambda L)^{-1}(I - N)$ is a compact operator, $\lambda = \frac{1}{\mu} \in \mathbb{R}_+$ and $g = (I + \lambda L)^{-1}f$. This it allows to introduce the coincidence degree for the pair (L, N), simply, as follows:

If $D \subset X$ is an open bounded set such that

$$N(u) + \lambda Lu \neq f \; \forall \; u \in \partial D, \; \lambda \in \mathbb{R}_+ \qquad (23)$$

then the coincidence degree of the pair (L, N) in D relatively to f is defined by

$$d_\lambda((L,N), D, f) := d_{LS}(I - M(\lambda), D, g), \qquad (24)$$

where d_{LS} denotes the Leray-Schauder degree. Of course, this degree has all the properties of Leray-Schauder degree, because of the above mentioned equivalence, and we can apply all classical results to (21). We point out here some in the linear case (see Mortici [4]):

Suppose that $N \in L(X)$. We say that $\lambda \in \mathbb{R}_+$ is a characteristic value for the pair (L, N) provided that $\ker(N + \lambda L) \neq \{\theta\}$ and we say that λ is a regular covalue of (L, N) if the resolvent $R(\lambda) := (N + \lambda L)^{-1}$ exists on X and it is continuous.

Since, from the above equivalence, we have

$$\ker(N + \lambda L) \equiv \ker(I - M(\lambda)) \qquad (25)$$

it results that each $\lambda > 0$ is a characteristic value or a regular covalue for (L, N).

Denote by $\mathcal{C}(L, N)$ the set of characteristic values of (L, N) and by L_α the spectral operator for (L, N)

$$L_\alpha := (N + \alpha L)^{-1} L \equiv R(\alpha) L, \qquad (26)$$

and observe that L_α is compact whenever L is compact. From the identity

$$N + \lambda L = (N + L)[I + (\lambda - \alpha) L_\alpha] \qquad (27)$$

we see that $N + \lambda A$ is invertible if and only if $\alpha - \lambda$ is not a characteristic value of L_α. Hence from Fredholm theory it results that $C(L, N)$ is at most a countable set with only one possible acumulation point at inønity, as we have seen in the theorem 1.

Suppose now that α is a øxed covalue of (L, N) and deøne the multiplicity $m(\lambda)$ of the characteristic value $\lambda \in C(L, N)$ to be the algebraic multiplicity of the eingenvalue $(\alpha - \lambda)^{-1}$ of the compact operator L_α.

Let $a \in D$ be an isolated solution of the equations (21) and choose an open ball $B(a, r)$ such that $R(\lambda) f \cap \bar{B}(a, r) = \{a\}$. By the excision property of the coincidence degree we can deøne the coincidence index of (L, N) in a with respect to f to be

$$i_\lambda((L, N), a, f) := d_\lambda((L, N), B(a, r), f) \qquad (28)$$

and, by the above equivalence, it is true the following tranversality property:

Proposition 1 (Leray-Shauder) *If* $[\lambda_1, \lambda_2] \cap C(L, N) = \{\lambda\}$, *then*

$$i_{\lambda_1}((L, N), a, f) = (-1)^{m(\lambda)} i_{\lambda_2}((L, N), a, f).$$

In the semilinear case it apperas naturally the following question: Are the well known results of Krosnoselskii and Rabinowitz, concerning bifurcation theory for compact operators, true for monotone type operators? The answer is Yes! and was given in [5] in the following way:

Let X be a reÆexive uniformly convex Banach space and X^* be its dual Banach space. As usually we denote by $\| \cdot \|$ the norm in both spaces, by $< \cdot, \cdot >$ the duality pairing and by $J : X \to X^*$, the duality map.

Consider the eingenvalue problem

$$Jx + \mu A x + R(\mu, x) = 0, \qquad (29)$$

where $A : X \to X^*$ is a linear continuous operator and $R(\mu, \cdot) : X \to X^*$ is a nonlinear perturbation such that $R(\mu, 0) = 0$, $\forall \mu \in \mathbb{R}$. Note that only J is monotone.

In this case $(\mu, 0)$ are solution of (29) for all $\mu \in \mathbb{R}$-named trivial solutions and the set of all trivial solutions are denoted by $\mathcal{C}$.

A point $(\mu_0, 0) \in \mathcal{C}$ is said to be a bifurcation point for (29) provided that there exists solutions (μ, x_μ), $x_\mu \neq 0$ in each neinghbourhood of $(\mu_0, 0)$. Let us denote by S_0 the set all of these nontrivial solutions and let $S := \bar{S}_0$ be its aderence in $\mathbb{R} \times X$.

The key step in our extension is the Browder-Ton theorem concerning the compact imbedding property for separable Banach spaces (e.g. D. Pascali and S. Sburlan [7,p.302]):

Theorem (Browder-Ton). Let X be a separable reflexive Banach space and let S be a countable subset of X. Then there exists a separable Hilbert space H and a compact one-to-one linear operator $\psi : H \to X$ such that $S \subset \psi(H)$ and $\psi(H)$ is dense in X.

Define now the adjoint operator $\phi : X^* \to H$ by using the inner product of H denoted by $(\cdot, \cdot)$:

$$<\phi u, v> = (u, \psi v), \forall u \in H, v \in X^*. \tag{30}$$

Then the operator $L := -\psi \phi A : X \to X$ is linear and compact as the composition of a continuous map with a compact one. Since the spectrum $\sigma(L)$ is discrete we can choose $\delta > 0$ such that

$$\sigma(L) \cap (I_1 \cup I_2) = \phi,$$

where $I_1 := (\varepsilon \mu_0 - \delta, \varepsilon \mu_0)$ and $I_2 := (\varepsilon \mu_0, \varepsilon \mu_0 + \delta)$ and $\mu_0 \in \sigma(L)$.

Let $\mu_1 \in I_1$ and $\mu_2 \in I_2$ be arbitrary fixed. Then the mappings

$$I + \tfrac{1}{\varepsilon}\mu_i \psi \phi A, \ i = 1, 2$$

are regular and thus there exists $M > 0$ such that

$$\left\|\left(I - \frac{\mu_i}{\varepsilon}L\right)x\right\| \geq M\|x\|, \forall x \in X, \ i = 1, 2. \tag{31}$$

Suppose that A is bounded from below in the sense

$$<Ax, x> \geq -\frac{\varepsilon}{\varepsilon \mu_0 + \delta}\|x\|^2, \forall X \in x \tag{i}$$

and the complementary part is asimptotical zero, i.e.,

$$Jx + R(\mu, x) = o(\|x\|) \tag{ii}$$

uniformly in μ on bounded sets. Using the following Leray-Schauder homotopy

$$h_t(x) := x + \frac{t}{\varepsilon}\psi\varphi Jx + \frac{1}{\varepsilon}\mu\psi\varphi Ax, \ t \in [0,1], \ x \in X, \tag{32}$$

we easily obtain

$$-\frac{\varepsilon}{\varepsilon\mu_0 + \delta} \cdot \|x\|^2 \leq <Ax, x> \leq -\frac{\varepsilon}{\mu} \cdot \|x\|^2$$

which is a contradiction, (C. Mortici and S. Sburlan [5]).

By homotopy invariance of the Leray-Schauder degree we can conclude that

$$d_{LS}\left(I + \frac{1}{\varepsilon}\mu\psi\phi A, B, 0\right) = d_{LS}\left(I + \frac{1}{\varepsilon}\psi\phi(J + \mu A), B, 0\right).$$

Eigenvalue Problem for Monotone Operators

Let now use the Berkovitz deønition of degree for (S_+) mappings (e.g. S. Sburlan and G. Moroanu [10])

$$d_S(J + \mu A, B, 0) = d_{LS}(I + \frac{1}{\varepsilon}\mu\psi\phi A, B, 0) \tag{33}$$

and consider the mapping $\varphi : I_1 \cup I_2 \to \mathbb{Z}$ deøned by

$$\varphi(\mu) := d_S(J + \mu A, B, 0), \quad \forall \mu \in I_1 \cup I_2.$$

Since L is a compact map and, by (33)

$$\varphi(\mu) = d_{LS}(I - \mu L, B, 0) = (-1)^{m(\mu)}$$

where $m(\mu)$ is the sum of algebraic multiplicity of all eingenvalues $\lambda > \frac{1}{\mu}$ of L. It then follows that $\varphi(\mu_1) = -\varphi(\mu_2)$ whenever the eingenvalue $\lambda_0 = \frac{1}{\varepsilon\mu_0}$ has odd algebraic multiplicity because it appears only in the expresion of $\varphi(\mu_2)$. Therefore in this case

$$d_S(J + \mu_1 A, B, 0) \neq d_S(J + \mu_2 A, B, 0). \tag{34}$$

Take now the Leray-Schauder homotopies $H_t^i : [0,1] \times X \to X$, $i = 1, 2$

$$H_t^i(x) = (I - \frac{\mu_i}{\varepsilon}L)x + \frac{t}{\varepsilon}\psi\phi(Jx + R(\mu_i, x)), \ t \in [0,1], \ x \in X,$$

and therefore,

$$d_{LS}(I - \frac{\mu_i}{\varepsilon}L, B, 0) = d_{LS}(I + \frac{1}{\varepsilon}(J + \mu_i A + R(\mu_i, \cdot)), B, 0), \ i = 1, 2.$$

Combine this equalitty with (34) to produce

$$d_S(J + \mu_1 A + R(\mu_1, \cdot), B, 0) \neq d_S(J + \mu_2 A + R(\mu_2, \cdot), B, 0).$$

Join linearly μ_1 with μ_2 by $\mu_t := (1-t)\mu_1 + t\mu_2$ and consider the (S_+) homotopy

$$\chi_t(x) := Jx + \mu_t Ax + R(\mu_t, x), \ t \in [0,1], \ x \in X.$$

Hence it could exist $\tau \in [0,1]$ and $x_\tau \in \partial B$ such that

$$Jx_\tau + \mu_\tau Ax_\tau + R(\mu_\tau, x_\tau) = 0,$$

that is the equation (29) has nontrivial solutions in any neighbourhood of $(\varepsilon\mu_0, 0) \in \mathbb{R} \times X$, and thus $(\varepsilon\mu_0, 0)$ is a bifurcation point for (29).

Thus we have proved an analogous of Krasnoselskii theorem for monotone type operators (e.g. S. Sburlan [8]).

Theorem 2. (Krasnoselskii). Let μ_0 be a characteristic value with odd algebraic multiplicity of the linear compact operator $L \in L(X)$.

If there exist $\varepsilon, \delta > 0$ such that (i)-(ii) hold, then $(\varepsilon\mu_0, 0) \in R \in X$ is a bifurcation point for (29).

Example 2. If consider $R(\mu, x) := \mu ||x||^2 Jx$, then
$$< R(\mu, x), x > = \mu ||x||^2 < Jx, x > = \mu ||x||^4.$$

The above results can be applied in Sobolev space $X := H^1(\Omega)$. The corresponding operator is $-\Delta u + \mu ||u|| \Delta u$ or p-laplacian operator in the general case $X := W^{1,p}(\Omega)$. ∎

Let us denote
$$i_- := d_S(J + \mu A, B, 0) = d_{LS}(I + \frac{1}{\varepsilon} \mu \psi \phi A, B, 0), \ \mu \in I_1,$$

$$i_+ := d_S(J + \mu A, B, 0) = d_{LS}(I + \frac{1}{\varepsilon} \mu \psi \phi A, B, 0), \ \mu \in I_2,$$

and observe that these degrees are constant in $\mu_1 \in I_1$ and $\mu_2 \in I_2$.

For any fixed $r > 0$ define the mapping $H_r : \mathbb{R}_+ \times X \to \mathbb{R}_+ \times X^*$ as follows
$$H_r(\mu, x) := (||x||^2 - r^2, Jx + \mu Ax + R(\mu, x)), \ \forall (\mu, x) \in \mathbb{R} \times X$$

and we have a formula similar to Ize's formula:
$$d_s(H_r, \mathcal{B}, 0) = i_- - i_+, \tag{35}$$

where $\mathcal{B} = \{(\nu, x) \in R \times X | \nu^2 + ||x||^2 < \delta^2 + r^2\}$.

Indeed, by definition of (S_+) degree (e.g., S. Sburlan and G. Moroanu [10]) we have
$$d_S(H_r, \mathcal{B}, 0) = d_{LS}(U_r, \mathcal{B}, 0),$$

where
$$U_r(\mu, x) := (||x||^2 - r^2, (I - \mu L)x - N(\mu, x)), \ \forall (\mu, x) \in \mathbb{R} \times X,$$

and
$$N(\mu, x) := \frac{1}{\varepsilon} \psi \phi (Jx + R(\mu, x)) = o(||x||).$$

We can now prove a global result concerning the bifurcation under monotonocity condition similar to those under compactness condition proved by Rabinowitz.

Theorem 3 (Rabinowitz). If $\mathcal{E}$ is a connected component containing the bifurcation point $(\varepsilon \mu_0, 0) \in \mathcal{S}$, then we have one of the following two posibilities:

(j) $\mathcal{E}$ is unbounded in $\mathbb{R} \times X$.

(jj) $\mathcal{E}$ contains a finite number of bifurcation points $(\varepsilon \mu_j, 0)$ where $\frac{1}{\varepsilon \mu_j} \in \sigma(L)$.

Moreover, the number of these points, including $(\varepsilon \mu_0, 0)$, is even.

Proof. Suppose that $\mathcal{E}$ is bounded in $\mathbb{R} \times X$ and let G be a bounded domain that contain $\mathcal{E}$ and a finite number of points $(\varepsilon \mu_0, 0)$. Moreover, suppose that there is no solution of (29) on ∂D. In this case $d(H_r, G, 0)$ is well defined and independent of $r > 0$. For r enough small
$$d_S(H_r, G, 0) = \sum_{j=1}^{k} d_S(H_r, G_j, 0),$$

where G_j are disjoint and each G_j contains only one point $(\varepsilon\mu_j, 0)$. It then follows from (35) that

$$d_S(H_r, G, 0) = \sum_{j=1}^{k}(i_-(j) - i_+(j))$$

where the sum has only an even nonvanishing terms because $i_-^*(j) - i_+^* \in \{0, \pm 2\}$. If m_j is the algebraic multiplicity of $\varepsilon\mu_j$, then

$$i_+(j) = (-1)^{m_j} i_-(j),$$

that is $i_-(j) \neq i_+(j)$, which is possible if and only if $\varepsilon\mu_j$ has an odd algebraic multiplicity. ∎

We conclude our work with the following

Example 3. Let $g : I \times \mathbb{R}^2 \to \mathbb{R}$ be a bounded continuous function

$$|g(t, \xi, \eta)| \leq M, \ \forall t \in I, \ (\xi, \eta) \in \mathbb{R}^2$$

and consider the eingenvalue problem

$$\begin{cases} -u''(t) + \lambda u(t) + g(t, u(t), u'(t)) = f(t), \ t \in I \\ (Bu)(t) = 0, \ t \in \partial I. \end{cases}$$

where $I := [0,1] \subset \mathbb{R}$, B denotes either Dirichlet boundary conditions

$$u(0) = u(1) = 0$$

or Neumann boundary conditions

$$u'(0) = u'(1) = 0,$$

or periodic boundary conditions

$$u(0) = u(1), \ u'(0) = u'(1).$$

Such problems was extensively studied in the last time (see e.g. Drbek [1]).

Take $X := L^2(I)$, $L : D(L) \to X$ with $D(L) := \{u \in \mathcal{C}^2(I); (Bu)(t) = 0, t \in \partial I\}$ deøned by $Lu := -u''$, $N(u) : X \to X$ the superposition operator $(N(u)(t) := g(t, u(t), u'(t))$ and $f \in \mathcal{C}(I)$. With these notations we arrive to the following equation in X.

$$Lu + \lambda u + N(u) = f$$

with $L \in L(X)$ symmetric and positive.

Since L is invertible with

$$(Au)(t) \equiv L^{-1}u(t) := \int_0^1 G(t,s)u(s)ds, \ u \in X, \ t \in [0,1]$$

where $G : [0,1]^2 \to \mathbb{R}$ is the continuous function

$$G(t,s) = \begin{cases} s(1-t), & 0 \leq s \leq t \leq 1 \\ t(1-s), & 0 \leq t \leq s \leq 1 \end{cases}$$

it results that A is compact and the above equation can be written as

$$u + \lambda Au + N(u) = Af.$$

Now we can apply the above results. ∎

REFERENCES

[1] P. Drbek, Solvability and Bifurcations of Nonlinear Equations, Pitman Res. Not. Math., 264, Longman, London, 1992.

[2] A. Haraux, Nonlinear Evolution Equations. Global Behaviour of Solutions, Lect. Notes Math., Vol. 841, Springer - Verlag, Berlin, 1981.

[3] G. Morosanu and S. Sburlan, Multiple Orthogonal Sequence Method and Applications, An. St. Univ.]Ovidius] Constanta, 2 (1994), 188-200.

[4] C. Mortici, Bifurcations for Semilinear Equations with Compact Nonlinearities, Bull. Appl. Com. Math., BAM-1714/'99XC-B, pp.265-272.

[5] C. Mortici and S. Sburlan, Bifurcations for Semilinear Equations of Monotone Type, An. Univ. Ovidius Constanta, vol. 7, (1998) fasc.2, pp.

[6] J. Neas, Les Mthods Discretes en Thorie des quations Elliptiques, Ed. Academia, Praque, 1967.

[7] D. Pascali and S. Sburlan, Nonlinear Mappings of Monotone Type, Sijhoœ & Noordhoœ Int. Publ., 1978.

[8] S. Sburlan, Gradul Topologic. Lecii asupra Ecuaiilor Neliniare, Ed. Academiei, Bucureti, 1983.

[9] S. Sburlan, Luminita Barbu and C. Mortici, Ecuaii Difereniale, Integrale i Sisteme Dinamice, Ex Ponto, Constanta, 1999.

[10] S. Sburlan and G. Morosanu, Monotonocity Methods for Partial Diœerential Equations, MB-11/PAMM, Budapest, 1999.

[11] E. Zeidler, Applied Functional Analysis, Appl. Math. Sci., 108, Springer-Verlag, Berlin, 1995.

Implied Volatility for American Options via Optimal Control and Fast Numerical Solutions of Obstacle Problems

Srdjan Stojanovic
Department of Mathematical Sciences
University of Cincinnati
Cincinnati, OH 45221-0025
U.S.A.
srdjan@math.uc.edu
http://math.uc.edu/~srdjan
http://CFMLab.com
December 2000

■ 1. Statement of the Problem

Let $S(t)$ denote the price S of a particular stock at the time t. We suppose that the price evolves on the stock market according to the stochastic differential equation

$$dS(t) = S(t)(a(t, S(t))\,dt + \sigma(t, S(t))\,dB(t))$$

where $a(t, S)$ is the appreciation rate, $\sigma(t, S)$ is the volatility, and $B(t)$ is the standard Brownian motion. An *American call option* is a contract that entitles the holder to buy the stock at the prescribed *strike* price, denoted k, up until the prescribed expiration time, denoted T. An *American put option* is a contract that entitles the holder to sell the stock at the prescribed *strike* price k, up until the prescribed expiration time T. Although an option does not have to be exercised, if it is, the time of the actual exercise is a stopping time $\tau \leq T$. We shall assume for simplicity that the underlying stock does not pay the dividend. It turns out that under such an assumption the optimal exercise time for the American call option is $\tau = T$ (or alternatively, option is let to expire). Therefore we concentrate our attention to American *put* options. Suppose the holder of the put option does not own the stock, and wishes to exercise at some time τ. That means he/she buys the underlying stock on the open market at the price $S(\tau)$ and then sells the same stock at the strike price k. The payoff for the holder is equal to $k - S(\tau)$. Moreover, since option does not have to be exercised, the payoff is never negative - i.e., the payoff is equal to $\phi(S(\tau)) = \text{Max}(0, k - S(\tau))$.

According to the celebrated Black-Scholes theory [2] (treating European options), the fair price of an American put option with fixed strike price k, and fixed expiration time T, as a function of the current running time t and the current price of the underlying stock S, is the unique solution of the (degenerate) *backward* parabolic obstacle problem:

$$\frac{\partial \psi(t, S)}{\partial t} + \frac{1}{2} S^2 \frac{\partial^2 \psi(t, S)}{\partial S^2} \sigma(t, S)^2 + rS \frac{\partial \psi(t, S)}{\partial S} - r\psi(t, S) = 0 \text{ in } N = \{(t, S); \psi(t, S) > \text{Max}(0, k - S)\}$$

$$\psi(t, S) \geq \text{Max}(0, k - S)$$

$$\psi(t, S) = \text{Max}(0, k - S) \text{ on } \Gamma = \partial N \cap \{(t, S); t < T, S > 0\}$$

$$\frac{\partial \psi(t, S)}{\partial S} = -1 \text{ on } \Gamma$$

together with the *terminal* condition

$$\psi(T, S) = \text{Max}(0, k - S)$$

where r is the interest rate, $\text{Max}(0, k - S)$ is the obstacle as well as the terminal condition, N is the non-coincidence set, $\Gamma = \partial N \cap \{(t, S); t < T, S > 0\}$ is the free boundary. Notice that no boundary condition is imposed, due to the degeneracy of the governing partial differential equation. The option trading significance of the free boundary is that if the price of the underlying stock drops below certain threshold (the free boundary Γ), then it does not make further sense to hold (or trade) the particular put option, and instead, such an option should be exercised.

On the other hand, according to the Dupire theory (see e.g. [3] for the case of European options), the fair price of an American put option with fixed underlying stock price S, at fixed time t, as a function of the expiration time T and the option strike price k, is the unique solution of the (degenerate) *forward* parabolic obstacle problem (written in the free boundary problem form):

$$-\frac{\partial V(T, k)}{\partial T} + \frac{1}{2} k^2 \frac{\partial^2 V(T, k)}{\partial k^2} \sigma(T, k)^2 - rk \frac{\partial V(T, k)}{\partial k} = 0 \text{ in } N = \{(t, S); V(T, k) > \text{Max}(0, k - S)\}$$

$$V(T, k) \geq \text{Max}(0, k - S)$$

$$V(T, k) = \text{Max}(0, k - S) \text{ on } \Gamma = \partial N \cap \{(T, k); t_0 < T, k > 0\}$$

$$\frac{\partial V(T, k)}{\partial k} = 1 \text{ on } \Gamma$$

together with the *initial* condition

$$V(t_0, k) = \text{Max}(0, k - S)$$

The trading significance of the free boundary now is that for the given current price of the underlying stock S, only options with strikes below the free boundary should be considered for trading. It is remarkable that the same volatility function σ, *with different arguments*, appears in both problems (as well as in the underlying stock price evolution SDE above). Also notice that in either one of the problems the underlying stock price appreciation rate $a(t, S)$ does not appear (but instead the *known* interest rate r does), while volatility σ is of the deciding importance (options with different underlying volatilities have significantly different prices). It is then of paramount importance to have a precise estimate of the volatility σ.

Volatility of the underlying stock price σ can be estimated from the historical statistical data accurately. Nevertheless, the volatility changes, and in the above equations it is the future volatility that appears and not the current or past. Moreover, knowing or having a good estimate of the future volatility may be useful in trading for other reasons, as well. Therefore it is of great importance in trading options or even stocks, to have a reliable and efficient technology for estimating future volatility σ based on the current prices of the whole variety of the corresponding call and put options. For example, for the underlying QQQ (Nasdaq-100 Index Tracking Stock), AS OF DEC 26, 2000 10:46:20 AM (E.T.), the collection of sufficiently liquid put options with expiration in January and

Implied Volatility for American Options

February of 2001 ($T = 1.05593, 1.13259$; time is measured in units of years, and we start at the year 2000), with different strike prices (the second column), had the following market prices (the third column):

$$\begin{pmatrix} 1.05593 & 48 & 0.78125 \\ 1.05593 & 50 & 1.03125 \\ 1.05593 & 51 & 1.1875 \\ 1.05593 & 52 & 1.375 \\ 1.05593 & 53 & 1.625 \\ 1.05593 & 54 & 1.875 \\ 1.05593 & 55 & 2.1875 \\ 1.05593 & 56 & 2.375 \\ 1.05593 & 57 & 2.75 \\ 1.05593 & 58 & 3.125 \\ 1.05593 & 59 & 3.6875 \\ 1.05593 & 60 & 4.3125 \\ 1.05593 & 61 & 4.6875 \\ 1.05593 & 62 & 5.1875 \\ 1.05593 & 63 & 5.8125 \\ 1.05593 & 64 & 6.375 \\ 1.05593 & 65 & 6.875 \\ 1.05593 & 66 & 7.625 \\ 1.05593 & 69 & 9.875 \\ 1.05593 & 70 & 10.75 \\ 1.05593 & 71 & 11.5 \\ 1.05593 & 75 & 14.875 \\ 1.05593 & 83 & 22.75 \\ 1.05593 & 86 & 25.75 \end{pmatrix}, \begin{pmatrix} 1.13259 & 48 & 1.625 \\ 1.13259 & 50 & 2.125 \\ 1.13259 & 52 & 2.625 \\ 1.13259 & 55 & 3.6875 \\ 1.13259 & 56 & 4.0625 \\ 1.13259 & 59 & 5.3125 \\ 1.13259 & 60 & 5.8125 \\ 1.13259 & 65 & 8.8125 \\ 1.13259 & 75 & 15.8125 \\ 1.13259 & 80 & 20.1875 \end{pmatrix}$$

(Notice, by the way, how much more trading activity is performed for options that expire sooner rather then later.) The question is what can be said about the (market consensus about the) future volatility σ of the underlying stock price based on these recorded prices. The present paper describes a methodology for answering that question based on the optimal control theory for obstacle problems, as well as on a fast numerical solution of the Dupire obstacle problems. Some alternative methods can be found in [3].

■ 2. Some Remarks on Obstacle Problems

We recall some known facts about obstacle problems, introduce some new (see also [6,8]), as well as a method for solving obstacle problems, employed in our fast numerical solution of the Dupire obstacle problem.

2.1. Various Formulations of Obstacle Problems

Consider, for the sake of simplicity, an obstacle problem for the Laplace's operator and with zero obstacle. Everything that follows can be generalized to the case of arbitrary, sufficiently regular obstacle, as well as to the case of an arbitrary elliptic (or parabolic) operator with smooth coefficients. Obstacle problem can be formulated in many different ways. Arguably, the most popular way is the *variational inequality* formulation. Let $\Omega \subset \mathbb{R}^n$ be a bounded domain with regular boundary, let $(-f) \in L^2(\Omega)$ be the right hand side, and let $g \in H^1(\Omega)$, $g \geq 0$ be the boundary value. Denote also $H^1_g(\Omega) = g + H^1_0(\Omega)$, and $K = \{v \in H^1_g(\Omega), v \geq 0\}$. The variational inequality formulation of such an obstacle problem is:

Find $v \in K$, such that

$$\int_\Omega (\nabla v \cdot \nabla (\varphi - v) + f(\varphi - v))\, dx \geq 0$$

for any $\varphi \in K$. It is worthwhile emphasizing that the obstacle appears explicitly as a *constraint* imposed on all functions considered as possible solutions.

As it is very well known, such a problem has an unique solution. The variational inequality formulation is useful because it allows an easy variational proof of existence of the weak solution, as well as its uniqueness. Furthermore, the higher regularity of the weak solution can be established posteriori via additional arguments. Indeed, under above assumptions $v \in H^2_{loc}(\Omega)$ (see e.g. [5]). Increasing the regularity of the right hand side yields higher regularity of the solution, but not beyond $W^{2,\infty}_{loc}(\Omega)$ (unless obstacle does not affect the solution). The solution v, being non-negative, determines two distinctive regions in Ω: the coincidence set $\Lambda = \{v = 0\} \cap \Omega$, and the non-coincidence set $N = \{v > 0\} \cap \Omega$. The boundary separating the two $\Gamma = \partial \{u > 0\} \cap \Omega$ is called *free boundary*. In general, free boundary is not a smooth surface no matter how smooth is the right hand side (see [4] for fundamental results ensuring smoothness of the free boundary under additional assumptions; see also [5]).

2.1.2. Semi-Linear PDE Formulations of Obstacle Problems

Here are couple, not so well known, formulations of the same problem, introduced and utilized by the author in [6,8].

Let $I_A(x)$ be a characteristic function of the set A, i.e., let $I_A(x) = \begin{cases} 1, & x \in A \\ 0, & x \notin A \end{cases}$. Also let $h^+ = \text{Max}(h, 0)$, $h^- = -\text{Min}(h, 0)$ be the positive and negative parts of h, so that $h = h^+ - h^-$. The above obstacle problem can be formulated as:

Find $v \in H^1_g(\Omega) \cap H^2_{loc}(\Omega)$ such that

$$-\Delta v + f\, I_{\{v>0\}} = 0$$
$$f^-\, I_{\{v=0\}} = 0$$

All the equalities are to be understood as equalities *almost* everywhere.

The above obstacle problem can also be formulated as:

Implied Volatility for American Options

Find *maximal* $v \in H_g^1(\Omega) \cap H_{loc}^2(\Omega)$ such that

$$-\Delta v + f\, I_{\{v>0\}} = 0$$

Alternatively, let

$$X_{1,\Omega,f,g} = \{w \in H_g^1(\Omega) \cap H_{loc}^2(\Omega),\ -\Delta v + f\, I_{\{v>0\}} = 0\}$$

then

$$v(x) = \text{Max}_{w \in X_{1,\Omega,f,g}}\, w(x).$$

$X_{1,\Omega,f,g}$ is typically a small set. All the elements of $X_{1,\Omega,f,g}$ are non-negative functions due to the maximum principle, and if $f \geq 0$ then $X_{1,\Omega,f,g}$ is a singleton consisting of the solution of the obstacle problem. Notice also that the non-negativity follows although no constraint is imposed explicitly in either of the above two formulations.

For example, let $\Omega = (0, 1)$, $g = 1$, $f(x) = -8\pi^2 \cos(4\pi x)$. Then $X_{1,\Omega,f,g}$ has two elements: $w_1(x) = \frac{1}{2}(\cos(4\pi x) + 1)$ and $w_2(x) = \frac{1}{2}(\cos(4\pi x) + 1)\, I_{\{x < \frac{1}{4}\} \cup \{x > \frac{3}{4}\}}$, the first one being larger, and therefore being the solution of the obstacle problem. Notice also $f^- I_{\{w_2=0\}} = -8\pi^2 \cos(4\pi x)\, I_{\{\frac{3}{4} < x < \frac{5}{4}\}}(x) \neq 0$.

■ 2.1.3. Classical Free Boundary Problem Formulation

Assume f is sufficiently regular. The free boundary problem can be formulated as:

Find an open set $N \subset \Omega$ and $v \in C^2(N) \cap C^1(\overline{N})$, such that

$$v > 0$$

and

$$-\Delta v + f = 0$$

in N, such that

$$v = g$$

on $\partial N \cap \partial \Omega$, and such that

$$v = 0$$

and

$$\nabla v = 0$$

on $\partial N \cap \Omega$. The last two conditions are called *free boundary conditions*.

Notice that if f is sufficiently regular, any element of $X_{1,\Omega,f,g}$ is a solution of the free boundary problem. This implies in particular that the free boundary problem formulation does not yield necessarily an unique solution. In the above example, both w_1 and w_2 are solutions of the free boundary problem.

■ 2.2. Maximum Boundary Value Formulation of the Obstacle Problem

For each open $N \subset \mathbb{R}^n$, let $\Lambda_N = \mathbb{R}^n - N$. Now let

$$H^1_{0,g,N}(\Omega) = \{w|_\Omega, w \in H^1_{loc}(\mathbb{R}^n), w|_N \in H^1_0(N), w|_{\partial\Omega \cap N} = g, w|_{\Lambda_N} = 0\}$$

Notice that $w \in H^1_{0,g,N}(\Omega)$ has more then one extension, i.e., representation in $H^1_{loc}(\mathbb{R}^n)$. Further, let w_N be the unique function in $H^1_{0,g,N}(\Omega)$ such that

$$-\Delta w_N + f = 0$$

in $N \cap \Omega$.

Notice that if $N \cap \Omega = \emptyset$ then $w_N = 0$. On the other hand, in order to see non-trivial examples of w_N, let $\Omega = (-1, 1)$, $g = 1$, $f(x) = -10 + 20\, I_{\{x>0\}}$ and let $N_1 = (-\infty, \frac{3}{4})$, $N_2 = (-\frac{3}{4}, \infty)$. Then

$$w_{N_1}(x) = \left(10\, I_{\{x>0\}} x^2 - 5x^2 - \frac{141 x}{28} + \frac{27}{28}\right) I_{\{x<\frac{3}{4}\}}$$

which looks like

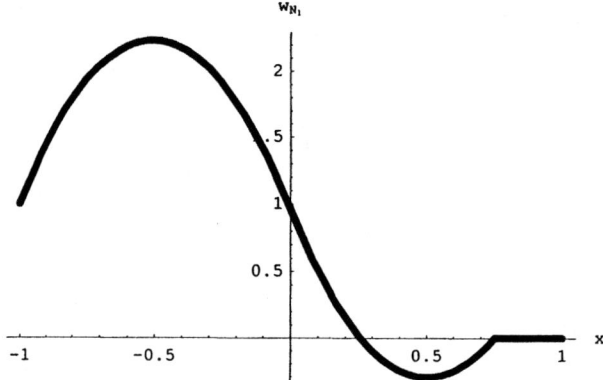

while $w_{N_2}(x) = (10\, I_{\{x>0\}} x^2 - 5x^2 - \frac{109 x}{28} - \frac{3}{28}) I_{\{x>-\frac{3}{4}\}}$, which looks like

Implied Volatility for American Options

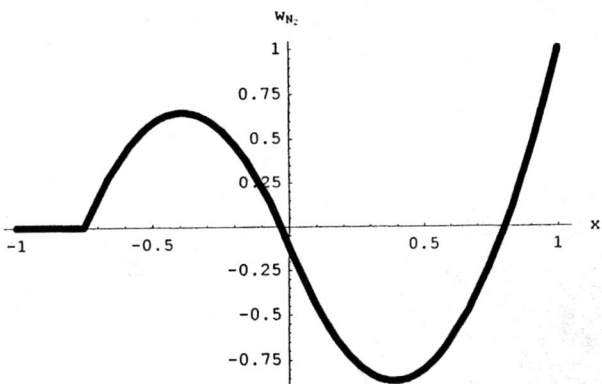

It turns out that the pointwise maximum of functions such as these, yields the solution of the obstacle problem, which by the way in this example is equal to

$$v(x) = \left(10\, I_{\{x>0\}}\, x^2 - 5x^2 - 2\left(-5 + \sqrt{55}\right)x - 2\sqrt{55} + 16\right) I_{\left\{x<-1+\sqrt{\frac{11}{5}}\right\}} + \left(5x^2 + 2\left(-5 + \sqrt{5}\right)x - 2\sqrt{5} + 6\right) I_{\left\{x>1-\frac{1}{\sqrt{5}}\right\}}$$

and looks like

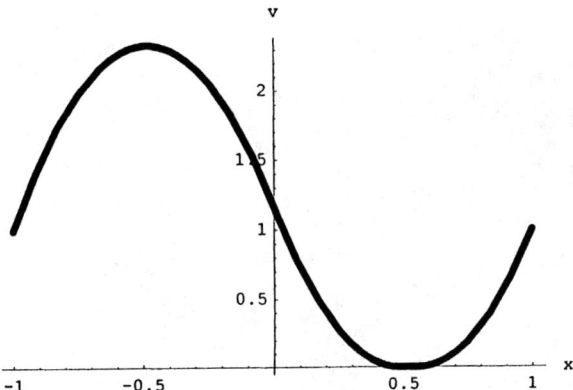

Notice also $v(x) = w_{\left\{x<-1+\sqrt{\frac{11}{5}}\right\} \cup \left\{x>1-\frac{1}{\sqrt{5}}\right\}}$.

So, formalizing, let

$$X_{\Omega,f,g} = \{w_N, N \subset \mathbf{R}^n\}$$

Compared to $X_{1,\Omega,f,g}$, $X_{\Omega,f,g}$ is much larger. Also, one should notice that, like in the examples above, depending on N, as opposed to $X_{1,\Omega,f,g}$ the elements of $X_{\Omega,f,g}$ take negative values, i.e., values below the obstacle, as well.

As announced, the obstacle problem can be formulated, and/or solved, simply as

$$v(x) = \text{Max}_{w \in X_{\Omega,f,g}}\, w(x).$$

Before proving this claim, it is instructive to visualize this maximization (on the same example), together with the solution, and its free boundary:

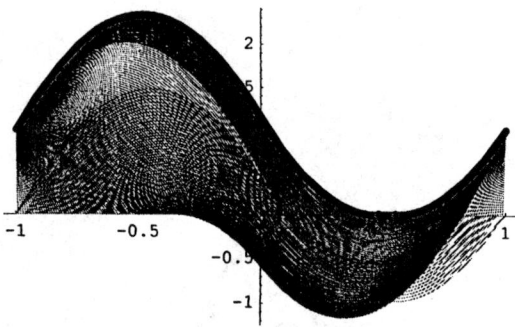

The following simple facts may have practical implications for numerical searching of the free boundary $\partial \{v > 0\} \cap \Omega$:

$$\{v > 0\} = \bigcup_{w \in X_{\Omega, f, g}} \{w > 0\}$$

or alternatively,

$$\{v = 0\} = \bigcap_{w \in X_{\Omega, f, g}} \{w \leq 0\}$$

and consequently, for any $Y \subset X_{\Omega, f, g}$

$$\{v > 0\} \supset \bigcup_{w \in Y} \{w > 0\}$$

and

$$\{v = 0\} \subset \bigcap_{w \in Y} \{w \leq 0\}$$

In particular, for any $Y \subset X_{\Omega, f, g}$, the set $\bigcap_{w \in Y} \{w \leq 0\}$ contains the free boundary $\partial \{v > 0\} \cap \Omega$. In order to prove the above claim, since $v \in X_{\Omega, f, g}$, we need to show only $\{v > 0\} \supset \bigcup_{w \in X_{\Omega, f, g}} \{w > 0\}$. Let $x \in \{w > 0\}$ for some $w \in X_{\Omega, f, g}$. This implies $v(x) \geq w(x) > 0$, and therefore $x \in \{v > 0\}$.

This formulation-solution of the obstacle problem can be compared - judged against the classical free boundary problem formulation. The free boundary problem can be viewed as the first order necessary condition for the above maximization problem. The computational efficiency of the present method lies in the fact that for each N, computing the corresponding w_N is a *linear* problem. Still, in general, especially in higher dimension, choosing proper N's, i.e., choosing sufficiently representative finite family $Y \subset X_{\Omega, f, g}$ may indeed be difficult, or at least additional issues will need to be addressed there. Nevertheless, in special cases such as the Dupire obstacle problem, when the geometry of the solution and of the free boundary is well understood a priori (in spite of being unknown), the above maximization can be done very efficiently.

Implied Volatility for American Options

■ 2.3. A Proof of the Equivalence

We shall prove that if

$$v(x) = \text{Max}_{w \in X_{\Omega, f, g}} w(x).$$

and if $z \in H_g^1(\Omega) \cap H_{\text{loc}}^2(\Omega)$ satisfies

$$-\Delta z + f \, I_{\{z>0\}} = 0$$
$$f^- \, I_{\{z=0\}} = 0$$

then $v = z$. Indeed, since $z \in X_{\Omega, f, g}$, all what is needed to prove is that if $w_N \in X_{\Omega, f, g}$, then $z \geq w_N$. To that end recall

$$\int_{\Omega \cap N} (\nabla w_N \cdot \nabla \varphi + f \varphi) \, dx = 0$$

for any $\varphi \in H_0^1(\Omega \cap N)$, and

$$\int_{\Omega} (\nabla z \cdot \nabla \varphi + f \, I_{\{z>0\}} \varphi) \, dx = 0$$

for any $\varphi \in H_0^1(\Omega)$. Consider $(w_N - z)^+ \in H_0^1(\Omega)$. Notice also $(w_N - z)^+ |_{\Omega \cap \Lambda_N} = 0$. Therefore

$$\int_{\Omega} (\nabla w_N \cdot \nabla (w_N - z)^+ + f \, I_N \, (w_N - z)^+) \, dx = 0$$

and

$$\int_{\Omega} (\nabla z \cdot \nabla (w_N - z)^+ + f \, I_{\{z>0\}} \, (w_N - z)^+) \, dx = 0$$

Subtracting

$$0 = \int_{\Omega} (|\nabla (w_N - z)^+|^2 + (f^+ - f^-)(I_N - I_{\{z>0\}})(w_N - z)^+) \, dx =$$
$$\int_{\Omega} (|\nabla (w_N - z)^+|^2 + f^+ (1 - I_{\{z>0\}})(w_N - z)^+ - f^- (I_N - 1)(w_N - z)^+) \, dx \geq \int_{\Omega} (|\nabla (w_N - z)^+|^2) \, dx$$

since $f^- \, I_{\{z>0\}} = f^- \, I_{\{z \geq 0\}} = f^-$, yielding $\nabla (w_N - z)^+ = 0$, and therefore $(w_N - z)^+ = 0$, and finally $z \geq w_N$.

■ 3. Fast Numerical Solution of the Dupire Obstacle Problem

Back to the Dupire obstacle problem (free boundary problem formulation):

$$-\frac{\partial V(T, k)}{\partial T} + \frac{1}{2} k^2 \frac{\partial^2 V(T, k)}{\partial k^2} \sigma(T, k)^2 - rk \frac{\partial V(T, k)}{\partial k} = 0 \quad \text{in } N = \{(t, S); V(T, k) > \text{Max}(0, k - S)\}$$

$$V(T, k) \geq \text{Max}(0, k - S)$$

$V(T, k) = \text{Max}(0, k - S)$ on $\Gamma = \partial N \cap \{(T, k); t_0 < T, k > 0\}$

$$\frac{\partial V(T, k)}{\partial k} = 1 \text{ on } \Gamma$$

together with the *initial* condition

$V(t_0, k) = \text{Max}(0, k - S)$

Under some theoretical assumptions, that we shall not be going into here, the Dupire obstacle problem admits an unique solution.

This problem is approximated by the time finite difference variant:

$$-\left(\frac{V(T, k) - V(T - dT, k)}{dT}\right) + \frac{1}{2} k^2 \frac{\partial^2 V(T, k)}{\partial k^2} \sigma(T, k)^2 - rk \frac{\partial V(T, k)}{\partial k} =$$
0 in $N = \{(t, S); V(T, k) > \text{Max}(0, k - S)\}$

$V(T, k) \geq \text{Max}(0, k - S)$

$V(T, k) = \text{Max}(0, k - S)$ on $\Gamma = \partial N \cap \{(T, k); t_0 < T, k > 0\}$

$$\frac{\partial V(T, k)}{\partial k} = 1 \text{ on } \Gamma$$

i.e., by the family of 1-dimensional elliptic obstacle problems, that are solved (numerically) in succession. For example, if $\sigma(T, k) = \sigma_{\text{stat}} = 0.531349$ (QQQ statistical historical volatility based on daily closse between March 10, 1999 and December 22, 2000), then it turns out that the corresponding solution looks like

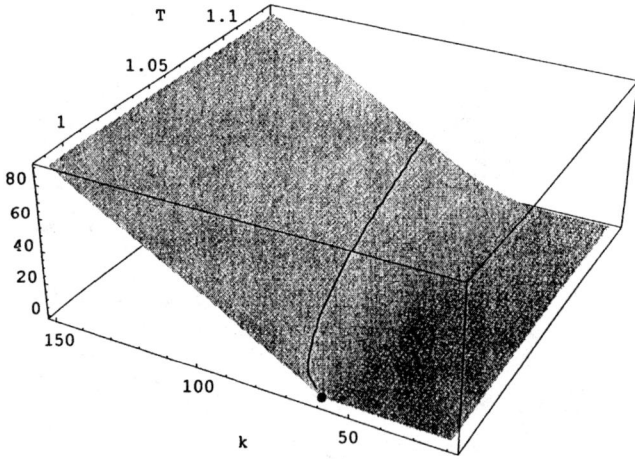

while, more precisely, its free boundary looks like

Implied Volatility for American Options

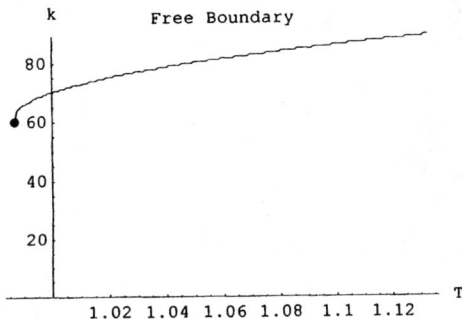

The geometry of the solution is very simple: the free boundary is a graph of a function, i.e., $\Gamma = \{\{T, g(T)\}, \{T, t_0, t_1\}\}$; moreover function g is strictly increasing (initiating at the price of QQQ AS OF DEC 26, 2000 10:46:20 AM (E.T.)) . Those two facts can be exploited in the search for the maximal boundary value solution of each 1-dimensional elliptic problem, like in 2.2.

■ 4. Optimal Control Problem

The above Dupire obstacle problem can be written more precisely, for example in its complementarity form

$$-\frac{\partial V(T, k)}{\partial T} + \frac{1}{2} k^2 \frac{\partial^2 V(T, k)}{\partial k^2} \sigma(T, k)^2 - rk \frac{\partial V(T, k)}{\partial k} \geq 0$$

$$V(T, k) \geq \text{Max}(0, k - S)$$

$$\left(-\frac{\partial V(T, k)}{\partial T} + \frac{1}{2} k^2 \frac{\partial^2 V(T, k)}{\partial k^2} \sigma(T, k)^2 - rk \frac{\partial V(T, k)}{\partial k}\right)(V(T, k) - \text{Max}(0, k - S)) = 0$$

together with the same initial condition above. Also, for the purpose of numerical solutions, we need to restrict the consideration on a finite domain, say $\{(T, k); t_0 < T < t_1, k_0 < k < k_1\}$, for some $0 < k_0 < k_1$, with some appropriate truncation boundary values at $k = k_0$ and $k = k_1$.

Let the class of considered volatilities be:

$$K = \left\{\sigma \in L^2((t_0, t_1) \times (k_0, k_1)); \sigma(T, k) \geq \sigma_{\min} > 0, \int_{k_0}^{k_1} \int_{t_0}^{t_1} \left(\left(\frac{\partial \sigma(T, k)}{\partial k}\right)^2 + \sigma(T, k)^2\right) dT\, dk < \infty\right\}$$

For each $\sigma \in K$ let $V(\sigma)$ denote the unique solution of the Dupire obstacle problem (this is formal, and motivated by the numerical scheme; for the sake of theoretical results it might be necessary to impose more restrictions on the set of admissible volatilities, which in turn would make the numerical scheme less natural). So, from the discrete option price observations, we construct the *target* functions $v_i(k)$, $i = 1, ..., n$, that the solution of the Dupire obstacle problem is supposed to match. So n is the number of different expiration times considered. We adopt the cost functional

$$J(\sigma) = \frac{1}{2} \sum_{i=1}^{n} \int_{k_0}^{k_1} (V(\sigma)(T_i, k) - v_i(k))^2 q_i(k)\, dk + \frac{1}{2} \int_{k_0}^{k_1} \int_{t_0}^{t_1} \left(\epsilon_1 \left(\frac{\partial \sigma(T, k)}{\partial k}\right)^2 + \epsilon_2 (\sigma(T, k) - \sigma_{\text{stat}})^2\right) dT\, dk$$

where ϵ_1, ϵ_2, $q_i(k) > 0$ are given and σ_{stat} is the statistical estimate of the constant volatility.

The optimal control problem is to find

$$\sigma_{opt} \in K$$

such that

$$J(\sigma_{opt}) = \text{Min}(J(\sigma), \sigma \in K)$$

■ 5. Derivation of the Minimization Algorithm

The state equation being what it is, the functional $J(\sigma)$ is not differentiable, but rather only Lipschitz continuous (in an appropriate sense). This is a very well known phenomenon in the theory of optimal control of variational inequalities (see e.g. [1,7],). Nevertheless from the practical point of view, from the point of view of designing a working numerical optimization algorithm, this is not a problem at all. Indeed, for a fixed $\sigma \in K$, such that J is differentiable at σ (see [7]) and $\overline{\sigma}$ such that $\sigma + \delta \overline{\sigma} \in K$, compute the directional derivative of J in the direction $\overline{\sigma}$:

$$J'(\sigma; \overline{\sigma}) = \lim_{\delta \to 0} \frac{J(\sigma + \delta \overline{\sigma}) - J(\sigma)}{\delta} = \sum_{i=1}^{n} \int_{k_0}^{k_1} (V(T_i, k) - v_i(s)) w(T_i, k) q_i(k) dk +$$

$$\int_{t_0}^{t_1} \int_{k_0}^{k_1} \left(\epsilon_1 \frac{\partial \sigma(T, k)}{\partial k} \frac{\partial \overline{\sigma}(T, k)}{\partial k} + \epsilon_2 (\sigma(T, k) - \sigma_{stat}) \overline{\sigma}(T, k) \right) dk \, dT$$

where

$$w(T, k) = \lim_{\delta \to 0} \frac{V(\sigma + \delta \overline{\sigma})(T, k) - V(\sigma)(T, k)}{\delta}$$

is the unique solution of the forward equation

$$-\frac{\partial w(T, k)}{\partial T} + \frac{1}{2} k^2 \frac{\partial^2 w(T, k)}{\partial k^2} \sigma(T, k)^2 - rk \frac{\partial w(T, k)}{\partial k} = -k^2 \frac{\partial^2 V(\sigma)(T, k)}{\partial k^2} \sigma(T, k) \overline{\sigma}(T, k)$$

in the non-coincidence set $N = \{(T, k); V(\sigma)(T, k) > \text{Max}(0, k - S)\} \cap \{(T, k); t_0 < T < t_1, k_0 < k < k_1\}$, and

$$w(T, k) = 0$$

in the coincidence set $\Lambda = \{(T, k); V(\sigma)(T, k) = \text{Max}(0, k - S)\}$. Let for any $1 \leq i \leq n$, $p_i(T, k)$ be the unique solution of the *backward equations* (not obstacle problems) in the non-coincidence region:

$$\frac{\partial p_i(T, k)}{\partial T} + \frac{1}{2} k^2 \frac{\partial^2 p_i(T, k)}{\partial k^2} \sigma(T, k)^2 + (2 \sigma(T, k)^2 + 2 k \sigma^{(0,1)}(T, k) \sigma(T, k) + r) k \frac{\partial p_i(T, k)}{\partial k} +$$
$$\left(\sigma(T, k)^2 + k (4 \sigma^{(0,1)}(T, k) + k \sigma^{(0,2)}(T, k)) \sigma(T, k) + k^2 \sigma^{(0,1)}(T, k)^2 + r \right) p_i(T, k) = 0$$

for $N \cap \{(T, k); t_0 < T < T_i\}$, with the *terminal* condition

$$p_i(T_i, k) = -(V(\sigma)(T_i, k) - v_i(k)) q_i(k)$$

and (non-cylindrical) boundary condition

Implied Volatility for American Options

$$p_i = 0$$

on the lateral boundary of $N \cap \{(T, k); t_0 < T < T_i\}$. Each p_i is furthermore extended as zero outside of the non-coincidence region. By means of integration by parts, this terminal-boundary value problem is equivalent to its weak formulation:

$$\iint_{N \cap \{(T,k); t_0 < T < T_i\}} \left(-\frac{\partial \phi(T, k)}{\partial T} + \frac{1}{2} k^2 \frac{\partial^2 \phi(T, k)}{\partial k^2} \sigma(T, k)^2 - rk \frac{\partial \phi(T, k)}{\partial k} \right) p_i(T, k) \, dk \, dT =$$

$$-\int_{k_0}^{k_1} \phi(T_i, k) \, p_i(T, k) \, dk = \int_{k_0}^{k_1} (V(\sigma)(T_i, k) - v_i(k)) \, q_i(k) \, \phi(T_i, k) \, dk$$

for any smooth test function ϕ such that $\phi = 0$ on the backward parabolic boundary of the non-coincidence set.

In particular, if $\phi = w$,

$$J'(\sigma; \overline{\sigma}) = \sum_{i=1}^{n} \int_{k_0}^{k_1} (V(\sigma)(T_i, k) - v_i(k)) \, w(T_i, k) \, q_i(k) \, dk +$$

$$\int_{t_0}^{t_1} \int_{k_0}^{k_1} \left(\epsilon_1 \frac{\partial \sigma(T, k)}{\partial k} \frac{\partial \overline{\sigma}(T, k)}{\partial k} + \epsilon_0 (\sigma(T, k) - \sigma_{\text{stat}}) \overline{\sigma}(T, k) \right) dk \, dT =$$

$$\sum_{i=1}^{n} \iint_{N \cap \{(T,k); t_0 < T < T_i\}} \left(-\frac{\partial w(T, k)}{\partial T} + \frac{1}{2} k^2 \frac{\partial^2 w(T, k)}{\partial k^2} \sigma(T, k)^2 - rk \frac{\partial w(T, k)}{\partial k} \right) p_i(T, k) \, dk \, dT +$$

$$\int_{t_0}^{t_1} \int_{k_0}^{k_1} \left(\epsilon_1 \frac{\partial \sigma(T, k)}{\partial k} \frac{\partial \overline{\sigma}(T, k)}{\partial k} + \epsilon_0 (\sigma(T, k) - \sigma_{\text{stat}}) \overline{\sigma}(T, k) \right) dk \, dT =$$

$$\sum_{i=1}^{n} \iint_{N \cap \{(T,k); t_0 < T < T_i\}} \left(-k^2 \frac{\partial^2 V(\sigma)(T, k)}{\partial k^2} \sigma(T, k) \, p_i(T, k) \right) \overline{\sigma}(T, k) \, dk \, dT +$$

$$\int_{t_0}^{t_1} \int_{k_0}^{k_1} \left(-\epsilon_1 \frac{\partial^2 \sigma(T, k)}{\partial k^2} + \epsilon_0 (\sigma(T, k) - \sigma_{\text{stat}}) \right) \overline{\sigma}(T, k) \, dk \, dT +$$

$$\epsilon_1 \left(\int_{t_0}^{t_1} \frac{\partial \sigma(T, k_1)}{\partial k} \overline{\sigma}(T, k_1) \, dT - \int_{t_0}^{t_1} \frac{\partial \sigma(T, k_0)}{\partial k} \overline{\sigma}(T, k_0) \, dT \right) =$$

$$\int_{t_0}^{t_1} \int_{k_0}^{k_1} \left(-k^2 \frac{\partial^2 V(\sigma)(T, k)}{\partial k^2} \sigma(T, k) \sum_{i=1}^{n} (p_i(T, k) \chi_{(t_0, T_i)}) + \left(-\epsilon_1 \frac{\partial^2 \sigma(T, k)}{\partial k^2} + \epsilon_0 (\sigma(T, k) - \sigma_{\text{stat}}) \right) \right)$$

$$\overline{\sigma}(T, k) \, dk \, dT + \epsilon_1 \left(\int_{t_0}^{t_1} \frac{\partial \sigma(T, k_1)}{\partial k} \overline{\sigma}(T, k_1) \, dT - \int_{t_0}^{t_1} \frac{\partial \sigma(T, k_0)}{\partial k} \overline{\sigma}(T, k_0) \, dT \right)$$

On the other hand, there exists z such that

$$\nabla J(\sigma) \cdot \overline{\sigma} = \int_{t_0}^{t_1} \int_{k_0}^{k_1} \left(\epsilon_1 \frac{\partial z(T, k)}{\partial k} \frac{\partial \overline{\sigma}(T, k)}{\partial k} + \epsilon_0 z(T, k) \overline{\sigma}(T, k) \right) dk \, dT =$$

$$\int_{t_0}^{t_1} \int_{k_0}^{k_1} \left(-\epsilon_1 \frac{\partial^2 z(T, k)}{\partial k^2} + \epsilon_0 z(T, k) \right) \overline{\sigma}(T, k) \, dk \, dT +$$

$$\epsilon_1 \left(\int_{t_0}^{t_1} \frac{\partial z(T, k_1)}{\partial k} \overline{\sigma}(T, k_1) \, dT - \int_{t_0}^{t_1} \frac{\partial z(T, k_0)}{\partial k} \overline{\sigma}(T, k_0) \, dT \right)$$

Since $J'(\sigma; \overline{\sigma}) = \nabla J_\epsilon(\sigma) \cdot \overline{\sigma}$ for any admissible $\overline{\sigma}$, we conclude that, for any T, $z(T, \cdot)$ is the solution of, and consequently can be computed as a unique solution of the boundary value problem for an ordinary differential equation (regularization equation):

$$-\epsilon_1 \frac{\partial^2 z(T, k)}{\partial k^2} + \epsilon_0 z(T, k) = -k^2 \frac{\partial^2 V(\sigma)(T, k)}{\partial k^2} \sigma(T, k) \sum_{i=1}^{n} (p_i(T, k) \chi_{(t_0, T_i)}) + \left(-\epsilon_1 \frac{\partial^2 \sigma(T, k)}{\partial k^2} + \epsilon_0 (\sigma(T, k) - \sigma_{\text{stat}})\right)$$

with boundary conditions

$$\frac{\partial z(T, k_0)}{\partial k} = \frac{\partial \sigma(T, k_0)}{\partial k}$$
$$\frac{\partial z(T, k_1)}{\partial k} = \frac{\partial \sigma(T, k_1)}{\partial k}$$

We summarize the *steepest descent* minimization algorithm. The single iterate is done in 4 steps:

1) for given $\sigma(T, k)$, $V(\sigma)(T, k)$ is computed as the unique solution of the Dupire obstacle problem;

2) $p_i(T, k)$, $i = 1, ..., n$ are computed using $V(\sigma)(T, k)$ as solutions of the adjoint equations in the non-coincidence region;

3) $V(\sigma)(T, k)$ and all of the $p_i(T, k)$'s are used in the regularization ODE's computing the gradient $z(T, k)$;

4) $\sigma_{\text{next}}(T, k) = \sigma(T, k) - \rho z(T, k)$, for some $\rho > 0$.

■ 6. Example: Put Implied Volatility for QQQ (Nasdaq-100 Index Tracking Stock)

We proceed with the data shown in Section 1. The first step is to construct the *target* functions $v_i(k)$, $i = 1, 2$, that the solution of the Dupire pde is supposed to match. They look like

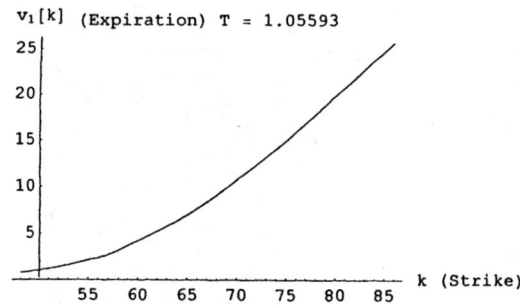

Implied Volatility for American Options

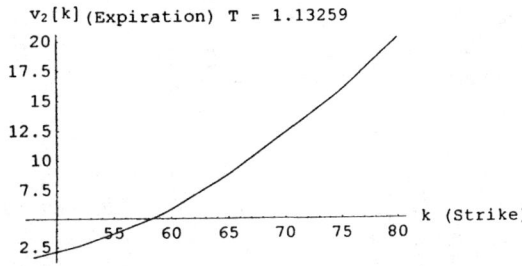

The steepest descent method is initialized by selecting the volatility function to be a statistical volatility $\sigma(T, k) = \sigma_{stat}$; see above. After a number of iterations, on two different grids, the optimal volatility function, as a function of strike and expiration time looks like

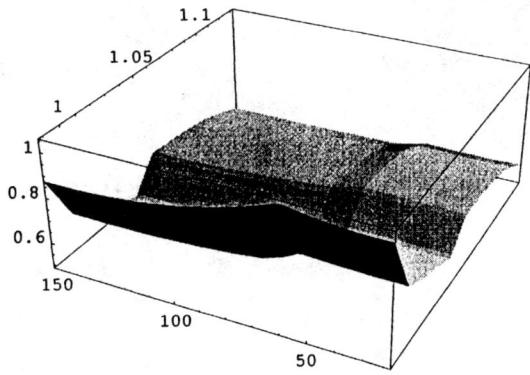

The corresponding solution of the Dupire obstacle problem, together with the free boundary, and together with the observed prices, looks like

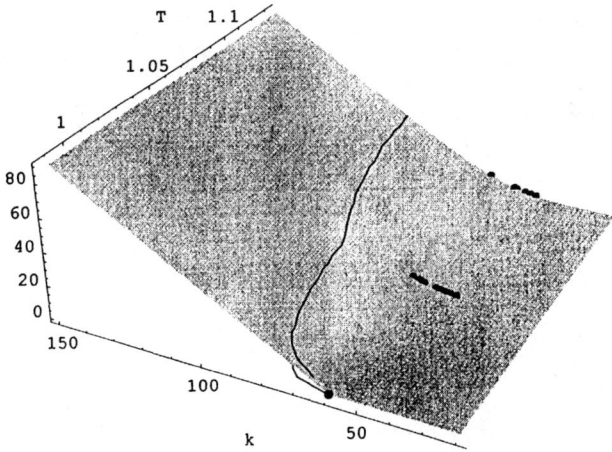

One can notice two features above. One is the reduced smoothness of the free boundary, as compared to the one presented before in the case of the constant volatility. The second feature is that the observed option market prices are all below the computed free boundary, as theory postulates: options with strikes above the free boundary need to be executed, and not be held or traded.

Proceeding, the corresponding gradient (the smallness of the gradient is the measure of the success of the minimization procedure; more iterations on the same grid, as well as going down on the finer grid and performing a number of iterations there, would reduce the gradient, and the precision of σ) was

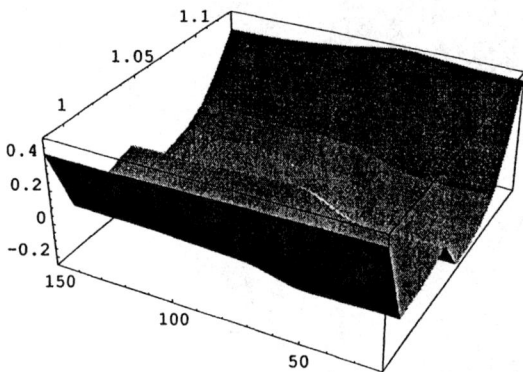

while the solution of the Dupire obstacle problem was matching the targets quite well:

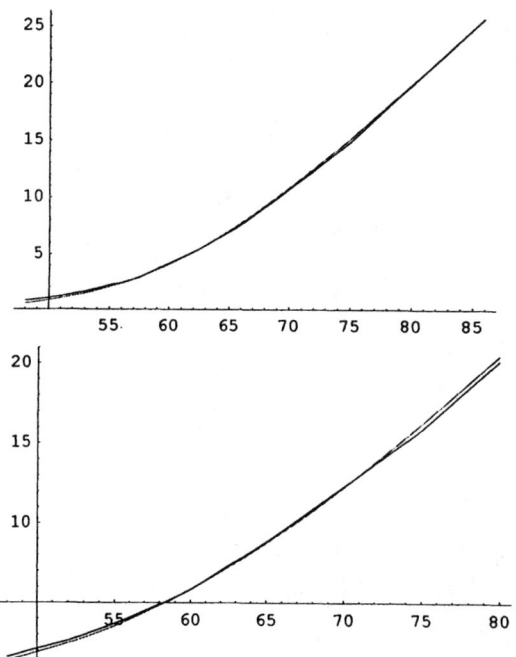

Implied Volatility for American Options

Finally, in order to concentrate our attention on time dependence of the implied volatility, we average the computed implied volatility for each time. The average, superimposed with points representing times of expiry (January and February 2001), looks like

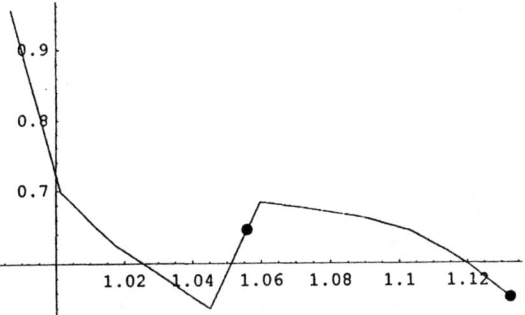

It is interesting to compare the term structure of the put implied volatility, with the one computed for calls (instead of the Dupire obstacle problem, Dupire partial differential *equation* and it's optimal control was used) on the same underlying (the constant statistical volatility is plotted, as well). The comparison, i.e. the mutual relationship between the two may suggest some insight about the option traders short term outlook of the (Nasdaq) market:

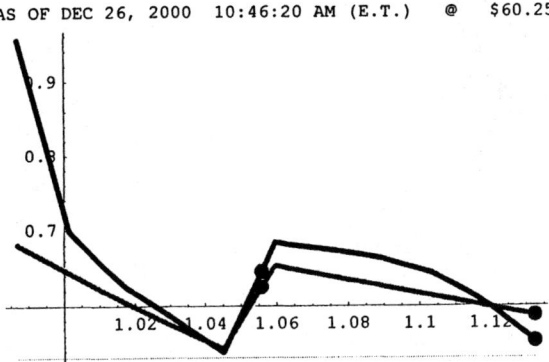

(Indeed, if an option trader expects QQQ price to increase, he/she would like to buy a call option (sell a put option). If many option traders share that opinion, the price of the call will go up, and price of the put (on the same underlying) will go down, and consequently its implied volatilities, as well. So, the difference in implied volatilities for calls and puts on the same underlying can be attributed to such a scenario, and consequently, can be exploited.) Although the market outlook sometimes changes suddenly, later that same day, the outlook did not look much different,

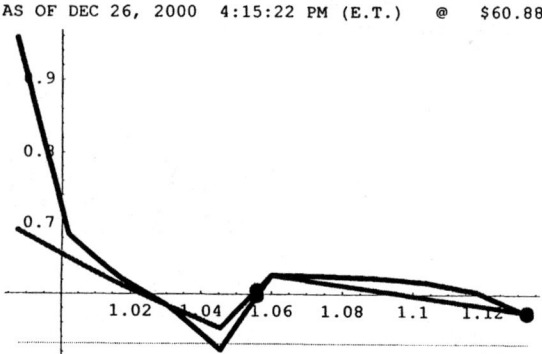

Many more details, including the complete implementation of the above algorithm, can be found in [9].

■ References

1. V. Barbu, *Optimal Control of Variational Inequalities*, Pitman, 1984.

2. F. Black and M. Scholes, The pricing of options and corporate liabilities, *J. Pol. Econ.* 81 (1973), 637-659.

3. I. Bouchouev and V. Isakov, Uniqueness, Stability, and Numerical Methods for the Inverse Problem that Arises in Financial Markets, *Inverse Problems*, 15 (1999), R1-R22

4. L. A. Caffarelli, Regularity of free boundaries in higher dimensions, *Acta Math.* 139 (1977), 155-184.

5. A. Friedman, *Variational Principles and Free Boundary problems*, Wiley, 1982.

6. S. Stojanovic, Remarks on $W^{2,p}$-solutions of bilateral obstacle problems, *IMA preprint* #1318, University of Minnesota (1995).

7. S. Stojanovic, Perturbation formula for regular free boundaries in elliptic and parabolic obstacle problems, *SIAM J. Control & Optimiz.*, 35 (1997), 2086-2100.

8. S. Stojanovic, Modeling and minimization of extinction in Volterra-Lotka type equations with free boundaries, *J. Differential Equations* 134 (1997), 320-342.

9. S. Stojanovic, *Computational Financial Mathematics*, (book and CD-ROM in preparation).

First Order Necessary Conditions of Optimality for Semilinear Optimal Control Problems

M.D. Voisei
Dept. of Math., Ohio University,
Athens, OH, 45701, USA

Abstract

This paper is concerned with the providing of necessary condition of optimality for optimal pairs (y^*, u^*) with respect to the cost functional $g(y) + h(u)$ subject to $Ay + Ly = Bu + f$, where A, B are linear and L is Lipschitz continuous. One example of applicatios of our necessary condition to a semilinear elliptic equation is presented in terms of Clarke's generalized gradient.

1. Preliminaries

In this article we investigate first order necessary conditions of optimality for optimal pairs of problem

(P) Minimize

$$g(y) + h(u)$$

on all $(y, u) \in V \times U$ subject to state equation

$$Ay + Ly = Bu + f.$$

Here $(U, (\cdot, \cdot)_U)$, $(H, (\cdot, \cdot)_H$ are Hilbert spaces, V is a Hilbert space which is not identified with its dual V^* such that $V \subset H \subset V^*$ algebraically and topologically with dense inclusions, V is compactly imbedded in H,

$g : H \to \mathbb{R}$ and $h : U \to \mathbb{R} \cup \{\infty\}$ are given functions, $L : H \to H$ is Lipschitz continuous, $A \in L(V, V^*)$, $B \in L(U, H)$ and $f \in H$.

We denote by $\|\cdot\|$ the norm in V, with $\langle \cdot, \cdot \rangle$ the pairing between V and V^*, with $|\cdot|_*$ the norm in V^*, with $|\cdot|_H$ the norm in H and with $|\cdot|_U$ the norm in U.

Assume (P) has an optimal pair $(y^*, u^*) \in V \times U$. The main idea is to linearize the state equation by introducing a new control variable and by penalizing the cost functional with an appropriate function we construct a sequence of approximative problems depending on (y^*, u^*) and a parameter $\varepsilon > 0$, given by

(P_ε) Minimize

$$g(y) + h(u) + \tfrac{1}{2\varepsilon}|v - Ly|_H^2 + \tfrac{1}{2}|u - u^*|_U^2 + \tfrac{1}{2}|y - y^*|_H^2,$$

on all $(y, u, v) \in V \times U \times H$ subject to state equation

$$Ay + v = Bu + f.$$

Under some suitable assumptions on g, h, B, A and L problem (P_ε) has at least one solution. We will derive the approximative necessary conditions of optimality and prove that these conditions converge in some sense, for $\varepsilon \to 0$, to the necessary optimality condition of (y^*, u^*).

Previous treatments of this problem for $L = \partial \varphi$ the subdifferential of a proper convex lower semicontinuous function φ, are due to Barbu [2] and use the penalization function $\varphi(y) + \varphi^*(v) - \langle y, v \rangle$ and the same state equation for the approximative problem (P_ε). Here we study the case of Lipschitz continuous L followed by an example governed by a semilinear elliptic equation.

Related results can be found in Aizicovici, Motreanu & Pavel [1] and Tiba [7].

In what follows, for a locally Lipschitz continuous function F defined in a Banach space X, we define its directional derivative by (see Clarke [4])

$$F^0(x; v) = \limsup_{\bar{x} \to x, t \downarrow 0} (1/t)(F(\bar{x} + tv) - F(\bar{x})), \; x, v \in X. \tag{1}$$

The Clarke generalized gradient $\partial F : X \to 2^{X^*}$ of F is defined by

$$y \in \partial F(x) \text{ iff } \langle y, w \rangle_{X^* \times X} \leq F^0(x; w), \forall w \in X, \tag{2}$$

where X^* is the dual of X and $\langle \cdot, \cdot \rangle_{X^* \times X}$ denotes the dual product between X and X^*.

When $(X, (\cdot, \cdot)_X)$ is a separable Hilbert space, we can construct the regularizations of F given by

$$F_\lambda(x) = \int_{\mathbb{R}^n} F(P_n x - \lambda \Lambda_n \tau) \rho_n(\tau) d\tau, \; x \in X, \tag{3}$$

where $\lambda > 0$, $n = [\lambda^{-1}]$, $\{e_m\}_{m=1}^\infty$ is an orthonormal basis in X, X_n is the finite dimensional space generated by $\{e_m\}_{m=1}^n$, P_n is the projection of X into X_n, that is, $P_n x = \sum_{m=1}^n (x, e_m)_X e_m$, $\Lambda_n : \mathbb{R}^n \to X_n$ is defined by $\Lambda_n(\tau) = \sum_{m=1}^n \tau_m e_m$, $\tau = (\tau_1, .., \tau_n)$ and $\rho_n \in C_0^\infty(\mathbb{R}^n)$, $\rho_n(\theta) = 0$ for $\|\theta\|_{\mathbb{R}^n} > 1$, $\rho_n \geq 0$, $\int \rho d\theta = 1$, $\rho_n(\theta) = \rho_n(-\theta)$, $\forall \theta \in \mathbb{R}^n$.

We recall that F_λ is Fréchet differentiable and if we denote its Fréchet derivative with ∇F_λ, then the sequence (∇F_λ) is uniformly bounded on bounded subsets of X, $\lim_{\lambda \to 0} F_\lambda(x) = F(x)$, $\forall x \in X$, and if $x_\lambda \to x$ strongly in X and $\nabla F_\lambda(x_\lambda) \to \xi$ weakly in X^*, then $\xi \in \partial F(x)$ (see Barbu [2]).

For a proper, convex, and lower semicontinuous function $h : U \to \mathbb{R} \cup \{\infty\}$ the directional derivative of h is given by

$$h'(u; \hat{u}) = \lim_{t \downarrow 0} (1/t)(h(u + t\hat{u}) - h(u)), \; u, \hat{u} \in U \tag{4}$$

and the subdifferential $\partial h : D(\partial h) \subset U \to 2^U$ is defined by

$$\eta \in \partial h(u) \text{ iff } (\eta, w - u)_U + h(u) \leq h(w), \forall w \in U. \tag{5}$$

We refer to Motreanu & Pavel [6] for results of convex analysis used in this paper.

We have considered the function $S^L : H \times H \to \mathbb{R}$ defined by

$$S^L(y, v) = \tfrac{1}{2}|v - Ly|^2, \; y, v \in H. \tag{6}$$

In the following Lemma we characterize its generalized gradient.

Lemma 1.1 Assume $L : H \to H$ is Lipschitz continuous. For $z \in H$ we define $f_z : H \to \mathbb{R}$ by

$$f_z(x) = -(z, Lx)_H, \; x \in H. \tag{7}$$

Then the following statements are true

i) S^L is locally Lipschitz continuous in $H \times H$ and f_z is Lipschitz in H for every $z \in H$,

ii) $S^{L\,0}(y, v; \overline{y}, \overline{v}) = (v - Ly, \overline{v})_H + f^0_{(v-Ly)}(y; \overline{y})$, $\forall y, v, \overline{y}, \overline{v} \in H$,

iii) $(w, z) \in \partial S^L(y, v)$ iff $z = v - Ly$ and $w \in \partial f_z(y)$,

iv) $|w|_H \leq K|z|_H$, $\forall (w, z) \in \partial S^L(y, v)$, (8)

where $K > 0$ is the Lipschitz constant of L.

v) If $(w_n, z_n) \in \partial S^L(y_n, v_n)$, $n \geq 1$ and

$$w_n \to w \text{ weakly in } H,$$
$$z_n \to z \text{ weakly in } H,$$
$$v_n \to v \text{ weakly in } H,$$

and

$$y_n \to y \text{ strongly in } H,$$

then $(w, z) \in \partial S^L(y, v)$.

Proof. Easy calculations show that

$$S^{L\,0}(y, v; \overline{y}, \overline{v}) = \limsup_{(\widetilde{y}, \widetilde{v}) \to (y, v), t \downarrow 0} \tfrac{1}{2t}(|\widetilde{v} + t\overline{v} - L(\widetilde{y} + t\overline{y})|^2_H - |\widetilde{v} - L\widetilde{y}|^2_H) =$$
$$= \limsup_{\widetilde{y} \to y, t \downarrow 0} (v - Ly, \overline{v} - \tfrac{1}{t}(L(\widetilde{y} + t\overline{y}) - L\widetilde{y}))_H = (v - Ly, \overline{v})_H + f^0_{(v-Ly)}(y; \overline{y}),$$

for every $y, v, \overline{y}, \overline{v} \in H$, so, ii) is true and it implies iii).

iv) For each $(w, z) \in \partial S^L(y, v)$ we have $z = v - Ly$ and $w \in \partial f_z(y)$ that is

$$(w, \overline{y})_H \leq \limsup_{\widetilde{y} \to y, t \downarrow 0} -\tfrac{1}{t}(z, L(\widetilde{y} + t\overline{y}) - L\widetilde{y}))_H \leq K|z|_H|\overline{y}|_H, \forall \overline{y} \in H,$$

or $|w|_H \leq K|z|_H$, where $K > 0$ is the Lipschitz constant of L.

v) We have $z_n = v_n - Ly_n$ and $w_n \in \partial f_{z_n}(y_n)$. Therefore the given hypotheses imply

$$z_n \to z = v - Ly \text{ weakly in } H.$$

From

$$(w_n, v)_H \leq \limsup_{y \to y_n, t \downarrow 0} -\tfrac{1}{t}(z_n, L(y + tv) - Ly)_H, \forall v \in H,$$

Semilinear Optimal Control Problems

we infer that for arbitrary fixed $v \in H$ there exist $(\overline{y}_n)$, (t_n) such that $|\overline{y}_n - y_n| < \frac{1}{n}$, $0 < t_n < \frac{1}{n}$ and

$$(w_n, v)_H \leq -\frac{1}{t_n}(z_n, L(\overline{y}_n + t_n v) - L\overline{y}_n)_H + \frac{1}{n}, \forall n \geq 1. \tag{9}$$

We can assume, by Mazur Theorem, that

$$\tilde{w}_n = \sum_{i \in I_n} \alpha_n^i w_{n_i} \to w \text{ strongly in } H \text{ and } \tilde{z}_n = \sum_{i \in I_n} \alpha_n^i z_{n_i} \to z \text{ strongly in } H$$

(Here I_n is a finite set of positive integers, $\alpha_n^i \geq 0$ and $\sum_{i \in I_n} \alpha_n^i = 1$).

Then (9) can be written as

$$(\tilde{w}_n, v)_H \leq -\frac{1}{t_n}(\tilde{z}_n, L(\overline{y}_n + t_n v) - L\overline{y}_n)_H + \frac{1}{n}, \forall n \geq 1. \tag{10}$$

If we let $n \to \infty$ in (10) then we get

$$(w, v)_H \leq \limsup_{n \to \infty} -\frac{1}{t_n}(\tilde{z}_n, L(\overline{y}_n + t_n v) - L\overline{y}_n)_H =$$
$$= \limsup_{n \to \infty} -\frac{1}{t_n}(z, L(\overline{y}_n + t_n v) - L\overline{y}_n)_H \leq$$
$$\leq \limsup_{\overline{y} \to y, t \downarrow 0} -\frac{1}{t}(z, L(\overline{y} + tv) - L\overline{y})_H,$$

because $\overline{y}_n \to y$ strongly in H.

Since v is arbitrarily chosen we conclude that $w \in f_z(y)$ and $(w, z) \in \partial S^L(y, v)$. ∎

Remark 1.1 Let us notice that in the previous Lemma we have proved the following

$$\left.\begin{array}{l} w_n \in f_{z_n}(y_n) \\ w_n \to w \text{ weakly in } H, \\ z_n \to z \text{ weakly in } H, \\ y_n \to y \text{ strongly in } H, \end{array}\right\} \Rightarrow w \in f_z(y). \tag{11}$$

2. The Main Result

Let us consider the assumptions

(A_1) V, U, H are Hilbert spaces, H is separable, V is dense in H and the inclusion of V into H is compact,

(A_2) g is locally Lipschitz continuous in H and $g(y) \geq -C|y|_H + D, \forall y \in H$, where $C > 0$, $D \in \mathbb{R}$,

(A_3) $h : U \to \mathbb{R} \cup \{\infty\}$ is proper convex and lower semicontinuous,

(A_4) $B \in L(U, H)$, $f \in H$,

(A_5) $L : H \to H$ is Lipschitz continuous, that is

$$|Ly_1 - Ly_2| \leq K|y_1 - y_2|_H, \forall y_1, y_2 \in H \ (K > 0), \tag{12}$$

(A_6) $A \in L(V, V^*)$ satisfies

$$\langle Av, v \rangle \geq \omega \|v\|^2, \forall v \in V \ (\omega > 0), \tag{13}$$

(A_7) A dominates L, i.e.,

$$\omega \lambda_1 > K,$$

where $\lambda_1 = \inf\{\frac{\|v\|^2}{|v|_H^2}; v \in V, v = 0\}$.

In the following we suppose that (A_1–A_7) are fulfilled.

Theorem 2.1 Let $(y^*, u^*) \in V \times U$ be a solution of (P). Then there exist $p \in V, \ell \in H$ such that

$$-A^*p - \ell \in \partial g(y^*), \ B^*p \in \partial h(u^*), \ \ell \in \partial f_{-p}(y^*).$$

Proof. Let us prove first that (P_ε) has at least one optimal solution $(y_\varepsilon, u_\varepsilon, v_\varepsilon)$. We define

$$A_H : D(A_H) = \{y \in V \ ; \ Ay \in H\} \subset H \to H \text{ by } A_H y = Ay, y \in D(A_H)$$

and, since A is a linear homeomorphism between V and V^*, we can consider $D(A_H)$ as a Hilbert space endowed with the inner product

$$(u, v)_{D(A_H)} = (Au, Av)_H, \forall u, v \in D(A_H).$$

Semilinear Optimal Control Problems

Then $(D(A_H), (\cdot, \cdot)_{D(A_H)})$ is compactly embedded in V and $A \in L(D(A_H), H)$.

The above assumption imply $\inf(P_\varepsilon) > -\infty$ and if (y_n, u_n, v_n) is a minimizing sequence, that is $Ay_n + v_n = Bu_n + f$ and

$$\inf(P_\varepsilon) \leq g(y_n) + h(u_n) + \tfrac{1}{2\varepsilon}|v_n - Ly_n|_H^2 + \tfrac{1}{2}|u_n - u^*|_U^2 + \tfrac{1}{2}|y_n - y^*|_H^2 \leq$$
$$\leq \inf(P_\varepsilon) + \tfrac{1}{n} \qquad (14)$$

then (y_n, u_n, v_n) is bounded in $H \times U \times H$.

The state equation $Ay_n + v_n = Bu_n + f$ shows, in fact, that (y_n) is bounded in $D(A_H)$ so, we can suppose that on a subsequence (denoted in the same way for simplicity) we have

$$y_n \to y \text{ strongly in } V,$$
$$v_n \to v \text{ weakly in } H,$$
$$u_n \to u \text{ weakly in } U,$$

and $Ay + v = Bu + f$.

If we let $n \to \infty$ in (14) we infer that $(y, u, v) \in D(A_H) \times U \times H =: \mathcal{V}$ is a solution of (P_ε). We denote this solution by $(y_\varepsilon, u_\varepsilon, v_\varepsilon)$.

Next, we show that

$$(y_\varepsilon, u_\varepsilon, v_\varepsilon) \to (y^*, u^*, Ly^*) \text{ strongly in } \mathcal{V} \qquad (15)$$

Since $(y_\varepsilon, u_\varepsilon, v_\varepsilon)$ is optimal for (P_ε) we have

$$g(y_\varepsilon) + h(u_\varepsilon) + \tfrac{1}{2\varepsilon}|v_\varepsilon - Ly_\varepsilon|_H^2 + \tfrac{1}{2}|u_\varepsilon - u^*|_U^2 + \tfrac{1}{2}|y_\varepsilon - y^*|_H^2 \leq$$
$$\leq g(y^*) + h(u^*) = \inf(P). \qquad (16)$$

Hence, $(y_\varepsilon, u_\varepsilon, v_\varepsilon)$ is bounded in $\mathcal{V}$ and on a subsequence it satisfies

$$(y_\varepsilon, u_\varepsilon, v_\varepsilon) \to (\overline{y}, \overline{u}, \overline{v}) \text{ weakly in } \mathcal{V},$$
$$y_\varepsilon \to \overline{y} \text{ strongly in } V,$$
$$h(u_\varepsilon) \to \liminf_{\varepsilon \downarrow 0} h(u_\varepsilon) \geq h(\overline{u}),$$

and $A\overline{y} + \overline{v} = B\overline{u} + f$.

By passing to limit in (16) we get

$$h(\overline{u}) \leq \liminf_{\varepsilon \downarrow 0} h(u_\varepsilon) \leq \inf(P) - g(\overline{y}) < \infty, \tag{17}$$

$$\limsup_{\varepsilon \downarrow 0} \tfrac{1}{2\varepsilon}|v_\varepsilon - Ly_\varepsilon|_H^2 \leq \inf(P) - g(\overline{y}) - h(\overline{u}) < \infty, \tag{18}$$

$$\lim_{\varepsilon \downarrow 0} |v_\varepsilon - Ly_\varepsilon|_H = 0, \ \overline{v} = L\overline{y}, \tag{19}$$

$$g(\overline{y}) + h(\overline{u}) + \tfrac{1}{2}\limsup_{\varepsilon \downarrow 0}\{|u_\varepsilon - u^*|_U^2 + \tfrac{1}{2}|y_\varepsilon - y^*|_H^2\} \leq \inf(P) \leq$$
$$\leq g(\overline{y}) + h(\overline{u}), \tag{20}$$

because $A\overline{y} + L\overline{y} = B\overline{u} + f$. This final inequality proves that

$$u_\varepsilon \to u^* \text{ strongly in } U,$$
$$y_\varepsilon \to y^* \text{ strongly in } H.$$

But (19) implies $v_\varepsilon \to Ly^*$ strongly in H, since $Ly_\varepsilon \to Ly^*$ strongly in H. Therefore

$$Ay_\varepsilon = Bu_\varepsilon + f - Ly_\varepsilon \to Bu^* + f - Ly^* = Ay^* \text{ strongly in } H,$$

i.e.

$$y_\varepsilon \to y^* \text{ strongly in } D(A_H),$$

so (15) is proved.

For $\lambda > 0$, we define

$$I_\lambda(y, u, v) = g_\lambda(y) + h(u) + \tfrac{1}{\varepsilon}S^L(y, v) + \tfrac{1}{2}|u - u^*|_U^2 + \tfrac{1}{2}|y - y^*|_H^2 +$$
$$+ \tfrac{1}{2}|u - u_\varepsilon|_U^2 + \tfrac{1}{2}|y - y_\varepsilon|_H^2,$$

where g_λ is the regularization of g in the separable Hilbert space H, given by (3).

Using a similar argument as above, for every $\lambda > 0$ problem
(R_λ) *Minimize*

$$I_\lambda(y, u, v)$$

on all $(y, u, v) \in V \times U \times H$ *subject to*

$$Ay + v = Bu + f,$$

Semilinear Optimal Control Problems

admits a solution $(y_\lambda, u_\lambda, v_\lambda) \in \mathcal{V}$ and we prove that (on a subsequence)
$$(y_\lambda, u_\lambda, v_\lambda) \to (y_\varepsilon, u_\varepsilon, v_\varepsilon) \text{ strongly in } V \times U \times V^* \qquad (21)$$
Indeed, since $Ay_\varepsilon + v_\varepsilon = Bu_\varepsilon + f$ and $(y_\lambda, u_\lambda, v_\lambda)$ is optimal for (R_λ), we have
$$\inf(R_\lambda) = I_\lambda(y_\lambda, u_\lambda, v_\lambda) \leq I_\lambda(y_\varepsilon, u_\varepsilon, v_\varepsilon) \leq C_\varepsilon < \infty \qquad (22)$$
and this provides us with the boundedness of $(y_\lambda, u_\lambda, v_\lambda)$ in $\mathcal{V}$.

Hence, on a subsequence we can assume that
$$(y_\lambda, u_\lambda, v_\lambda) \to (\tilde{y}, \tilde{u}, \tilde{v}) \text{ weakly in } \mathcal{V}$$
$$y_\lambda \to \tilde{y} \text{ strongly in } V, \; v_\lambda \to \tilde{v} \text{ strongly in } V^*, \; A\tilde{y} + \tilde{v} = B\tilde{u} + f$$
$$g_\lambda(y_\lambda) \to g(\tilde{y}), \; h(u_\lambda) \to \liminf_{\lambda \downarrow 0} h(u_\lambda) \geq h(\tilde{u})$$
$$S^L(y_\lambda, v_\lambda) \to \liminf_{\lambda \downarrow 0} S^L(y_\lambda, v_\lambda) \geq S^L(\tilde{y}, \tilde{v}).$$

Then, by passing to lim sup in (22) we get
$$g(\tilde{y}) + h(\tilde{u}) + \tfrac{1}{\varepsilon} S^L(\tilde{y}, \tilde{v}) +$$
$$+ \tfrac{1}{2} \limsup_{\lambda \downarrow 0} \{|u_\lambda - u^*|_U^2 + |y_\lambda - y^*|_H^2 + |u_\lambda - u_\varepsilon|_H^2 + |y_\lambda - y_\varepsilon|_H^2\} \leq$$
$$\leq \limsup_{\lambda \downarrow 0} I_\lambda(y_\lambda, u_\lambda, v_\lambda) \leq \lim_{\lambda \downarrow 0} I_\lambda(y_\varepsilon, u_\varepsilon, v_\varepsilon) = \inf(R_\lambda) \leq$$
$$\leq g(\tilde{y}) + h(\tilde{u}) + \tfrac{1}{\varepsilon} S^L(\tilde{y}, \tilde{v}) + \tfrac{1}{2}|\tilde{u} - u^*|_U^2 + \tfrac{1}{2}|\tilde{y} - y^*|_H^2,$$
and so
$$\limsup_{\lambda \downarrow 0} \{|u_\lambda - u_\varepsilon|_U^2 + |y_\lambda - y_\varepsilon|_H^2\} \leq 0. \qquad (23)$$

This implies $u_\lambda \to u_\varepsilon = \tilde{u}$ strongly in U, $\tilde{y} = y_\varepsilon$, $\tilde{v} = v_\varepsilon$ and
$$Ay_\lambda \to Ay_\varepsilon = A\tilde{y} \text{ weakly in } H$$
since $v_\lambda \to v_\varepsilon$ weakly in H.

Hence $y_\lambda \to y_\varepsilon$ weakly in $D(A_H)$ and eventually on a subsequence $y_\lambda \to y_\varepsilon$ strongly in V, thereby (21) is proved.

The necessary condition of optimality for $(y_\lambda, u_\lambda, v_\lambda)$ can be easily deducted according to the principle that if $F : X \to \mathbb{R}$ is a given function defined on a Banach space X and $F(z) = \inf_{x \in X} F(x)$ then $F^o(z, v) \geq 0$ for every $v \in X$.

In our case, for problem (R_λ) we find

$$(\nabla g_\lambda(y_\lambda), \overline{y})_H + h'(u_\lambda, \overline{u}) + \tfrac{1}{\varepsilon}(S_y^L(y_\lambda, v_\lambda), \overline{y})_H + \tfrac{1}{\varepsilon}(S_v^L(y_\lambda, v_\lambda), \overline{v})_H +$$
$$+ (2u_\lambda - u^* - u_\varepsilon, \overline{u})_U + (2y_\lambda - y^* - y_\varepsilon, \overline{y})_H \geq 0 \qquad (24)$$

for every $(\overline{y}, \overline{u}, \overline{v}) \in V \times U \times H$ such that $A\overline{y} + \overline{v} = B\overline{u}$, where

$$(S_y^L(y_\lambda, v_\lambda), S_v^L(y_\lambda, v_\lambda)) \in \partial S^L(y_\lambda, v_\lambda) \subset H \times H.$$

According to Lemma 1.1, we have $S_v^L(y_\lambda, v_\lambda) = v_\lambda - Ly_\lambda$, $S_y^L(y_\lambda, v_\lambda) \in \partial f_{(v_\lambda - Ly_\lambda)}(y_\lambda)$ and $|S_y^L(y_\lambda, v_\lambda)|_H \leq K|S_v^L(y_\lambda, v_\lambda)|_H$.
Relation (24) can be equivalently written as

$$(\nabla g_\lambda(y_\lambda) + \tfrac{1}{\varepsilon} S_y^L(y_\lambda, v_\lambda) + 2y_\lambda - y^* - y_\varepsilon, \overline{y})_H - \tfrac{1}{\varepsilon}(v_\lambda - Ly_\lambda, A\overline{y})_H \geq 0, \qquad (25)$$

for every $\overline{y} \in D(A_H)$, and

$$h'(u_\lambda, \overline{u}) + (\tfrac{1}{\varepsilon} B^*(v_\lambda - Ly_\lambda) + 2u_\lambda - u^* - u_\varepsilon, \overline{u})_U \geq 0, \forall \overline{u} \in U,$$

or,

$$\partial h(u_\lambda) \ni B^* p_\lambda - 2u_\lambda + u^* + u_\varepsilon, \qquad (26)$$

where $p_\lambda = -\tfrac{1}{\varepsilon}(v_\lambda - Ly_\lambda) \in H$.
We have

$$p_\lambda \to p_\varepsilon := -\tfrac{1}{\varepsilon}(v_\varepsilon - Ly_\varepsilon) \text{ weakly in } H,$$
$$S_y^L(y_\lambda, v_\lambda) \to w_\varepsilon \text{ weakly in } H,$$

and since $S_y^L(y_\lambda, v_\lambda) \in \partial f_{(v_\lambda - Ly_\lambda)}(y_\lambda)$ then by Remark 1.1 we know that

$$w_\varepsilon \in \partial f_{(v_\varepsilon - Ly_\varepsilon)}(y_\varepsilon).$$

If we let $\lambda \to 0$ in (25), (26) then we get

$$(\tau_\varepsilon + \tfrac{1}{\varepsilon} w_\varepsilon + y_\varepsilon - y^*, \overline{y})_H - \tfrac{1}{\varepsilon}(v_\varepsilon - Ly_\varepsilon, A\overline{y})_H \geq 0, \forall \overline{y} \in D(A_H), \qquad (27)$$

where $\tau_\varepsilon \in \partial g(y_\varepsilon)$, and

$$\partial h(u_\varepsilon) \ni B^* p_\varepsilon + u^* - u_\varepsilon. \qquad (28)$$

Let $\eta_\varepsilon \in D(A_H) = \{y \in V; A^* y \in H\}$ be such that

Semilinear Optimal Control Problems

$$A^*\eta_\varepsilon = \tau_\varepsilon + \tfrac{1}{\varepsilon}w_\varepsilon + y_\varepsilon - y^*$$

Then relation (27) can be stated as

$$(A^*\eta_\varepsilon, \overline{y})_H = -(p_\varepsilon, A\overline{y})_H, \quad \forall \overline{y} \in D(A_H) \tag{29}$$

But

$$(A^*\eta_\varepsilon, \overline{y})_H = \langle A^*\eta_\varepsilon, \overline{y}\rangle = \langle \eta_\varepsilon, A\overline{y}\rangle = (\eta_\varepsilon, A\overline{y})_H.$$

Therefore (29) is equivalent to $\eta_\varepsilon = -p_\varepsilon$.

In this way we proved that $p_\varepsilon \in D(A_H^*)$ and

$$-A^*p_\varepsilon - \tfrac{1}{\varepsilon}w_\varepsilon + y^* - y_\varepsilon \in \partial g(y_\varepsilon). \tag{30}$$

If we multiply the last relation with $-p_\varepsilon$ then we get, according to (12), that

$$\omega\|p_\varepsilon\|^2 - \|p_\varepsilon\|(C_1 + |\tfrac{1}{\varepsilon}w_\varepsilon|_*) \le 0, \ (C_1 < \infty),$$

since $\{\cup \partial g(y_\varepsilon)\}$ is bounded in H.

Hence

$$\omega\sqrt{\lambda_1}|p_\varepsilon|_H \le \omega\|p_\varepsilon\| \le |\tfrac{1}{\varepsilon}w_\varepsilon|_* + C_1 \le \tfrac{1}{\sqrt{\lambda_1}}|\tfrac{1}{\varepsilon}w_\varepsilon|_* + C_1 \le \tfrac{K}{\sqrt{\lambda_1}}|\tfrac{1}{\varepsilon}w_\varepsilon|_* + C_1.$$

We have $\omega\lambda_1 > K$, therefore, (p_ε) and $(\tfrac{1}{\varepsilon}w_\varepsilon)$ are bounded in H so, eventually on a subsequence we infer that

$$p_\varepsilon \to p \text{ weakly in } H, \ \tfrac{1}{\varepsilon}w_\varepsilon \to \ell \text{ weakly in } H.$$

We can pass to limit in (29) and (30) because A^* is weakly-strongly closed in H and $B^* \in L(H, U)$ to get

$$-A^*p - \ell \in \partial g(y^*), \ B^*p \in \partial h(u^*). \tag{31}$$

In order to determine a relation between p and ℓ we start recalling that $w_\varepsilon \in \partial f_{(v_\varepsilon - Ly_\varepsilon)}(y_\varepsilon)$, i.e.,

$$(w_\varepsilon, h)_H \le \limsup_{y \to y_\varepsilon, t\downarrow 0} -\tfrac{1}{t}(v_\varepsilon - Ly_\varepsilon, L(y + th) - Ly)_H, \forall h \in H,$$

and so for arbitrary but fixed $h \in H$ we have

$$(\tfrac{1}{\varepsilon}w_\varepsilon, h)_H \leq \limsup_{y\to y_\varepsilon, t\downarrow 0} \tfrac{1}{t}(p_\varepsilon, L(y+th) - Ly)_H.$$

Therefore for every $\varepsilon > 0$, there exist $y'_\varepsilon \in H$, $t_\varepsilon \in (0,\varepsilon)$ such that $|y'_\varepsilon - y_\varepsilon|_H < \varepsilon$ and

$$(\tfrac{1}{\varepsilon}w_\varepsilon, h)_H \leq \tfrac{1}{t_\varepsilon}(p_\varepsilon, L(y'_\varepsilon + t_\varepsilon h) - Ly'_\varepsilon) + \varepsilon.$$

Because $\tfrac{1}{\varepsilon}w_\varepsilon =: \ell_\varepsilon \to \ell$ weakly in H and $p_\varepsilon \to p$ weakly in H we may assume by the Mazur theorem that on a convex combination we have

$$\ell_n = \sum_{i\in I_n} \alpha_n^i \ell_{\varepsilon_i} \to \ell \text{ strongly in } H \text{ and } p_n = \sum_{i\in I_n} \alpha_n^i p_{\varepsilon_i} \to p \text{ strongly in } H$$

(Here I_n is a finite set of positive integers, $\alpha_n^i \geq 0$ and $\sum_{i\in I_n} \alpha_n^i = 1$).

This allows us to find $(\ell_n, h)_H \leq \tfrac{1}{t_\varepsilon}(p_n, L(y'_\varepsilon + t_\varepsilon h) - Ly'_\varepsilon)_H + \varepsilon$, and passing to limit with $\varepsilon \to 0$ ($n \to \infty$) we obtain

$$(\ell, h)_H \leq \limsup_{\varepsilon \to 0} \tfrac{1}{t_\varepsilon}(p, L(y'_\varepsilon + t_\varepsilon h) - Ly'_\varepsilon)_H \leq$$

$$\leq \limsup_{y \to y^*, t\downarrow 0} \tfrac{1}{t}(p_\varepsilon, L(y+th) - Ly)_H, \forall h \in H.$$

because $t_\varepsilon \downarrow 0$, $y'_\varepsilon \to y^*$ strongly in H. This final inequality shows that $\ell \in \partial f_{-p}(y^*)$. The proof is complete. ∎

3. An Example to Optimal Control of a Semilinear Elliptic Equation

We investigate the following particular case of (P)

(P_1) *Minimize*

$$g(y) + h(u),$$

on all $(y, u) \in H_0^1(\Omega) \times U$, *subject to*

$$(SE) \quad \begin{cases} -\Delta y + \beta(y) = Bu + f & \text{in } \Omega, \\ y = 0 & \text{in } \Omega, \end{cases}$$

where Ω is a bounded open domain of class C^2 in $\mathbb{R}^n$, $V = H_0^1(\Omega)$, U is a given Hilbert space, $H = L^2(\Omega)$, g, h, B, f satisfy hypotheses (A_2), (A_3), (A_4) and $\beta : \mathbb{R} \to \mathbb{R}$ is a Lipschitz continuous function, i.e.,

Semilinear Optimal Control Problems

$$|\beta(r_1) - \beta(r_2)| \leq K|r_1 - r_2|, \forall r_1, r_2 \in \mathbb{R} \ (K > 0). \tag{32}$$

We may see (SE) as $Ay + Ly = Bu + f$, where $A = -\Delta : H_0^1(\Omega) \to H^{-1}(\Omega)$ and $L : L^2(\Omega) \to L^2(\Omega)$ is given by

$$Ly = \beta(y), \ y \in L^2(\Omega).$$

Theorem 3.1 Suppose, in addition to the above hypotheses, that (y^*, u^*) is a solution of (P_1) and

$$\lambda_1 > K \tag{33}$$

where $\lambda = \inf\{|\nabla v|_{L^2(\Omega)}/|v|_{L^2(\Omega)}; v = 0\}$ is the first eigenvalue of $-\Delta$ in Ω.

Then there exist $p \in H_0^1(\Omega)$, $\ell \in L^2(\Omega)$ such that
(i) $\partial g(y^*) + \ell \ni \Delta p$,
(ii) $B^* p \in \partial h(u^*)$,
(iii) $\ell(x) \in p(x)\partial\beta(y^*(x))$ a.e. $x \in \Omega$.

Proof. Since all the hypotheses are fulfilled, we may apply Theorem 2.1 to find $p \in H_0^1(\Omega)$, $\ell \in L^2(\Omega)$ such that

$$\partial g(y^*) + \ell \ni -A^* p, \ B^* p \in \partial h(u^*), \ \ell \in \partial f_{-p}(y^*)$$

The final relation is equivalent to

$$(\ell, h)_H \leq \limsup_{y \to y^*, t \downarrow 0} (1/t) \int (\beta(y + th) - \beta(y))p \, dx, \ \forall v \in H_0^1(\Omega),$$

and it can be described as $\ell \in \partial G(y^*)$ where $G : L^2(\Omega) \to \mathbb{R}$ is defined by

$$G(\xi) = \int \beta(\xi) p \, dx, \ \xi \in L^2(\Omega).$$

Thus $\ell(x) \in p(x)\partial\beta(y^*(x))$ a.e. $x \in \Omega$ (see Clarke [4] or Ioffe and Levin [5]). Since $-A^* = \Delta$, the proof is complete. ∎

References

1. Aizicovici, S., Motreanu D. & Pavel N.H., *Nonlinear Programming Problems Associated with Closed Range Operators*, Appl. Math. Optim., Vol. 40, pp. 211-228, 1999,

2. BARBU, V., *Analysis and Control of Nonlinear Infinite Dimensional Systems*, Academic Press, Boston, 1993,

3. BARBU, V., *Partial Differential Equations and Boundary Value Problems*, Kluwer Academic Publishers, Dordrecht, Boston, 1998,

4. CLARKE, F. H., *Optimization and Nonsmooth Analysis*, John Wiley and Sons, New-York, 1983,

5. IOFFE, A. D. and LEVIN, V. I., *Subdifferential of Convex Functions*, Trudy Moskov. Mat. Obshch. (translated in Transaction of the Moscow Mathematical Society), Vol. 26, pp. 3-13, 1972,

6. Motreanu D., Pavel N.H., *Tangency, Flow Invariance for Differential Equations and Optimization Problems*, Monographs and Textbooks in Pure and Applied Mathematics, Vol. 219, Marcel Dekker, New-York - Basel, 1999,

7. TIBA, D., *Lectures on the Optimal Control of Elliptic Equations*, The 5th International Summer School Jyväskylä, Lecture Notes 32, Finland, 1995.

Lyapunov Equation and the Stability of Nonautonomous Evolution Equations in Hilbert Spaces

Quoc-Phong Vu
Department of Mathematics
Ohio University
Athens, OH 45701, U.S.A.
and
Siu Pang Yung
Department of Mathematics
University of Hong Kong
Hong Kong

Abstract. We apply the method of Lyapunov equations $A^*P + PA = -I$ to the question of exponential stability of the differential equation $u'(t) = A(t)u(t), t \geq 0$, in a Hilbert space E. Under some suitable conditions, we show that the solutions are exponentially stable provided that $A(t)$ generate exponentially stable semigroups with exponential types $\leq \sigma < 0$, and are slowly varying in some sense. Estimates are also given for the rates of convergence of the solutions to zero.

1. Introduction

Let $A(t)$ be closed linear operators on a Banach space E such that the differential equation

$$u'(t) = A(t)u(t), t \geq 0, \tag{1}$$

is well-posed, i.e., for each x_0 from a dense subset $\mathcal{D} \subset E$, Eq.(1) has a unique solution with $u(0) = x_0$, which depends continuously on the initial value x_0. In this paper, we are concerned with conditions which imply that solutions to Eq.(1) are exponentially stable, i.e. $\|u(t)\| \leq Ne^{-\sigma t}\|u(0)\|$ for every solution $u(t)$ of Eq.(1), where N and σ are positive constants.

Note that if $\dim E < \infty$ and Eq.(1) is autonomous, i.e. $A(t) \equiv A$ are independent of t, then its solutions are exponentially stable if and only if all eigenvalues of A have

negative real parts, i.e., there exists a positive number σ such that
$$\sigma(A) \subset \{\lambda \in \mathbb{C} : \operatorname{Re} \lambda \leq -\sigma\},$$
where $\sigma(A)$ is the set of eigenvalues (the spectrum) of A.

However, it is well known that a similar statement does not holds for nonautonomous equations. Indeed, assume that $A(t) = U^{-1}(t)A_0 U(t)$, where
$$A_0 = \begin{pmatrix} -1 & -5 \\ 0 & -1 \end{pmatrix}, \quad U(t) = \begin{pmatrix} \cos t & \sin t \\ -\sin t & \cos t \end{pmatrix}.$$
Then
$$\sigma(A(t)) = \{-1\}, \text{ for all } t \geq 0,$$
but the solution matrix is
$$X(t) = \begin{pmatrix} e^t(\cos t + \tfrac{1}{2}\sin t) & e^{-3t}(\cos t - \tfrac{1}{2}\sin t) \\ e^t(\sin t - \tfrac{1}{2}\cos t) & e^{-3t}(\sin t + \tfrac{1}{2}\cos t) \end{pmatrix},$$
and therefore is unstable (see [2]).

Various conditions have been obtained for equations (1) in a finite dimensional space E, that, together with the condition of uniform exponential stability of the matrices $A(t)$, i.e.
$$\|e^{sA(t)}\| \leq Ne^{-\sigma s}, \quad (\sigma > 0, N \geq 1 \text{ are independent of } t), \tag{2}$$
imply the exponential stability of solutions of Eq.(1) (see [1,3,5,7,10]). All these conditions express the property that $A(t)$ are slowly varying matrices in a suitable sense. To our knowledge, analogous results have not been obtained for Eq.(1) in an infinite dimensional space.

In this paper, we consider the problem of exponential stability of Eq.(1) in infinite dimensional Hilbert space, and we extend some results of the above mentioned papers to this case. Namely, we prove that if $A(t)$ are uniformly bounded (i.e. $\sup_{t\geq 0} \|A(t)\| < \infty$) and slowly varying in an appropriate sense, and if (2) holds, then solutions to Eq.(1) are exponentially stable. We also obtain estimates of the rate of convergence of the solutions $u(t)$ to zero (Theorems 7 and 8).

The method we use to obtain these results is based on the Lyapunov's equation $A^*(t)P + PA(t) = -I$ with the variable operators $A(t)$. This method also allows us to extend the results to some cases involving unbounded operator coefficients. For other recent applications of Lyapunov equations in the stability theory of evolution equations in infinite dimensional spaces see [13-15].

Throughout the paper $\mathcal{D}(A)$ denotes the domain of an operator A, and $\sigma(A)$ denotes the spectrum of A.

2. Main Results.

Assume that $A(t)$ are closed linear operators on a Hilbert space E which satisfy the following conditions:

H1. $A(t) = iA_0 + A_1(t)$, where $A_0 : E \to E$ is a self-adjoint, generally unbounded, operator, $A_1(t)$ are bounded, and A_0 commutes with $A_1(t)$ for all t, i.e. from $x \in \mathcal{D}(A_0)$ we have $A_1(t)x \in \mathcal{D}(A_0)$ and $A_0 A_1(t)x = A_1(t)A_0 x$, for all t.

H2. $\sup_{t \geq 0} \|A_1(t)\| < \infty$, and
$$\|e^{sA_1(t)}\| \leq Ne^{-\sigma s}, \quad (\sigma > 0, N \geq 1 \text{ are independent of } t).$$

H3. For every $x \in \mathcal{D} := \mathcal{D}(A_0)$, the function $t \mapsto A_1(t)x$ is differentiable and $A_1'(t)x$ is a bounded operator from $\mathcal{D}$ to E, so that it can be extended by continuity to a bounded operator on E.

Note that conditions H1-H3 include the case of Eq.(1) with bounded operators $A(t)$, in an infinite dimensional Hilbert space (set $A_0 = 0$). It also follows from the theory of evolution equations (see e.g. [8], Chapter 5, Theorem 2.3 and 3.1) that under these conditions there exists an evolution system $U(t,s)(0 \leq s \leq t < \infty)$ associated with Eq.(1), such that solution of Eq.(1) with the initial value $u(0) = x$ is given by $u(t) = U(t,0)x, t \geq 0$.

Our method of investigation of the asymptotic behavior of Eq.(1) is based on the operator equation of the following form
$$PA + BP = C, \tag{3}$$
where A and B are closed linear operators on E and $C : E \to E$ is a bounded linear operator. A bounded linear operator $P : E \to E$ is called *solution* of Eq.(3) if $P\mathcal{D}(A) \subset \mathcal{D}(B)$ and $PAx + BPx = Cx$ for all $x \in \mathcal{D}(A)$. The following proposition is well known (see, e.g. [9], [12]).

Proposition 1. Let A and B be generators of exponentially stable C_0-semigroups $\{e^{sA}\}_{s \geq 0}$ and $\{e^{sB}\}_{s \geq 0}$, respectively. Then the integral
$$P = -\int_0^\infty e^{sB} C e^{sA} ds \tag{4}$$
converges in the uniform operator topology and is the unique solution of Eq.(2).

Now let $A(t)$ satisfy conditions H1-H3. We will need the following lemma.

Lemma 2. An operator $P(t)$ is a solution to equation

$$P(t)A(t) + A^*(t)P(t) = -I \tag{5}$$

if and only if it is a solution of

$$P(t)A_1(t) + A_1^*(t)P(t) = -I. \tag{6}$$

Proof. Assume that $P(t)$ is a solution to (5). By Proposition 1, and since A_0 and commutes with $A_1(t)$ as well as with $A_1(t)^*$, we have

$$P(t) = -\int_0^\infty e^{sA_0^* + sA_1(t)^*} e^{sA_0 + sA_1(t)} ds = -\int_0^\infty e^{sA_1(t)^*} e^{sA_1(t)} ds,$$

hence $P(t)$ is a solution to (6). The converse is proved analogously. $\square$

Thus, for each $t \geq 0$,

$$P(t) := \int_0^\infty e^{sA_1(t)^*} e^{sA_1(t)} ds \tag{7}$$

is a bounded solution of Eqs(5)-(6).

Proposition 3. There exist $\alpha, \beta > 0$, independent of t, such that

$$\beta \|x\|^2 \leq \langle P(t)x, x \rangle \leq \alpha \|x\|^2, \text{ for all } x \in E. \tag{8}$$

Moreover, one can choose $\alpha = \frac{N}{\sqrt{2\sigma}}$ and $\beta = \frac{1}{2M}$.

Proof. From (7) and condition H2 it follows that

$$\|P(t)x\|^2 \leq \int_0^\infty N^2 e^{-2\sigma s} ds \|x\|^2 = \alpha^2 \|x\|^2,$$

where $\alpha = \frac{N}{\sqrt{2\sigma}}$. Hence $\langle P(t)x, x \rangle \leq \alpha \|x\|^2$. On the other hand

$$\langle P(t)x, x \rangle = \int_0^\infty \|e^{sA_1(t)}x\|^2 ds \geq \int_0^\infty (e^{-s\|A_1(t)\|})^2 ds \|x\|^2 \geq$$
$$\int_0^\infty e^{-2sM} ds \|x\|^2 = \frac{1}{2M} \|x\|^2 = \beta \|x\|^2. \quad \square$$

Proposition 4. The operator function $P(t)$ is differentiable and satisfies

$$\|P'(t)\| \leq K \|A_1'(t)\|,$$

where $K = \frac{N^3}{\sqrt{2\sigma^3}}$.

Proof. That $P(t)x$ is differentiable for every $x \in E$ follows from (7). Since

$$P(t)A_1(t)x + A_1^*(t)P(t)x = -x,$$

it follows that

$$P'(t)A_1(t)x + A_1(t)^*P'(t)x = -[P(t)A_1'(t)x + A_1'(t)^*P(t)x].$$

Since by H2 $A_1'(t)$ is bounded, and generates an exponentially stable semigroup, we have by Proposition 1

$$P'(t)x = \int_0^\infty e^{sA_1(t)^*}[-P(t)A_1'(t) - A_1'(t)^*P(t)]e^{sA_1(t)}x\,ds.$$

Therefore, from (8) we have

$$\|P'(t)\| \leq \int_0^\infty N^2 e^{-2\sigma s}[2\alpha\|A_1'(t)\|]ds \leq K\|A_1'(t)\|,$$

where $K = \frac{\alpha N^2}{\sigma} = \frac{N^3}{\sqrt{2\sigma^3}}$. Box

Consider the Cauchy problem

$$\begin{cases} u'(t) = A(t)u(t) \\ u(0) = x_0 \end{cases} \tag{9}$$

where the operators $A(t)$ satisfy conditions H1-H3. We need the following simple lemma.

Lemma 5. Let $\varphi(t)$ and $f(t)$ be real functions on $[0,\infty)$ such that $\varphi(t) > 0$ for all $t \geq 0$, is differentiable, satisfies $\varphi'(t) \leq f(t)$ for all $t \geq 0$. Then

$$\varphi(t) \leq \varphi(0)e^{\int_0^t f(s)ds}.$$

Proof. The required inequality follows from

$$\int_0^t \frac{d}{ds}\{\ln(\varphi(t))\}ds \leq \int_0^t f(s)ds. \quad \Box$$

Proposition 6. Let $u(t)$ be a solution of Eq.(9) and $P(t)$ be solutions of Eqs.(5)-(6). Then the following estimate holds.

$$\langle P(t)u(t), u(t)\rangle \leq \exp\left\{-\left[\frac{t}{\alpha} - 2MK\int_0^t \|A_1'(s)\|ds\right]\right\}\langle P(0)x_0, x_0\rangle, \text{ for all } t \geq 0. \tag{10}$$

Proof. Since Eq.(1) is well posed, if $u(t_0) = 0$ for some t_0 then $u(t) = 0$ for all $t \geq 0$. Hence, we can assume that $u(t) \neq 0$ for all $t \geq 0$. By Proposition 3, $\langle P(t)u(t), u(t)\rangle > 0$ for all $t \geq 0$. We have

$$\frac{d}{dt}\langle P(t)u(t), u(t)\rangle = \langle P'(t)u(t), u(t)\rangle + \langle P(t)u'(t), u(t)\rangle + \langle P(t)u(t), u'(t)\rangle.$$

On the other hand,

$$\langle P(t)u'(t), u(t)\rangle + \langle P(t)u(t), u'(t)\rangle =$$
$$\langle P(t)A(t)u(t), u(t)\rangle + \langle P(t)u(t), A(t)u(t)\rangle =$$
$$\langle P(t)A(t)u(t), u(t)\rangle + \langle A^*(t)P(t)u(t), u(t)\rangle =$$
$$-\langle u(t), u(t)\rangle \leq -\frac{1}{\alpha}\langle P(t)u(t), u(t)\rangle.$$

Moreover, from Propositions 3-4 it follows

$$\langle P'(t)u(t), u(t)\rangle \leq \|P'(t)\|\|u(t)\|^2 \leq K\|A'_1(t)\|\|u(t)\|^2 \leq$$
$$\frac{K}{\beta}\|A'_1(t)\|\langle P(t)u(t), u(t)\rangle = 2MK\|A'_1(t)\|\langle P(t)u(t), u(t)\rangle.$$

Therefore

$$\frac{d}{dt}\langle P(t)u(t), u(t)\rangle \leq \left[2MK\|A'_1(t)\| - \frac{1}{\alpha}\right]\langle P(t)u(t), u(t)\rangle,$$

and (10) follows from Lemma 5. □

We remark that Proposition 6 gives better estimates than those in [4]. From Propositions 3 and 6 we obtain the following result.

Theorem 7. Let $u(t)$ be a solution of Eq.(9). Then

$$\|u(t)\|^2 \leq 2\alpha M \exp\left\{-\left[\frac{t}{\alpha} - 2MK\int_0^t \|A'_1(s)\|ds\right]\right\}\|x_0\|^2,$$

where α and K are constants defined in Propositions 3-4.

From Theorem 7 we obtain the following stability result which are extensions of results in [1-5], [7-10].

Theorem 8. Assume that $A(t)$ satisfy conditions H1-H3. In addition, assume that $A_1(t)$ satisfy one of the following conditions:

$$\gamma := \sup_{s \geq 0}\|A'_1(s)\| < \frac{1}{2\alpha MK}, \text{ or} \tag{11}$$

$$\lim_{t \to \infty}\frac{1}{t}\int_0^t \|A'_1(s)\|ds = 0. \tag{12}$$

Then Eq.(1) is exponentially stable, i.e. $\|u(t)\| \leq Le^{-\omega t}\|u(0)\|$ for every solution $u(t)$ and some positive constants L, ω.

Proof. If (11) holds, then

$$\|u(t)\|^2 \leq 2\alpha M e^{-t\sigma}\|x_0\|^2, \text{ where } \sigma = \frac{1}{\alpha} - 2MK\gamma > 0.$$

Assume now that (12) holds. Choose $\varepsilon > 0$ such that

$$\sigma := \frac{2\alpha M}{K} - \varepsilon > 0.$$

By (12), there exists t_0 such that, for every $t \geq t_0$, we have

$$\int_0^t \|A_1'(s)\|ds < \varepsilon t.$$

Therefore,

$$\|u(t)\|^2 \leq 2\alpha M e^{-t\sigma}, \text{ for all } t \geq t_0. \quad \square$$

We remark that the conclusions in Theorems 7 and 8 remain valid for mild solutions, i.e., for functions $u(t) = U(t, 0)x$, thanks to the well-posedness of Eq.(1) in the considered situation. In other words, there exist positive constants L, σ such that

$$\|U(t,s)\| \leq Le^{-\sigma(t-s)}\|, \text{ for all } t \geq s.$$

We also remark that the following condition (considered by Cesari for ordinary differential equations)

$$\int_0^\infty \|A_1'(t)\|dt < \infty \quad (13)$$

implies that (12) holds. On the other hand, if (13) holds, then, for every $x \in \mathcal{D}$, the limit $\lim_{t \to \infty} A_1(t)x$ exists. In fact,

$$\lim_{t \to \infty} A_1(t)x = \lim_{t \to \infty} \int_0^t A_1'(s)xds + A_1(0)x = \int_0^\infty A_1'(s)xds + A_1(0)x, \text{ for all } x \in \mathcal{D}.$$

Hence $A(t)x$ converges as $t \to \infty$, for all $x \in \mathcal{D}$. However, even in this case Theorem 8.1 of [8, p.173] is not applicable since $A(t)$ do not generate, in general, analytic semigroups.

Examples.

1. Let $E := L^2(\mathbb{R})$ and $b(t)$ be a bounded differentiable scalar valued function such that $b(t) \leq -\sigma$ for some $\sigma > 0$, and assume that one of the following conditions is fulfilled:

(a) $\sup_{t>0} |b'(t)|$ is sufficiently small;
(b) $\int_0^\infty |b'(t)| dt < \infty$;
(c) $\lim_{t\to\infty} \frac{1}{t} \int_0^t |b'(s)| ds = 0$.

Define operators $A(t)$ on E by

$$A(t)u(x) := i\frac{\partial^2 u}{\partial x^2} + b(t)u(x), \ u(t) \in E.$$

Then, as is easily seen, $A(t)$ satisfy conditions H1-H3 with $A_0 := \frac{\partial^2}{\partial x^2}$ and $A_1(t)u := b(t)u$. Consequently, solutions of the equation

$$u_t(t,x) = i\frac{\partial^2}{\partial x^2}u(t,x) + b(t)u(t,x), t \geq 0, x \in \mathbb{R},$$

converge to zero exponentially (in $L^2(\mathbb{R})$-norm).

2. We generalize the above example by putting $A_0 := \Delta$, where Δ is a Laplace operator on $E := L^2(\mathbb{R}^3)$, or $E := L^2(\Omega)$ where Ω is a bounded domain in $\mathbb{R}^3$ with smooth boundary, and Δ is defined with an appropriate boundary condition. It is well known that A_0 is a self-adjoint operator (see e.g. [6, Chapter 5]). Furthermore, let $A_1(t)$ be a uniformly bounded family of operators on E such that

$$\|e^{sA_1(t)}\| \leq Ne^{-\sigma s}, \ (\sigma > 0, N \geq 1 \text{ are independent of } t),$$

and assume that $A_1(t)$ satisfies (11) or (12). Then solutions to the equation

$$u_t(t,x) = i\Delta u(t,x) + A_1(t)u(t,x), \ t \geq 0,$$

are exponentially stable.

References

1. L. Cesari, *Un nouvo criterio di stabilità per soluzioni delle equazioni differenziali lineari,* Ann. Scoula Norm. Sup. Pisa **(2)9** (1940), 163–186.

2. W.A. Copell, *Dichotomies in Stability Theory,* Lecture Notes in Mathematics, vol. 629, Springer, Berlin, 1978.

3. J.K. Hale and A.P. Stokes, *Conditions for the stability of nonautonomous differential equations*, J. Math. Anal. Appl. **3** (1961), 50–69.

4. D. Henry, *Geometric Theory of Semilinear Parabolic Equations*, Lecture Notes in Mathematics, vol. 840, Springer, Berlin, 1981.

5. C.S. Kahane, *On the stability of solutions of linear differential systems with slowly varying coefficients*, Czechoslovak Math. J. **42 (117)** (1992), 715–726.

6. T. Kato, *Perturbation Theory for Linear Operators*, Springer, Berlin, 1966.

7. L. Markus and H. Yamabe, *Global stability criteria for differential systems*, Osaka Math. J. **12** (1960), 305–317.

8. A. Pazy, *Semigroups of Linear Operators and Applications to Partial Differential Equations*, Springer, New York, 1983.

9. C.R. Putnam, *Commutation Properties of Hilbert Space Operators and Related Topics*, Springer, Berlin, 1967.

10. G. Sansone and R. Conti, *Nonlinear Differential Equations*, McMillan, New York, 1964.

11. H. Tanabe, *Evolution Equations*, Pitman, London, 1979.

12. Vu Quoc Phong, *The operator equation $AX - XB = C$ with unbounded operators A and B and related abstract Cauchy problems*, Math. Z. **208** (1991), 567–588.

13. Vu Quoc Phong, *On the exponential stability and dichotomy of C_0-semigroups*, Studia Math. **132** (1999), 141–149.

14. Vu Quoc Phong and E. Schüler, *the operator equation $AX - XB = C$, admissibility, and asymptotic behavior of differential equations*, J. Differential Equations **145** (1998), 394–419.

15. E. Schüler and Vu Quoc Phong, *The operator equation $AX - X\mathcal{D}^2 = -\delta_0$ and second order differential equations in Banach spaces*, Semigroups of operators: theory and applications (Newport Beach, CA, 1998), 352–363, Progr. Nonlinear Differential Equations Appl., **42**, Birkhauser, Basel, 2000.

Least Action for *N*-Body Problems with Quasihomogeneous Potentials

Shih-liang Wen
Department of Mathematics, Ohio University, Ohio 45701, USA
Shiqing Zhang
Mathematical Department, Chongqing University, Chongqing 400044, China

Abstract Using variational minimization methods, we study the existence of a noncollision or a generalized periodic solution for N-body problems with quasihomogeneous potentials, specially for N–body and N+1–body problems in R^{2k} (k $\geq$ 1), we study the geometric characterization for variational minimization solutions.

1. Introduction and Main Results

N-body problems with quasihomogeneous potentials ([6], [12]-[14], [22]) are related with the motion of N point masses $m_1, \cdots, m_N$ in R^K (K$\geq$1) under the action of the potential $-W(q)$ given by

$$W(q) = U(q) + V(q) = \frac{1}{2}\sum_{1\leq i\neq j\leq N} W_{ij}(q_i - q_j) \qquad (1.1)$$

Where
$$U(q) = \frac{a}{2}\sum_{1\leq i\neq j\leq N}\frac{m_i m_j}{|q_i - q_j|^\alpha} \qquad (1.2)$$

$$V(q) = \frac{b}{2}\sum_{1\leq i\neq j\leq N}\frac{m_i m_j}{|q_i - q_j|^\beta} \qquad (1.3)$$

Where $a, b \geq 0, a^2 + b^2 \neq 0, \alpha, \beta \geq 0, \alpha^2 + \beta^2 \neq 0, q = (q_1, \cdots, q_N), q_i \in R^K$.

The equations of the motion for the N–body problems with a potential $-W(q)$ are given by

$$m_i \ddot{q}_i = \frac{\partial W(q)}{\partial q_i}, i = 1, \cdots, N \qquad (1.4)$$

Note that $-W(q)$ is the classical Newtonian potential when $a=0, \beta=1$, or $b=0, \alpha=1$ or $\alpha=\beta=1$, and is a homogeneous potential when $a=0$ or $b=0$ or $\alpha=\beta$.

References [1]–[5], [8], [10]–[12], [15], [19]–[23] used variational methods to study the periodic solutions of N–body problems.

The true motions of the celestial bodies should obey the Least Action Principle of Fermat-Maupertuis in some sense. Hence, we use variational minimizing methods to study the periodic solutions of N-body problems.

Serra-Terracini ([15]) used variational minimizing methods to study the existence of a noncollision periodic solution for 3-body problems with the classical Newtonian potential and a radial perturbation potential in R^3.

Long and Zhang ([10–12]) and Zhang-Zhou ([22]) and Chenciner and Desolneux ([3]) studied the shape of the solution minimizing the Lagrangian action integral on the T/2-antiperiodic or zero mean loop of class $W^{1,2}(R/TZ, R^k)$ for some N-body problems.

Zhang and Zhou ([23]) studied the variational characterization of Lagrangian elliptical solutions with equilateral triangle configurations for planar Newtonian 3-body problems, they proved that the regular minimizers of the Lagrangian action integral restricted on the periodic orbit space with three nonzero relative winding numbers are exactly the Lagrangian elliptical solutions.

In this paper, we study the existence of a noncollision or a generalized periodic solution for the systems (1.1)–(1.4).

For (1.1)-(1.4) in R^{2k} ($k \geq 1$), we also study the shape of the orbit minimizing its Lagrangian action integral defined on the periodic orbit space with the same integral mean for each body during one period.

Definition 1.1 ([2]、[4]) Given T>0, Let $q_i \in W^{1,2}([0,T], R^K)$, we say $q = (q_1, \cdots, q_N)$ is a generalized periodic solution (or a noncollision periodic solution) of (1.1)—(1.4), if there hold:

(i) $S(q) = \{t \in [0,T] | \exists 1 \leq i \neq j \leq N, s.t., q_i(t) = q_j(t)\}$ has zero Lebesgue measure (or is an empty set)

(ii) For all $t \in [0,T] \setminus S(q)$ (or $t \in [0,T]$), $q(t)$ satisfies (1.4).

(iii) For all $t \in [0,T] \setminus S(q)$ (or $t \in [0,T]$), there holds
$$\frac{1}{2}\sum_{i=1}^{N} m_i |\dot{q}_i(t)|^2 - U(q) \equiv Const = h$$

Theorem 1.1 For any given T>0,

If (i) $\alpha \geq 2, \beta \geq 2$

or (ii) $\alpha \geq 2, 0 \leq \beta \leq 2$

or (iii) $\beta \geq 2, 0 \leq \alpha \leq 2$

holds, then the system (1.1)–(1.4) has a noncollision periodic solution.

If (iv) $0 \leq \alpha < 2, 0 \leq \beta < 2, \alpha^2 + \beta^2 \neq 0$ Then the system (1.1)—(1.4) has a generalized T-periodic solution.

Theorem 1.2 Given energy $h \in R$, we consider N-body problems (1.1)-(1.4) and
$$\frac{1}{2}\sum_{i=1}^{N} m_i |\dot{q}_i|^2 - W(q) = h \tag{1.5}$$

If one of the following three conditions holds:

(i) $\alpha \geq 2$, $\beta \geq 2$, $(\alpha, \beta) \neq (2, 2)$

(ii) $\alpha \geq 2$, $0 \leq \beta \leq 2$, $(\alpha, \beta) \neq (2, 2)$

(iii) $\beta \geq 2$, $0 \leq \alpha \leq 2$, $(\alpha, \beta) \neq (2, 2)$

Then there is a T>0 such that (1.1)–(1.4) has a noncollision T-periodic solution with energy h.

If the following condition holds

(iv) $0 \leq \alpha < 2$, $0 \leq \beta < 2$, $\alpha^2 + \beta^2 \neq 0$

Then there is a T>0 such that (1.1)–(1.4) has a generalized T-periodic solution with energy h.

Theorem 1.3 For N–body problem (1.4) with a quasihomogeneous potential (1.1) in $R^{2k}(k \geq 1)$, we define

$$\tilde{\Lambda} = \left\{ \begin{array}{l} q = (q_1, q_2, \cdots, q_N) \mid q_i \in W^{1,2}(R/TZ, R^{2k}), \int_0^T q_1(t)dt = \cdots = \int_0^T q_N(t)dt \\ q_i(t) \neq q_j(t), 1 \leq i \neq j \leq N \end{array} \right\} \quad (1.5)$$

$$f(q) = \frac{1}{2} \sum_{i=1}^{N} \int_0^T m_i |\dot{q}_i|^2 dt + \int_0^T W(q) dt, q \in \tilde{\Lambda} \quad (1.6)$$

Then the global minimum point $q(t)=(q_1(t), q_2(t), \cdots, q_N(t))$ of $f(q)$ exists and must be the non-collision relative equilibrium T-periodic solution with minimal period T of the N-body problem (1.4) in R^{2k} ($k \geq 1$) whose configuration is the central configuration minimizing all configurations with potential U and V at any moment t and whose mass points rotate along circular orbits around the center (the common integral mean) with the same constant angular velocity on a fixed plane.

Theorem 1.4 Let $m_1 = m_2 = \cdots = m_N > 0$, $m_{N+1} > 0$,

$$S = \left\{ q = (q_1, \cdots, q_{N+1}), q_i \in W^{1,2}(R/TZ, R^{2k}), [q_i] = \frac{1}{T} \int_0^T q_i(t)dt = 0, q_{n+1}(t) = 0, \right.$$

$$\left. \sum_{i=1}^{n} m_i q_i(t) = 0, q_i(t) \neq q_j(t), 1 \leq i \neq j \leq n+ \right\} \quad (1.7)$$

$$I(q) = \frac{1}{2} \int_0^T \sum_{i=1}^{N+1} m_i |\dot{q}_i|^2 dt + a \int_0^T \sum_{1 \leq i < j \leq N+1} \frac{m_i m_j}{|q_i - q_j|^\alpha} dt$$

$$+ b \int_0^T \sum_{1 \leq i < j \leq N+1} \frac{m_i m_j}{|q_i - q_j|^\beta} dt \quad (1.8)$$

Then the minimizer $\bar{q} = (\bar{q}_1(t), \cdots, \bar{q}_N(t), \bar{q}_{N+1}(t))$ of $I(q)$ on $\bar{S}$ is a classical T-periodic solution of

$$m_i \ddot{q}_i = \frac{\partial \widetilde{W}(q)}{\partial q_i}, \quad i = 1, \cdots, N+1 \tag{1.9}$$

Where $\quad \widetilde{W}(q) = \widetilde{U}(q) + \widetilde{V}(q) \tag{1.10}$

$$\widetilde{U}(q) = \frac{a}{2} \sum_{1 \leq i \neq j \leq N+1} \frac{m_i m_j}{|q_i - q_j|^\alpha} \tag{1.11}$$

$$\widetilde{V}(q) = \frac{b}{2} \sum_{1 \leq i \neq j \leq N+1} \frac{m_i m_j}{|q_i - q_j|^\beta} \tag{1.12}$$

and satisfies that $\bar{q}_i (i=1,\cdots,N)$ moves on the circle centered at $\bar{q}_{N+1} = 0$ with a fixed angular velocity and a fixed motion plane and the configuration of $m_1, \cdots, m_N$ is the central configuration minimizing all configurations with potential U and V at any moment.

2. The Proof of Theorem 1.1

Given T>0, Let $\Lambda = \{q = (q_1, \cdots, q_N) | q_i \in W^{1,2}(R/TZ, R^K) \; ; \; q_i(t) \neq q_j(t) ,$
$\forall 1 \leq i \neq j \leq N, t \in [0,T]; \int_0^T q_i(t) dt = 0 \}$ (2.1)

$$f(q) = \frac{1}{2} \int_0^T \sum_{i=1}^N m_i |\dot{q}_i(t)|^2 dt + \int_0^T W(q) dt \tag{2.2}$$

Lemma 2.1 The functional $f(q)$ in (2.2) attains its global minimum value on the closure $\overline{\Lambda}$ of Λ, and the minimizer $q(t) = (q_1(t), \cdots, q_N(t))$ is a generalized T-periodic solution for the system (1.1)-(1.4)

Proof Since $f(q)$ is coercive and weakly lower-semi-continuous on Λ, so f(q) attains the global minimum value on $\overline{\Lambda}$, furthermore, similar to the proof of [1], [2], [4], [8], we know that the minimum point q(t) is a generalized solution for the system (1.1)–(1.4).

Lemma 2.2 In the systems (1.1)–(1.4),
if (i) $\alpha \geq 2, \beta \geq 2$
or (ii) $\alpha \geq 2, 0 \leq \beta \leq 2$
or(iii) $0 \leq \alpha \leq 2, \beta \geq 2$

Then there holds the Gordon's strong force condition ([8]): there exists a function $G \in C^1(R^K - \{0\}, R)$ and a neighborhood N of 0 in R^k such that

$$\begin{cases} \lim G(\xi) = -\infty, \xi \to 0 \\ W_{ij}(\xi) \geq |\nabla G(\xi)|^2, \forall \xi \in N - \{0\} \end{cases}$$

Lemma 2.3 ([8] [2])Let $\{q^n\}$ be a sequence in Λ and $q^n \to q \in \partial \Lambda$, then $f(q^n) \to +\infty$.

Now we can prove Theorem 1.1:

Under the assumption (i) or (ii) or (iii) of Theorem 1.1, by Lemma 2.2, $W_{ij}(\xi)$ satisfies the Gordon's strong force condition, then Lemma 2.3 implies the generalized solution q(t) obtained by Lemma 2.1 is a noncollision periodic solution, otherwise, then there is a minimizing sequence q^n such that $f(q^n) \to f(q) = +\infty$, this is contradiction with $\inf\{f(q)|q \in \Lambda\} < +\infty$.

3. The Proof of Theorem 1.2

We define

$$g(x) = \frac{1}{2}\int \sum_{i=1}^{N} m_i |\dot{x}_i|^2 dt \cdot \int (h + W(x)) dt, \forall x \in M \tag{3.1}$$

Where
$$M = \left\{ x = (x_1, \cdots, x_N) \in \Gamma, \int [-W(x) - \frac{1}{2}W'(x) \cdot x] dt = h \right\} \tag{3.2}$$

$$\Gamma = \{ x = (x_1, \cdots, x_N) \in W^{1,2}([0,1], (R^k)^N), x_i(t) \neq x_j(t),$$
$$\forall 1 \leq i \neq j \leq N, \forall t \in [0,1], \int x_i(t) dt = 0 \} \tag{3.3}$$

Similar to the proof in [1], we have

Lemma 3.1 If x is a non-constant critical point for g(x) on M, then we can define T>0:

$$T = \left[\frac{\frac{1}{2}\int \sum_{i=1}^{N} m_i |\dot{x}_i(t)|^2 dt}{-\int W'(x)x dt} \right]^{1/2} \tag{3.4}$$

and q(t)=x(t/T) is a T-periodic solution of (1.1)-(1.5)。

Lemma 3.2 ([12]) For $\alpha \geq 0 \quad \beta \geq 0; \alpha^2 + \beta^2 \neq 0, h \in R$, if g has a non-constant critical point then we have

(i) if $(\alpha, \beta) \in [2, +\infty)^2 \setminus \{(2,2)\}$ then h>0,

(ii) if $(\alpha, \beta) = (2, 2)$ then $h = 0$,

(iii) if $(\alpha, \beta) \in [0, 2]^2 \setminus \{(2,2), (0,0)\}$ then h<0.

Lemma 3.3 For any integer k>0, $N \geq 2$ and positive masses $m_1, \cdots m_N$, power index $\alpha > 0$

and $q = (q_1, \cdots, q_N) \in (R^k)^N$ the following homogeneous function with degree 0:

$$h(q) = \left(\frac{C}{2} \sum_{1 \leq i \neq j \leq N} \frac{m_i m_j}{|q_i - q_j|^\alpha}\right) \left(\sum_{i=1}^N m_i |q_i|^2\right)^{-\alpha/2} \tag{3.5}$$

attains global minimum on $(R^k)^N$, we denote this minimum by $D_\alpha(N, m, C)$.

Lemma 3.4 (Wirtinger's inequality) For any $m_i > 0$ and $x_i \in W^{1,2}([0,1], R^k)$, ($i=1, \cdots, N$) if

$$\int_0^1 x_i(t)dt = 0, \; i = 1, \cdots, N \tag{3.6}$$

then

$$\int_0^1 \sum_{i=1}^N m_i |\dot{x}_i|^2 dt \geq (2\pi)^2 \int_0^1 \sum_{i=1}^N m_i |x_i|^2 dt \tag{3.7}$$

In the following, we'll prove Theorem 1.2:

(i) $\alpha \geq 2$, $\beta \geq 2$, but α, β can't be 2 simultaneously.

By Lemma 3.2, we have h>0. Hence

$$g(x) \geq \frac{1}{2} h \int_0^1 \sum_{i=1}^N m_i |\dot{x}_i|^2 dt$$

By Lemma 3.4, We know that g(x) is coercive on M.

(ii) $0 \leq \alpha < 2$, $\beta \geq 0$ or $0 \leq \beta < 2$, $\alpha \geq 0$, but $\alpha^2 + \beta^2 \neq 0$ or $\alpha = 2$, $0 \leq \beta < 2$ or $\beta = 2$, $0 \leq \alpha < 2$, by Lemma 3.2, h<0, but

$$g(x) = \frac{1}{2} \int_0^1 \sum_{i=1}^N m_i |\dot{x}_i|^2 dt \int_0^1 (-\frac{1}{2} W'(x)x) dt = \frac{1}{2} \int_0^1 \sum_{i=1}^N m_i |\dot{x}_i|^2 dt \int_0^1 (\frac{\alpha}{2} U(x) + \frac{\beta}{2} V(x)) dt$$

Hence by Lemma 3.3, We have

$$g(x) \geq \left(\frac{1}{2} \int_0^1 \sum_{i=1}^N m_i |\dot{x}_i|^2 dt\right) \int_0^1 \left[\frac{a\alpha}{2} D_\alpha(N, m) \left(\sum_{i=1}^N m_i |x_i|^2\right)^{-\alpha/2} \right.$$

$$\left. + \frac{b\beta}{2} D_\beta(N, m) \left(\sum_{i=1}^N m_i |x_i|^2\right)^{-\beta/2}\right] dt$$

By Wirtinger's inequality (Lemma 3.4) we have

$$g(x) \geq \left[\frac{a\alpha}{4} D_\alpha(N, m)(2\pi)^\alpha \left(\int_0^1 \sum_{i=1}^N m_i |\dot{x}_i|^2 dt\right)^{1-\frac{\alpha}{2}}\right.$$

$$\left. + \frac{b\beta}{4} D_\beta(N, m)(2\pi)^\beta \left(\int_0^1 \sum_{i=1}^N m_i |\dot{x}_i|^2 dt\right)^{1-\frac{\beta}{2}}\right]$$

Hence g(x) is coercive on M. Under the above two cases (i) and (ii), It's easy to know

Least Action for N-Body Problems

that g(x)>0 and g(x) is weakly lower semi-continuous on M, and $\overline{M}$ is a weakly closed subset. Hence g(x) attains the global minimum on $\overline{M}$. The minimum value point is the required periodic solution.

4. The Proof of Theorem 1.3

In order to prove Theorem 1.4, we also note that $f(q)$ is invariant under translations. Assume $q=(q_1,q_2,\cdots,q_N)\in \widetilde{\Lambda}$, Let

$$[q_i]=\frac{1}{T}\int_0^T q_i(t)dt, \quad x_i(t)=q_i(t)-[q_i], i=1,2,\cdots,N \tag{4.1}$$

Then $[x_i]=0$ and $x=(x_1, x_2,\cdots,x_N)$ satisfies that $x_i(t)\neq x_j(t)$ ($1\le i\neq j\le N$) if and only if $q_i(t)\neq q_j(t)$ ($1\le i\neq j\le N$).

By Wirtinger's inequality and Coti Zelati's type inequality (Lemma 3.3) and Jensen's inequality we have

$$f(q)=f(x)\ge \frac{1}{2}(\frac{2\pi}{T})^2\int_0^T \sum_{i=1}^N m_i|x_i|^2\,dt + D_\alpha(N,m,a)\int_0^T (\sum_{i=1}^N m_i|x_i|^2)^{-\alpha/2}dt$$
$$+ D_\beta(N,m,b)\int_0^T (\sum_{i=1}^N m_i|x_i|^2)^{-\beta/2}dt \tag{4.2}$$

$$\ge \frac{1}{2}(\frac{2\pi}{T})^2\int_0^T \sum_{i=1}^N m_i|x_i|^2\,dt + D_\alpha(N,m,a)T^{1+\alpha/2}(\int_0^T (\sum_{i=1}^N m_i|x_i|^2)dt)^{-\alpha/2}$$
$$+ D_\beta(N,m,b)T^{1+\beta/2}(\int_0^T \sum_{i=1}^N m_i|x_i|^2\,dt)^{-\beta/2} \tag{4.3}$$

Let
$$s=\left(\int_0^T \sum_{i=1}^N m_i|x_i|^2\,dt\right)^{1/2} \tag{4.4}$$

Then
$$f(q)\ge \inf\{\varphi(s), s>0\} \tag{4.5}$$

Where
$$\varphi(s)=\frac{1}{2}(\frac{2\pi}{T})^2\cdot s^2 + D_\alpha(N,m,a)T^{1+\alpha/2}\cdot s^{-\alpha} + D_\beta(N,m,b)T^{1+\beta/2}\cdot s^{-\beta} \tag{4.6}$$

We note that $\varphi(s)$ is a strictly convex smooth function on $s>0$ and $\varphi(s)\to +\infty$, as $s\to 0^+$ or $s\to +\infty$, so it has a unique global minimizing point $s=s_0>0$, hence $f(x)$ attaining its infimum if and only if the inequalities of Wirtinger and Coti Zelati's type and Jensen take equalities simultaneously. Hence we have

(i) $x_i(t)=a_i\cos\frac{2\pi}{T}t + b_i\sin\frac{2\pi}{T}t, a_i, b_i \in R^{2k}, \forall t\in R, i=1,2,\cdots,N,$

(ii) $U(x)\left(\sum_{i=1}^N m_i|x_i|^2\right)^{-\alpha/2}$ and $V(x)\left(\sum_{i=1}^N m_i|x_i|^2\right)^{-\beta/2}$ attains their minimum simultaneously,

(iii) $\sum_{i=1}^{N} m_i |x_i(t)|^2 \equiv \frac{1}{T} s_0^2$

Similar to the proof of [10, 11, 22], we deduce that the orbits $x_i(t)$ ($i=1, 2, \cdots, N$) locate the same fixed plane and must be circles whose centers are at the origin by (i) and (ii) and (iii).

Hence $q_i(t)$ ($i=1, 2, \cdots, N$) have the properties of Theorem 1.1 by (4.1).

5. The Proof of Theorem 1.4

Lemma 5.1 Let $q_{N+1}=0$, then
$$\sum_{i=1}^{N} \frac{m_i m_{N+1}}{|q_i|^\nu} \geq \left(\sum_{i=1}^{N} m_i\right)^{1+\frac{\nu}{2}} \cdot m_{N+1} \left(\sum_{i=1}^{N} m_i |q_i|^2\right)^{-\nu/2} \quad (5.1)$$

and inequality (5.1) takes equality if and only if

$$|q_i| = |q_l|, \quad 1 \leq i \neq l \leq n \quad (5.2)$$

Now we prove Theorem 1.4:

By the inequalities of Wirtinger and Lemma 3.3 and Lemma 5.1 we have

$$I(q) = \frac{1}{2} \int_0^T \sum_{i=1}^{N} m_i |\dot{q}_i|^2 dt + \int_0^T \frac{a}{2} \sum_{1 \leq i \neq j \leq N} \frac{m_i m_j}{|q_i - q_j|^\alpha} dt + a \int_0^T \sum_{i=1}^{N} \frac{m_{N+1} m_i}{|q_i|^\alpha} dt$$

$$+ \int_0^T \frac{b}{2} \sum_{1 \leq i \neq j \leq N} \frac{m_i m_j}{|q_i - q_j|^\beta} dt + b \int_0^T \sum_{i=1}^{N} \frac{m_{N+1} m_i}{|q_i|^\beta} dt$$

$$\geq \left(\frac{2\pi}{T}\right)^2 \int_0^T \sum_{i=1}^{N} m_i |q_i|^2 dt + D_\alpha(N, m, a) \cdot \int_0^T \left(\sum_{i=1}^{N} m_i |q_i|^2\right)^{-\alpha/2} dt$$

$$+ a \left(\sum_{i=1}^{N} m_i\right)^{1+\alpha/2} m_{N+1} \int_0^T \left(\sum_{i=1}^{N} m_i |q_i|^2\right)^{-\alpha/2} dt$$

$$+ D_\beta(N, m, b) \int_0^T \left(\sum_{i=1}^{N} m_i |q_i|^2\right)^{-\beta/2} dt$$

$$+ b \left(\sum_{i=1}^{N} m_i\right)^{1+\beta/2} m_{N+1} \int_0^T \left(\sum_{i=1}^{N} m_i |q_i|^2\right)^{-\beta/2} dt$$

By Jensen's inequality we have

$$I(q) \geq \left(\frac{2\pi}{T}\right)^2 S^2 + \left[D_\alpha(N, m, a) + a \left(\sum_{i=1}^{N} m_i\right)^{1+\alpha/2} m_{N+1}\right] T^{1+\alpha/2} \cdot S^{-\alpha}$$

$$+ \left[D_\beta(N, m, b) + b \left(\sum_{i=1}^{N} m_i\right)^{1+\beta/2} m_{N+1}\right] T^{1+\beta/2} \cdot S^{-\beta} \equiv \tilde{\varphi}(S)$$

Where note that $\tilde{\varphi}(S)$ is strictly convex smooth function on $S>0$ and $\tilde{\varphi}(S) \to +\infty$ as

$S \to +\infty$, hence $\widetilde{\varphi}(S)$ has a unique global minimizing Point $S=S_0>0$ and $I(q)$ attains its infimum if and only if all inequalities which we have used take equalities simultaneously, hence we have

(i) $q_i(t) = a_i \cos\dfrac{2\pi}{T}t + b_i \sin\dfrac{2\pi}{T}t, a_i, b_i \in R^{2k}, \forall t \in R, i=1,2,\cdots N,$

(ii) $U(q)\left(\sum\limits_{i=1}^{N} m_i |q_i|^2\right)^{-\alpha/2}$ and

$V(q)\left(\sum\limits_{i=1}^{N} m_i |q_i|^2\right)^{-\beta/2}$

attains their minimum simultaneously,

(iii) $|q_1(t)| = \cdots = |q_N(t)|$

(iv) $\sum\limits_{i=1}^{N} m_i |q_i|^2 = const = \dfrac{1}{T}S_0^2$

References

1. Ambrosetti A. and Coti Zelati V., Periodic solutions of singular Lagrangian systems, Birkhauser, Basel, 1993.

2. Bahri A. and Rabinowitz P., Periodic solutions of Hamiltonian systems of three-body type, Ann. IHP. Anal non lineaire 8(1991), 561-649.

3. Chenciner A. and Desolneux N., Minima de l'intégrale d'action et équilibres relatifs de n corps, C. R. Acad. Sci. Paris, Sr. I. Math. 326(1998), 1209-1212.

4. Coti Zelati V., The periodic solutions of N-body type problems, Ann. IHP Anal. non linéaire 7(1990), 477-492.

5. Degiovanni M. and Giannoni F., Dynamical systems with Newtonian type potentials, Ann. Scuola Norm. Super. Pisa 15(1989), 467-494.

6. Diacu, F., Near-collision dynamics for particles systems with quasihomo- geneous potentials, J. Diff. Equ. 128(1996), 58-77.

7. Euler, L., De motu rectilineo trium corpörum se mutuo attrahentium, Novi. Comm. Acad. Sci. Imp. PetropII (1767), 145-151.

8. Gordon W, A minimizing property of Keplerian orbits, Amer J.Math. 99(1977), 961-971.

9. Lagrange J., Essai sur le probléme des trois corps, Ouvres 3(1783), 229-331.

10. Long Y. and Zhang S. Q., Geometric characterization for variational minimization

solutions of the 3-body problem, Abstract, Chinese Science Bulletin 44(1999), 1653-1654. Acta Math Sinica, to appear.

11. Long Y. and Zhang S. Q., Geometric characterizations for variational minimization solutions of the 3-body problem with fixed energy, J. Diff Equ, to appear.

12. Long Y. and Zhang S. Q., Variational minimization solutions of 3-body problem with quasi-homogeneous potential and fixed energy, preprint.

13. Manev, G., La gravitation et l'énergie au zéro, Comptes Rendus 178(1924), 2159-2161.

14. Pérez-Chavela E. and Vela-Arévalo, L.V. , Triple collision in the quasi-homogeneous collinear three-body problem, J. Diff. Equ. 148(1998), 186-211.

15. Serra E. and Terracini S. , Collisionless periodic solutions to some three-body problems, Arch. Rational Mech. Anal. 120(1992), 305-325.

16. Siegel C. and Moser J., Lectures on celestial mechanics, Springer, Berlin, 1971.

17. K. F. Sundman, Memoire sur le problem des trois corps, Acta Math. 36(1913), 105-179.

18. Wintner A., Analytical foundations of celestial mechanics, Princeton University Press, Princeton, 1941.

19. Zhang S. Q., Variational minimizing properties for N-body problems, in Proceedings in memory of Professor Liao Santao, World Scientific, to appear.

20. Zhang S. Q., Periodic solutions for N-body problems. Proceedings in honor of the 60th birthday of Professor Rabinowitz, World Scientific, to appear.

21. Zhang S. Q. and Zhou Qing, Symmetric periodic noncollision solutions for N-body-type problems, Acta Math Sinica, New Series 11(1995), 37-43.

22. Zhang S. Q.and Zhou Qing, Geometric characterization for the least Lagrangian action of N-body problems, Chinese Science, to appear.

23. Zhang S.Q. and Zhou Qing, A minimizing property of Lagrangian solutions, Acta Math.Sinica, to appear.